fu ther

advanced physics

further
advanced physics

DAVID BRODIE

JOHN MURRAY

Titles in this series:
Introduction to Advanced Biology 0 7195 7671 7
Introduction to Advanced Chemistry 0 7195 8587 2
Introduction to Advanced Physics 0 7195 8588 0

Further Advanced Chemistry 0 7195 8608 9
Further Advanced Physics 0 7195 8609 7

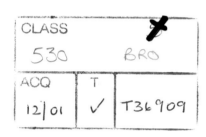
© David Brodie 2001

First published in 2001
by John Murray (Publishers) Ltd
50 Albemarle Street
London W1S 4BD

Layouts by Eric Drewery
Illustrations by Oxford Designers & Illustrators Ltd

Typeset in 10/12pt Gill Sans by Wearset, Boldon, Tyne & Wear
Printed and bound by G Canale, Italy

A CIP catalogue entry for this title is available from the British Library

ISBN 0 7195 8609 7

Contents

CONTENTS

Introduction

Part A of this book, *Further Advanced Topics* (Chapters 1 to 8), builds on the first book, *Introduction to Advanced Physics*, and completes coverage of specification content required for A2 assessment at Advanced level. It deals with those topics which are only found in A2 specifications (see the matrix below).

As in *Introduction to Advanced Physics*, work in these chapters is contextualised through introductory *Background* sections and through extensive *Comprehension and application* exercises. The latter provide opportunities, not only for thoughtful consideration and application of chapter content, but also for practice at general physics skills, including use of language and of mathematical techniques. In addition to these sections, an exceptionally large number of practice questions are provided, some integrated with the explanatory text and images, and some in the form of past examination questions.

Further support for student and teacher is offered by *Extra skills tasks*, so that physics content and issues can be used for practice and preparation of evidence in Communication, Application of Number and Information Technology.

Part B, *Themes and Applications* (Chapters 9 to 16), primarily provides support for preparation for synoptic assessment, which demands that students can apply ideas from the specification as a whole. In that these chapters expect the student to consider material from across the specification, they also have general revision value. Questions in these shorter chapters require the student to refer to earlier work as necessary. Examination questions are again provided, where appropriate.

The unifying themes and general topics that are dealt with by each of these Part B chapters are listed at the start of each one. One chapter addresses the specific skill of use of logarithms, illustrating this with examples from different specification content areas. Two chapters look at pervasive concepts: rate of change and visual representations relating to energy. Other chapters in Part B provide opportunities for students to work with areas of the specification that might otherwise seem unrelated, and so these later chapters stress the unity of physics.

Specification-matching matrix

Board	AQA (NEAB)					AQA (AEB)				Edexcel						OCR				WJEC			
Module*	1	2	3	4	5	1	2	4	5	1	2	3C	4	5	6	A	B	C1	D	PH1	PH2	PH4	PH5
Chapter 1				✓				✓					✓						✓			✓	
Chapter 2				✓				✓					✓						✓			✓	
Chapter 3				✓					✓				✓	✓					✓				✓
Chapter 4				✓				✓					✓	✓					✓			✓	
Chapter 5				✓					✓				✓						✓				✓
Chapter 6				✓					✓				✓						✓			✓	✓
Chapter 7				✓	✓				✓					✓					✓				✓
Chapter 8													✓										

* Modules in the tinted columns make up the A2 part of the course

Note: Chapters 9 to 16 are synoptic chapters that draw together ideas from the whole of the physics course.

Acknowledgements

I would like to thank the editorial team Katie Mackenzie Stuart, Jane Roth and Geoff Amor; Thelma Gilbert for researching excellent photographs, Ronald Trevillion for his advice on the questions, John Gregson and Malcolm Parry for their valuable comments, and my nearest and dearest for remaining so despite my attachment to the computer keyboard.

David Brodie

Examination questions have been reproduced with kind permission of the following examination boards:

Assessment and Qualifications Alliance (AQA): the Associated Examining Board (AEB) and the Northern Examinations and Assessment Board (NEAB)
Edexel Foundation (London Examinations)
International Baccalaureate (IB) Organisation
OCR, incorporating the former UCLES and UODLE

The answers in this book have not been provided by or approved by the examining boards. Their accuracy and the method of working are the sole responsibility of the author.

Thanks are due to the following for permission to reproduce copyright photographs:

Cover David Nunuk/SPL; **p.2** Fig. 1.1 (*l*) J. Townson/Creation, (*r*) Science Photo Library/Françoise Sauze; **p.16** Fig. 2.1 Science Photo Library/Dr G. Owan Bredbeig; **p.20** Fig. 2.7 University of Washington Library Special Collections Division; Fig. 2.8 J. Townson/Creation; Fig. 2.9 Redfern/Patrick Ford; Fig. 2.10 Science & Society Picture Library; **p.30** Fig. 2.25 Auto Express; **p.41** Fig. 3.1 Action Plus; **p.43** Fig. 3.4 Action Plus; **p.46** Fig. 3.10 Science Photo Library/NASA; **p.48** Fig. 3.15 Science Photo Library/NASA; Fig. 3.16 Robert Harding Pictures; **p.52** Fig. 3.24 Science Photo Library/US Geological Surveys; **p.54** Fig. 3.27 Science Photo Library/Colin Cuthbert; **p.58** Fig. 3.33 Science Photo Library; **p.60** Fig. 3.35 Science Photo Library/European Space Agency; **p.65** Fig. 3.41 Corbis; **p.66** Fig. 3.43 Science Photo Library; **p.73** Fig. 4.1 Stock Market; **p.75** Fig. 4.5 Science Photo Library/Maximilian Stock; **p.80** Fig. 4.14 Andrew Lambert; **p.89** Fig. 4.29 Science Photo Library; Fig. 4.30 Allsport/Doug Pensinger; **p.93** Fig. 5.1 Science Photo Library/Jack Finch; Fig. 5.2 Science & Society Picture Library; **p.102** Fig. 5.16 Science Photo Library; **p.110** Fig. 5.27 Science Photo Library/John Gretin; **p.115** Fig. 6.1 Science Photo Library/European Southern Observatory; Fig. 6.2 J. Townson/Creation; **p.122** Fig. 6.13 Science Photo Library; **p.129** Fig. 6.26 Colorific; **p.133** Fig. 6.31 J. Townson/Creation; **p.139** Fig. 6.41 (*l & r*) J. Townson/Creation; **p.140** Fig. 6.42 Science Photo Library; **p.145** Fig. 7.1 Advertising Archives; Fig. 7.2 Historical Newspapers Loan Service; **p.149** Fig. 7.7 Science Photo Library; **p.150** Fig. 7.8 Science Photo Library/Philippe Plailly/Eurelios; **p.154** Fig. 7.14 Science Photo Library/US Dept of Energy; **p.158** Fig. 7.21 Science Photo Library/Los Alamos National Laboratory; Fig. 7.22 Rex Features; **p.161** Fig. 7.25 Science Photo Library/Mere Words; **p.162** Fig. 7.26 Science Photo Library/US Dept of Energy; **p.164** Fig. 7.28 Science Photo Library; **p.168** Fig. 8.1 Science Photo Library/NASA; **p.169** Fig. 8.2 Science Photo Library/Kaptyn Laboratarium; **p.170** Fig. 8.4 Science Photo Library/NASA; Fig. 8.5 Science Photo Library/Space Telescope Science Institute/NASA; **p.172** Fig. 8.9 Science Photo Library/Gordon Gerrad; **p.173** Fig. 8.10 Science Photo Library/Royal Observatory, Edinburgh; **p.179** Fig. 8.16 Science Photo Library; Fig. 8.17 Science Photo Library/Space Telescope Science Institute; **p.181** Fig. 8.21 Popperfoto; Fig. 8.22 Historical Newspapers Loan Service; **p.183** Fig. 8.23 Science Photo Library/Space Telescope Science Institute; **p.184** Fig. 8.25 Science Photo Library; **p.186** Fig. 8.29 Science Photo Library/Nina Lampen; **p.187** Fig. 8.30 GeoScience Features; Fig. 8.31 Collections; **p.211** Fig. 11.1 Science Photo Library; Fig. 11.2 Science Photo Library/P. Dumas/Eurelios; **p.215** Fig. 11.6 Science Photo Library/Dr Jeremy Burgess; Fig. 11.7 Dr Dimitri A. Rusakov & Dr

Attila Gulyas, National Institute of Medical Research; **p.216** Fig. 11.8 Science Photo Library/Dr Mitsuo Ohtsuki; Fig. 11.9 British Geological Survey IPR/19-1C, © NERC; **p.221** Fig. 12.6 Jodrell Bank Science Centre; **p.227** Fig. 13.1 Science Photo Library; **p.229** Fig. 13.2 Science Photo Library/Lawrence Berkeley; Fig. 13.04 Science Photo Library/CC Studio; **p.231** Fig. 13.6 Science Photo Library/CERN; **p.232** Fig. 13.8 Science Photo Library/David Parker; **p.235** Fig. 14.1 J.G. Armitage/St Andrews Physics Trust; **p.237** Fig. 14.5 Science Photo Library; **p.241** Fig. 15.3 Science Photo Library/Philippe Plailly; **p.243** Fig. 15.6 Science Photo Library/Taskeshi Takahara; **p.245** Fig. 15.9 Science Photo Library/STU; **p.246** Fig. 15.11 Ace Photo Library; **p.251** Fig. 16.4 Science Photo Library/Dept of Clinical Radiology, Salisbury District Hospital.

l = left, *r* = right

The publishers have made every effort to contact copyright holders. If any have been inadvertently overlooked the publishers will be pleased to make the necessary arrangements at the earliest opportunity.

PART A

FURTHER ADVANCED TOPICS

Circular motion

THE BIG QUESTION

● Circular motion is common in natural and in manufactured environments, but how can we model circular motion graphically and mathematically?

KEY VOCABULARY

angular displacement   angular velocity   camber (also known as banking and super-elevation)
centripetal acceleration   centripetal force   Coriolis effect   geostationary orbit
period   uniform circular motion

BACKGROUND

Figure 1.1
A satellite dish fixed to the side of a house (left), pointing at a geostationary satellite (right).

A satellite dish on the side of a house stays aimed at a point in the sky (Figure 1.1); at that point high above the Earth's atmosphere a satellite is in orbit, rotating around the Earth but staying above the same point on the surface below (Figure 1.2). The orbit is **geostationary** – the satellite's orbit is synchronised with the Earth's own spin. The dish on the wall receives radio signals from space without having to track across the sky. The motion of the satellite is described by the mathematics of circular motion.

Figure 1.2
Geostationary orbit.

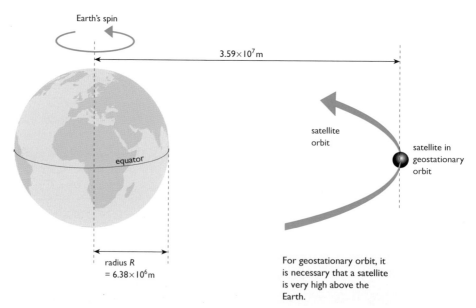

Earth's spin

3.59×10^7 m

satellite orbit

satellite in geostationary orbit

equator

radius R = 6.38×10^6 m

For geostationary orbit, it is necessary that a satellite is very high above the Earth.

Speed and velocity in circular motion

Figure 1.3
Changing velocity in
uniform circular motion.

When we think about circular motion the distinction between scalar speed and vector velocity is highlighted. A satellite in circular orbit and the seats on a big wheel fairground ride move at constant speed, travelling the same distance in each unit of time. But the direction of the motion changes continuously, which means that velocity changes continuously (Figure 1.3). So bodies in circular motion at constant speed are accelerating all of the time. This chapter is, for the most part, about the circular motion of bodies of fixed mass at constant speed and radius of motion, which we can describe as **uniform circular motion**.

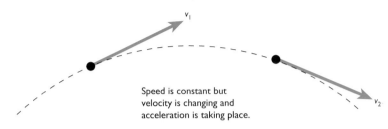

Speed is constant but
velocity is changing and
acceleration is taking place.

Angular displacement and angular velocity

When a body moves in a circle, the radius that joins its position to the centre of the circle also moves. **Angular displacement** is the angle, θ, through which this radius has turned, away from its initial position.

The rate of change of angular displacement is **angular velocity**, ω (Figure 1.4a):

$$\omega = \frac{d\theta}{dt}$$

Figure 1.4 (below)
a Angular displacement θ
and angular velocity ω, and
b the radian.

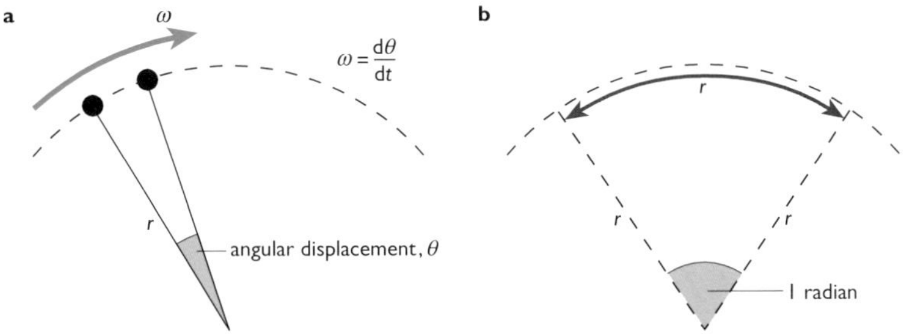

a

$\omega = \frac{d\theta}{dt}$

— angular displacement, θ

b

1 radian

The angle θ is normally measured in radians, and so the unit of angular velocity is the radian per second, rad s^{-1}. Remember that one radian is the angle created at the centre of a circle by an arc of length equal to the radius (Figure 1.4b). So there are 2π radians in a complete revolution, as there are 2π radii in one circumference:

$$2\pi\,\text{rad} = 1 \text{ revolution} \qquad = 360°$$
$$1\,\text{rad} \ = 1/2\pi \text{ revolutions} = 57.3°$$

Note that a body in
uniform circular motion
has changing velocity, but
constant speed and
constant angular velocity.

For a small but finite angle $\delta\theta$,

$$\sin\delta\theta \approx \delta\theta$$

For example, if $\delta\theta = 0.01$ radians (equal to little more than half of one degree), then $\sin\delta\theta - 0.00999$. As $\delta\theta$ continues to become smaller, the approximation becomes nearer and nearer to absolute truth. Also, the trigonometry (Figure 1.5) tells us that, for small $\delta\theta$ and a corresponding arc of length δx,

$$\sin\delta\theta = \frac{\text{opposite}}{\text{hypotenuse}} \approx \frac{\delta x}{r}$$

where r is the radius of the circle.

So,

$$\omega = \frac{d\theta}{dt} \approx \frac{\delta\theta}{\delta t} \approx \frac{\delta x/r}{\delta t} \approx \frac{1}{r}\frac{dx}{dt} = \frac{v}{r} \quad \text{where } v \text{ is the linear velocity.}$$

When $\delta\theta$, δx and δt are vanishingly small, then, again, the approximations become equalities. Then,

$$\omega = \frac{v}{r} \quad \text{and} \quad v = \omega r$$

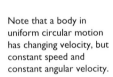
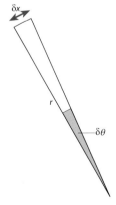

Figure 1.5 (above)
For small $\delta\theta$, $\sin\delta\theta \approx \dfrac{\delta x}{r}$.

So we have a relationship between angular and linear velocities.

Note that the word 'velocity' should be taken to mean linear velocity unless it is specified otherwise.

Acceleration – the story so far

To consider the acceleration of a body in circular motion we need to remember that the definition of acceleration is rate of change of velocity, which we write as

$$a = \frac{dv}{dt}$$

This is the general calculus formula. It is the formula that applies in *all* circumstances. In other *particular* circumstances we can also use the following formulae:

$$a = \frac{v - u}{t}$$

and

$$a = \frac{\Delta v}{\Delta t}$$

A variation on this that is sometimes used when the changes in velocity and time are very small is

$$a = \frac{\delta v}{\delta t}$$

Figure 1.6 shows when each formula is appropriate.

a

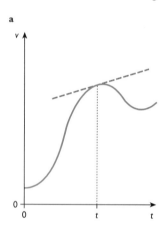

In general, acceleration is given by the formula

$$a = \frac{dv}{dt}$$

which is equal to the gradient of the *v*–*t* graph at any instant, *t*.

Figure 1.6
Graphical and mathematical representations of acceleration, and the situations in which each is appropriate.

b

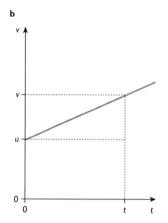

For uniform acceleration, for the period from time 0 to time *t*,

uniform acceleration
$$= \frac{v - u}{t}$$

c

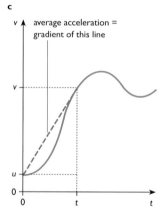

average acceleration = gradient of this line

For the period from time 0 to time *t*,

average acceleration
$$= \frac{v - u}{t}$$

d

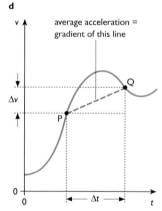

average acceleration = gradient of this line

For an arbitrary portion of a journey,

average acceleration $= \dfrac{\Delta v}{\Delta t}$

$\left(\text{or } \dfrac{\delta v}{\delta t} \text{ for small values}\right)$

This becomes equal to $\dfrac{dv}{dt}$ when δt and δv are vanishingly small, and the points P and Q are an infinitesimal distance apart.

Acceleration in uniform circular motion

Acceleration of a body in uniform circular motion is related to change in direction of velocity rather than change in magnitude. To see what change, δv, takes place in the velocity of a body during a short period of time δt, we need to think about the velocity vectors at the start and finish (Figure 1.7a).

Figure 1.7
Change in velocity, δv, in uniform circular motion.

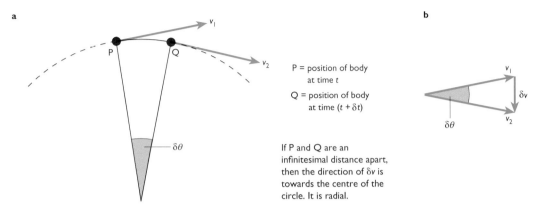

a

b

P = position of body
at time t

Q = position of body
at time $(t + \delta t)$

If P and Q are an infinitesimal distance apart, then the direction of δv is towards the centre of the circle. It is radial.

In Figure 1.7, $\delta \theta$ is the angle through which the velocity vector has turned. It is also the angle made at the centre of the circle by the radii that correspond to the body's initial and final positions.

The change in velocity is shown by the third side of a triangle made by the initial and final velocities (Figure 1.7b). If $\delta \theta$ is very small then the triangle approximates to a right-angled triangle; we can write $v_1 = v_2 = v$ and then say that

$$\sin \delta \theta \approx \frac{\delta v}{v}$$

Also, for small values of angle $\delta \theta$, measured in radians,

$$\delta \theta \approx \sin \delta \theta$$

So, we get

$$\delta \theta \approx \frac{\delta v}{v}$$

and dividing both sides by δt and rearranging,

$$v\frac{\delta \theta}{\delta t} \approx \frac{\delta v}{\delta t}$$

For vanishingly small (infinitesimal) values of $\delta \theta$, δv and δt, the approximations that we have used above become equalities. And in this case we can write

$$v\frac{d\theta}{dt} = \frac{dv}{dt}$$

or

$$v\omega = a$$

Since we have already seen the relationship between linear velocity and angular velocity, $\omega = v/r$, we can write the above as

$$a = \frac{v^2}{r}$$

and

$$a = \omega^2 r$$

Figure 1.8
The relative directions of *v* and *a* for a body in uniform circular motion

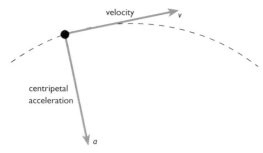

For a body in circular motion at constant speed, velocity and acceleration are always perpendicular to each other.

The acceleration of a body in uniform circular motion is called **centripetal acceleration**. The relative directions of the velocity and the acceleration are important (Figure 1.8). Velocity always acts in a direction parallel to the circumference of the circle – along a tangent. Centripetal acceleration always acts towards the centre of the circle, and can be described as radial.

1 What are the values of sin θ when
  a $\theta = 1°$
  b $\theta = 0.01°$
  c $\theta = 0.02$ radians
  d $\theta = 0.002$ radians?

2 Compare the validity of the approximation sin $\theta = y/x$ for the two situations shown in Figure 1.9.

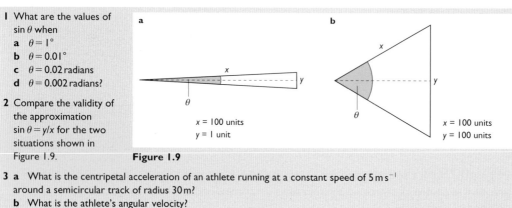

Figure 1.9

3 a What is the centripetal acceleration of an athlete running at a constant speed of $5\,\mathrm{m\,s^{-1}}$ around a semicircular track of radius 30 m?
  b What is the athlete's angular velocity?

Centripetal force

Since a body moving in circular motion is always accelerating, there must be a force producing this acceleration. This force is called **centripetal force**.

The equation $F = ma$ tells us that force and acceleration are in the same direction in all circumstances, including circular motion.

$$F = ma$$
$$\text{centripetal force} = m \times \text{centripetal acceleration}$$

From page 5,

$$F = \frac{mv^2}{r}$$
$$= m\omega^2 r$$

Centripetal force and Newton's First Law

Every body continues in a state of constant velocity unless acted upon by an external force. That is Newton's First Law. The 'constant velocity' may equal zero. Or it may not, in which case to be constant it must be always in the same direction, as well as always having the same size. Newton's First Law tells us that force is needed for changes in direction. For circular motion this force is the centripetal force.

In the absence of a force, a body continues with no change in its motion. (People often find this surprising. It is 'counter-intuitive' in that everyday experience seems to show us that bodies slow down unless we exert a force. The slowing down is, of course, due to the presence of a net resistive force.) The tendency of any body with mass to continue with unchanged motion is described as being due to the inertia of the body. You can feel this in a vehicle that's turning a corner – you can feel your own body's tendency to continue in a straight line while friction with the seat provides the necessary centripetal force to give you an acceleration that's perpendicular to your velocity.

Origins of centripetal force

Centripetal force can have many origins – for an electron in an atom it is electric force, for the Earth in orbit around the Sun it is gravitational force, for a car on a bend with a horizontal surface it is friction with the surface, and for a conker whirled around on a string it is the tension in the string. Some other examples are shown in Figure 1.10.

Figure 1.10
Examples of centripetal forces on passengers.

centripetal force
(provided by friction)

centripetal force provided by handrail and by friction with the floor

centripetal force provided by neighbour

For circular motion at constant speed, centripetal force given by

$$F = \frac{mv^2}{r}$$

is essential. The interaction between inertia and centripetal force can become a matter of personal experience.

If a body is moving in uniform circular motion and the centripetal force is suddenly removed, then it reverts immediately to straight line motion – it flies off at a tangent (Figure 1.11).

Figure 1.11
Motion with and without centripetal force.

If you lose centripetal force then straight away you revert to straight line motion.

Forces in non-uniform circular motion

Figure 1.12
Examples of non-uniform circular motion.

For *uniform* circular motion, all quantities in the relationship $F = mv^2/r$ are constant.

Here the mass of a horse and the radius of its motion are constant, but the speed may not be constant. The change in speed must be compensated by a change in centripetal force.

Here again, mass and radius are constant but other variables are not.

acceleration due to normal reaction

acceleration due to gravity

Circular motion may be *non-uniform* due to a combination of:

- change in the centripetal force (such as can be achieved when spinning a mass on a string)
- change in mass of the body
- change in radius of the motion
- change in speed

If one of these four quantities changes, then if uniform circular motion is to be maintained at least one of the others must also change. Change in speed may also be the result of the action of a tangential force, such as a frictional force that opposes motion. Figure 1.12 shows two examples of non-uniform circular motion.

While for uniform circular motion velocity and acceleration are always perpendicular, this is not necessarily so for non-uniform acceleration. (Force and acceleration, linked by $F = ma$, are *always* parallel to each other.)

For non-uniform circular motion with constant radius and constant mass:
- centripetal acceleration is not constant
- speed is not constant
- angular velocity is not constant
- acceleration has tangential and radial components.

4 a Estimate the force necessary to provide the centripetal acceleration of the athlete in question 3.
   b What is the origin of the force?
5 What is the origin of the centripetal force for:
   a a satellite in orbit round the Earth
   b a speedboat going round in a circle?
6 Why aren't car seats made of smooth material?
7 If a centripetal force of 220 N acts on a cycle and cyclist, of total mass 110 kg, when moving in a circle of radius 25 m, what is the speed?

8 Use centripetal force $= \dfrac{mv^2}{r}$ to explain

   a what will happen to a conker being swung round and round on a string if the force is increased
   b what will happen (there is more than one possibility) if the string is fed through the hand so that the radius of the motion increases.

9 Use centripetal force $= \dfrac{mv^2}{r}$ to explain whether a spanner released by a space-walking astronaut in uniform circular orbit will stay in orbit with the astronaut or move to a different orbital radius and/or speed. (Remember that the origin of the centripetal force is gravitational, and is given by $F = mg$, where g is the local gravitational field strength.)

10 Bodies can experience changes in radius of their circular motion. Two possibilities are that they do this at constant speed v or at constant angular velocity ω.
   a Use $v = \omega r$ to show what happens to angular velocity when radius changes at constant speed.
   b Use $v = \omega r$ to show what happens to speed when radius changes at constant angular velocity.
   c Give a real-world example of both of these kinds of change.
   d Use $F = \dfrac{mv^2}{r}$ and $F = m\omega^2 r$ to sketch graphs of centripetal force against radius for changes in radius at constant speed and at constant angular velocity (and constant mass in both cases).
11 A washing machine drum of radius 0.2 m rotates with a constant angular velocity of 200 rad s⁻¹.
   a What are the angular and linear velocities of a sock when it is
   i at the rim of the drum
   ii half-way between the rim and the centre of the drum?
   b Weight is one possible origin of the centripetal force in this example.
   i What is the other?
   ii Estimate the ratio of the weight of the sock and the centripetal force acting on it when at the rim.

Period

Period is the time taken for one complete revolution, and is often given the symbol T. Note that it is the time that has elapsed when distance travelled is one circumference, and the total angle through which the radius of the motion has turned is 360°, which is 2π radians:

$$\text{average speed for one revolution} = \frac{x}{t} = \frac{2\pi r}{T}$$

Note that we use the word 'speed' and not 'velocity'. The average linear velocity for one complete revolution is zero, since total displacement after one revolution is zero. (Average linear speed after one revolution ≠ average linear velocity after one revolution. It is true, however, that at any particular instant speed and velocity have the same magnitudes.)

Note that, although the linear quantities of speed and velocity have very different meanings and mean values, the distinction between angular speed and angular velocity has no meaning:

$$\text{average angular velocity for one revolution} = \frac{\theta}{t} = \frac{2\pi}{T}$$

A body in uniform circular motion has a constant angular velocity. So instantaneous angular velocity is equal to the average angular velocity. That is,

$$\omega = \frac{d\theta}{dt} = \frac{\theta}{t}$$

We can thus say that

$$\omega = \frac{2\pi}{T}$$

Period is related to angular velocity by

$$T = \frac{2\pi}{\omega}$$

Frequency

The frequency of circular motion is the number of revolutions in each unit of time. Its unit is revolutions (or cycles) per second, which can be abbreviated to $\text{rev}\,s^{-1}$. Since the number of revolutions is itself a number with no units (it is dimensionless), this can be written as s^{-1}. The unit s^{-1} is more often known as the hertz, Hz. (Note, however, that in the context of radioactivity the unit s^{-1} is called the becquerel, Bq.)

Since period is the number of units of time for each revolution, period and frequency are simple inverses of each other:

$$\text{frequency} = \frac{1}{\text{period}}$$

$$f = \frac{1}{T}$$

But $T = 2\pi/\omega$, so

$$f = \frac{\omega}{2\pi}$$

or

$$\omega = 2\pi f$$

12 For the washing machine in question 11, what is the frequency in
  a revolutions per second (revs⁻¹)
  b revolutions per minute (rpm)?

13 The big end of a piston rod is connected to the rotating crankshaft in a car engine (Figure 1.13). If the radius of the motion of the big end is 10 cm, what is its linear velocity when the crankshaft rotates at 2000 rpm?

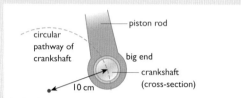

Figure 1.13

14 a What is the period of the spin of the Earth around its own axis:
  i in hours
  ii in seconds?
  b What is the frequency of the spin of the Earth in hertz?

c What is the angular velocity of the Earth's spin around its own axis?
d What is the linear velocity of a point on the Earth's equator? (Radius of the Earth = 6.4 × 10⁶ m.)
e At what latitude would you have a linear velocity that is half of that at the equator?
f The origin of centripetal force for geostationary orbit is gravitational, for which

$$F = \frac{km}{r^2}$$

where $k = 3.99 \times 10^{14}\,\mathrm{N\,m^2\,kg^{-1}}$, m is the mass of the body in orbit, and r is the radius of the orbit. Use this
  i to obtain a relationship between the radius of the orbit and angular velocity
  ii to calculate the radius required for geostationary orbit.

15 a What is the period of the Earth's motion around the Sun?
  b Assuming the Earth's orbit to be circular with a radius of 1.49 × 10¹¹ m, what is its
  i linear velocity
  ii angular velocity
  due to its motion around the Sun?

● ●

● **Comprehension and application**

The Coriolis effect

A roundabout in a children's playground has an angular velocity that can change only slowly. It can change, for example, by being pushed to accelerate it or due to the effects of friction. Over short periods of time we can think of the roundabout's angular velocity as unchanging.

A child near the centre of a turning roundabout who decides to move to the edge will experience no change in angular velocity. But as radius increases, the child's linear velocity must increase (Figure 1.14), in order to remain in motion with the roundabout. This requires an accelerating force. Likewise, a child moving towards the centre must be subject to a reduction in speed and a tangential decelerating force, in order to stay in contact with the roundabout. The force is provided by friction, or by the handles on the roundabout.

Figure 1.14
Position and velocity on a roundabout.

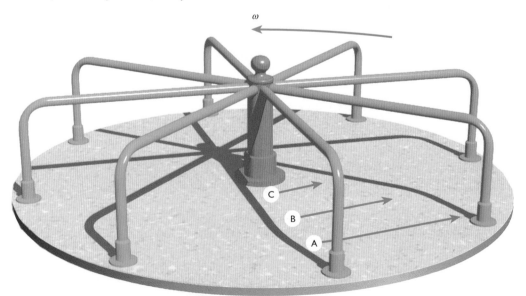

For a roundabout with constant angular velocity, the linear velocity (and hence speed) of a child depends on radial position, such as A, B, or C.

A ball rolled inwards on a roundabout has less frictional contact with the roundabout, and it *does* tend to retain its original speed as it moves inwards. The result is that it moves forwards relative to the roundabout (Figure 1.15). Likewise when it moves outwards it fails to increase its velocity to match that of the outer parts of the roundabout, and tends to get left behind. This relative motion is a manifestation of the **Coriolis effect**.

Figure 1.15
Origin of the Coriolis effect for a ball on a roundabout.

a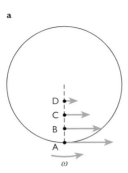

b ball released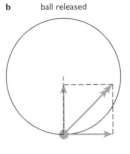

c a short time later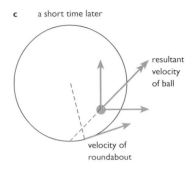

Points A, B, C and D on the roundabout have the same angular velocity but different linear velocities.

A ball rolled towards the centre of the roundabout by a person riding on the roundabout has two components of velocity, which remain unchanged in the absence of friction.

At a position closer to the centre of the roundabout, the ball has the same components of velocity, but its velocity is significantly bigger than that of the roundabout over which it is rolling.

Masses of air in the atmosphere behave more like the ball than the child. They experience only low levels of friction with the ground or with other air. The consequence is that such masses of air apparently have lateral (eastwards or westwards) components to their motions relative to the globe (Figure 1.16). Thus the Coriolis effect has a huge influence on climate – giving Northwestern Europe, for example, a predominantly west to east flow of air.

Figure 1.16
Global air flows.

Where air masses have a tendency to move northwards in the Northern Hemisphere, the Coriolis effect creates a movement across the globe from west to east.

16 Show mathematically that as radius increases for a child moving out towards the edge of a roundabout but retaining the same angular velocity, the child's linear velocity must increase.

17 Why does a ball behave differently from the child?

18 Sketch the pathway of a mass of air that moves away from the Earth's North Pole.

19 A mass of air sits above the equator and moves with the same speed as the surface below. It then moves southwards from the equator. It experiences only a little friction with the Earth's surface and with other air, and so its linear velocity decreases only slightly and it now has linear velocity that is higher than that of the surface below. Draw sketches to show the relative movement of the air mass and the Earth when seen

a from above the South Pole

b from above the point on the equator from which it moved.

● **Comprehension and application**

The Bubble Wheel

Eltom Waters is an adventure park that is considering introducing new rides. An idea has been suggested. The Bubble Wheel (Figure 1.17) is a variation on a big wheel fairground ride.

Figure 1.17
The Bubble Wheel, and its specifications.

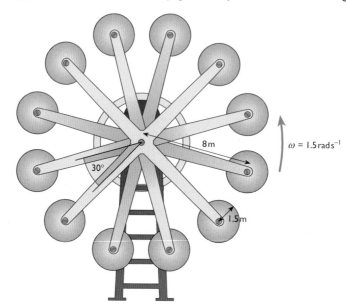

Bubble Wheel – technical specifications:
• radius, wheel centre to bubble centre $= 8\,m$
• operating angular velocity $= 1.5\,rad\,s^{-1}$
• angular separation of bubbles $= 30°$
• individual bubble radius $= 1.5\,m$
• smooth inner surface, no seats – passengers lie on the see-through plastic surface.

20 What is the linear velocity of a bubble centre?
21 a When a Bubble Wheel passenger has rotated through an angle of 0.1 radians, what change in velocity have they experienced?
  b What is the direction of this change in velocity?
22 a If two passengers have masses of 50 kg and 90 kg, estimate the centripetal forces acting on each of them when the Bubble Wheel has its full angular velocity.
  b If the two passengers are on diametrically opposite bubbles of the Bubble Wheel, what is the net force they exert on the centre of the wheel when:
  i the wheel is stationary and the passengers are at opposite ends of a horizontal diameter
  ii the wheel is stationary and the passengers are at opposite ends of a vertical diameter
  iii in the same position as i with the wheel rotating at full angular velocity
  iv in the same position as ii with the wheel rotating at full angular velocity?
23 Suppose that an acrobatic child attempts to stand up inside a bubble, feet pointing outwards. Calculate the centripetal force acting on each kilogram of
  a the child's head
  b the child's feet.
24 Comment on the following aspects of the Bubble Wheel:
  a likely customer satisfaction
  b the safety considerations.

Passengers are in individual transparent plastic bubbles that are fitted with horizontal axes about which they are free to rotate. The bubbles have no seats – the passengers are in a lying position and can slide around against the smooth walls. The horizontal pivots are fixed to the framework of the wheel.

The Bubble Wheel reaches a constant angular velocity, and as it does so the bubbles rotate so that the radius of the motion of each passenger increases to the maximum possible value.

● **Comprehension and application**

Cornering

A road may look like a dull flat surface, but there is more to it than that. A road surface can exert a force on another surface such as a bicycle tyre in two ways – through normal reaction and through friction. Friction between surfaces in relative motion dissipates energy and produces wear. On a sharp curve in a road, such as a roundabout, the road surface is given a slope perpendicular to its direction. This slope, called **camber** or banking (or sometimes super-elevation), cuts down wear on the road surface and on the tyres that run across it, cuts down noise and aids safety. The camber is there to provide centripetal force, and so to reduce the need for friction between road and tyre to produce centripetal force.

Figure 1.18
The forces involved in cornering.

a

On a flat road, centripetal force must be provided by friction between tyre and road surface.

b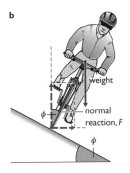

On a road with an ideal camber, the vertical component of normal reaction balances weight, while the horizontal component provides the centripetal force without the need for a sideways frictional force:

$$mg = F\cos\phi$$
$$\frac{mv^2}{r} = F\sin\phi$$

Suppose that a cyclist is travelling around a roundabout in flat countryside. This requires the presence of a centripetal force, which acts horizontally. If the road surface itself is horizontal, then the normal reaction is vertical. So the surface must exert a centripetal force on a vehicle by means of friction (Figure 1.18a). But a camber on the road can give the normal reaction both vertical and horizontal components. The horizontal component can contribute to the necessary centripetal force (Figure 1.18b).

We can work out the vertical and horizontal components of the normal reaction for the cyclist going around the roundabout with a camber:

vertical component of the normal reaction $= mg = F\cos\phi$

so

$$F = \frac{mg}{\cos\phi}$$

where ϕ is the angle of the camber from the horizontal and F is the total normal reaction that acts perpendicularly to the surface;

horizontal component of the normal reaction $= F\sin\phi$

$$= \frac{mg}{\cos\phi}\sin\phi$$
$$= mg\frac{\sin\phi}{\cos\phi}$$
$$= mg\tan\phi$$

In an ideal situation, all of the centripetal force can be applied by the horizontal component of the normal reaction. Then:

$$mg\tan\phi = \frac{mv^2}{r}$$

and the ideal angle of camber is given by

$$\tan\phi = \frac{v^2}{rg}$$

25 For a car going round a roundabout with a horizontal (no camber) surface, what is the origin of the necessary centripetal force?

26 Express the 'ideal' angle of banking in terms of angular velocity. In what sense is this an 'ideal' banking?

27 You might occasionally see a road sign saying: 'Caution, Adverse Camber'. For an adverse camber, the normal reaction acting on the car has a horizontal component in the opposite direction to that of the required centripetal force.
a Draw a sketch to show a car travelling around an adverse camber, with the forces of friction, F (parallel to the surface), and reaction, R (perpendicular to the surface).
b Calculate the necessary contribution of friction to centripetal force for a car of mass 1000 kg travelling at 15 m s^{-1} on a bend of radius 40 m with
i an appropriate camber of 5°
and with
ii an adverse camber of 5°.

• **Extra skills task**   Information Technology and Communication

1 Sketch the shape of the graph of $\tan\phi$, where ϕ is the ideal angle of camber, against the inverse of radius, $1/r$, for vehicles moving at 15 m s^{-1}.

2 What is the gradient of the graph?

3 How does the graph change when the value of v is doubled?

4 Use a spreadsheet program to predict values of ideal camber angle, ϕ, for speeds in the range 10 m s^{-1} to 30 m s^{-1} and for circles of radius from 5 m to 40 m.

5 Spreadsheet programs are useful for many purposes. Use your work on camber angle as the basis for a leaflet to illustrate the general usefulness of spreadsheet programs and to provide instruction, for an intelligent beginner, in how to use them. Use a publishing package to edit and design your leaflet.

Examination questions

1 A car is travelling round a bend in a road at a constant speed of $22\,\mathrm{m\,s^{-1}}$. The driver moves along a circular path of radius 25 m.
  a Explain why, although the speed of the driver is constant, his velocity is not constant. (1)
  b The diagram illustrates the position of the car and the driver at an instant during this motion.

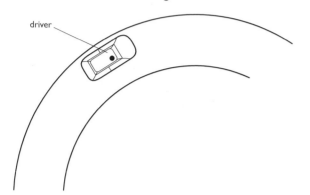

Show the directions of the velocity and of the acceleration of the driver at this instant. (2)
  c Calculate the magnitude of the driver's acceleration. (2)
  d Suggest what provides a force to cause this acceleration. (2)

UCLES, AS/A level, 4831, June 1998

2 The diagram shows a compact disc (CD).

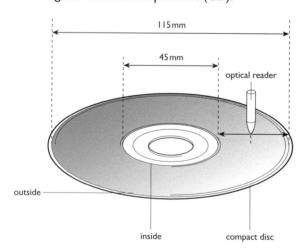

The information is recorded along a path which spirals around the surface of the disc from the inside to the outside. The path passes under an optical reader. The distance of the optical reader from the edge of the disc can be changed.

  a The disc spins at 500 revolutions per minute (r.p.m.) when the optical reader is at the inside of the disc. Calculate
    i the angular speed of the disc, in $\mathrm{rad\,s^{-1}}$
    ii the linear speed of a point on the inside of the disc which passes under the reader. (4)
  b The path passes under the reader at a fixed linear speed, regardless of its position on the disc. Use your answer to **a** to show that the rate at which the disc spins when the reader is at the outside of the disc is about 200 r.p.m. (2)

UCLES, A level, 4833, March 1998

3 a A satellite orbits the Earth once every 120 minutes. Calculate the satellite's angular speed. (2)
  b Draw a free-body force diagram for the satellite. (1)
  c The satellite is in a state of free fall. What is meant by the term *free fall*? How can the height of the satellite stay constant if the satellite is in free fall? (3)

London, AS/A level, Module Test PH1, June 1996

4 a State the period of the Earth about the Sun. Use this value to calculate the angular speed of the Earth about the Sun in $\mathrm{rad\,s^{-1}}$. (2)
  b The mass of the Earth is $5.98 \times 10^{24}\,\mathrm{kg}$ and its average distance from the Sun is $1.50 \times 10^{11}\,\mathrm{m}$. Calculate the centripetal force acting on the Earth. (2)
  c What provides this centripetal force? (1)

London, AS/A level, Module Test PH1, June 1997

5 A stone on a string is whirled in a vertical circle of radius 80 cm at a constant angular speed of 16 radians per second.
  a Calculate the speed of the stone along its circular path. (2)
  b Calculate its centripetal acceleration when the string is horizontal. (2)
  c Calculate the resultant acceleration of the stone at the same point. (3)
  d Explain why the string is most likely to break when the stone is nearest the ground. (2)

London, AS/A level, Module Test PH1, January 1997

6 A child sits on the edge of a roundabout of diameter 3.2 m. The roundabout rotates at a constant rate of one revolution every 3.5 s.
  a Calculate, for the child,
    i the speed,
    ii the angular speed,
    iii the magnitude of the acceleration and state its direction. (5)

b The child is holding a football which she releases from a position above the edge of the roundabout. With reference to horizontal and vertical components, describe the subsequent motion of the ball as it falls to the ground. (3)

NEAB, AS/A level, Module Test PH01, June 1998

7 The path of the Earth around the Sun can be treated as a circle of radius 1.50×10^8 km.

a By considering the distance travelled in one year, show that the average speed of the Earth around the Sun is approximately 3.0×10^4 m s^{-1}.

b Use your answer to part **a** to calculate the centripetal force of the Sun on the Earth.
Mass of the Earth $= 6.00 \times 10^{24}$ kg. (4)

NEAB, AS/A level, Module Test PH03, March 1998 (part)

8 A grinding wheel of diameter 0.12 m spins horizontally about a vertical axis, as shown in the diagram. P is a typical grinding particle bonded to the edge of the wheel.

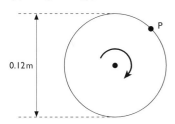

0.12 m

P

a If the rate of rotation is 1200 revolutions per minute, calculate
i the angular velocity,
ii the acceleration of P,
iii the magnitude of the force acting on P if its mass is 1.0×10^{-4} kg. (5)

b The maximum radial force at which P remains bonded to the wheel is 2.5 N.
i Calculate the angular velocity at which P will leave the wheel if its rate of rotation is increased.
ii If the wheel exceeds this maximum rate of rotation, what will be the speed and direction of motion of particle P immediately after it leaves the wheel? (4)

NEAB, AS/A level, Module Test PH01, June 1997

2 Oscillations

BACKGROUND

Oscillation is periodic motion. A bungee jumper coming to rest provides an example of oscillation in a straight line, and so does a piston in a car engine. Oscillating bodies, some more complex than others, include sources of sound like a guitar string, the layers of air in a wind instrument, a ruler twanged on the edge of a table, and the cone of a loudspeaker. The eardrum and the tiny hairs of the inner ear (Figure 2.1) that create responses in our brains are also oscillating systems.

Figure 2.1
Hair cells on the cochlea as imaged by an electron microscope. Your two cochleae, one in each ear, and their hair cells are physical systems that can produce different responses to different frequencies of oscillation.

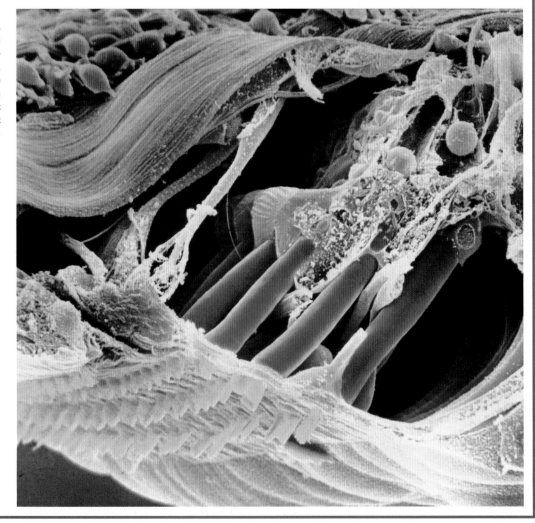

Travelling or progressive waves

Energy can be given to a rope by oscillation of one end. Then a wave travels down the rope. The same forces that hold the rope together ensure that oscillation of one section of the rope is passed to the next section. It takes time for the energy to pass from section to section, and the wave travels with a finite speed. The nature of these forces and the quantity of material involved – the density and the thickness of the rope – determine the flexibility of the rope and the speed of the wave.

The wave that travels along the rope, called a progressive wave, is carried by transverse oscillations – the direction of oscillation and the direction of travel are perpendicular. (See *Introduction to Advanced Physics*, Chapter 2.)

Transverse waves require shear forces between particles, and they can also be called shear waves (see Figure 2.2), or **S-waves**.

Travelling waves that carry sound

If a metal railing is hit with a hard object, then the particles within the railing and the object are moved by the impact. They are pulled back towards their equilibrium positions by the action of **restoring forces**, so that they vibrate around these positions. It is these interparticle forces that transmit oscillation from particle to particle, at a finite speed, just as for the wave on the rope. There is a difference, though, between the rope waves and the waves travelling within the body of the metal. In the metal the oscillation of the particles is partially parallel to the direction of travel of the wave. The oscillations have longitudinal as well as transverse components (Figure 2.2).

Figure 2.2
A solid can transmit both longitudinal (pressure) and transverse (shear) waves. Oscillation around an equilibrium position and transmission of oscillation from particle to particle are due to interparticle forces. In a gas, pressure waves are possible but shear waves are not.

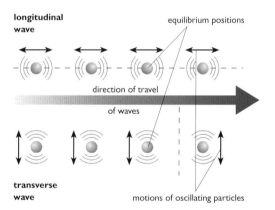

longitudinal wave

equilibrium positions

direction of travel

of waves

transverse wave

motions of oscillating particles

For a *longitudinal* wave, the force between neighbouring particles acts along the line shown (dashed line), creating pressure differences that move through the material. The oscillations are *parallel* to the direction of travel.

For a *transverse* wave, oscillation passes from particle to particle as a result of shear force, acting in the direction shown (dashed line). The oscillations are *perpendicular* to the direction of travel.

Shear forces can only be very small in a gas because of the very nearly independent behaviour of particles. Gases cannot transmit transverse waves to any significant extent. Sound waves in air are therefore longitudinal.

In a small volume of a gas that is carrying a longitudinal wave, the particles can be thought of as all oscillating together. Neighbouring volumes move further apart and closer together. This makes the gas denser in some places and less dense in others – it creates regions of compression and regions of rarefaction (Figure 2.3). These compressions and rarefactions travel, resulting in transmission of a sound wave through a gas. The wave is a pressure wave, or **P-wave**.

Figure 2.3
Sound waves in a gas are travelling compressions and rarefactions of the medium through which they travel.

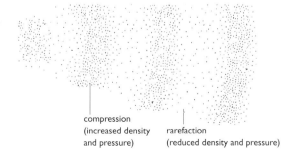

compression (increased density and pressure)

rarefaction (reduced density and pressure)

Seismic waves – transverse and longitudinal

Figure 2.4
The spread of waves from a source that is experiencing one-dimensional oscillation. Transverse waves subject the material to varying shear, and longitudinal waves produce varying pressure.

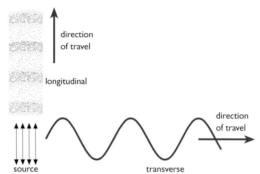

An earthquake releases energy stored by rock under very high tension, at a point (or epicentre) within the Earth's crust. It is possible to imagine a very simple energy release that results in a one-dimensional oscillation of material. Then transverse waves will spread strongly in directions perpendicular to this oscillation, and longitudinal waves will spread in a parallel direction (Figure 2.4). These waves due to earthquakes are **seismic waves**.

Of course, a real earthquake can never be so simple, and transverse and longitudinal waves spread from the epicentre in all directions. These are S-waves (shear waves – which are transverse) and P-waves (pressure waves – which are longitudinal). When oscillations reach the Earth's surface, then a third kind of wave travels along the surface – these are also transverse waves, called **L-waves** (or Love waves, after the scientist who studied them). It is the L-waves that cause most death and destruction.

The fact that S-waves and P-waves travel at different average speeds through the body of the Earth allows geologists to find the location of an epicentre by calculation using data from different monitoring stations (Figure 2.5).

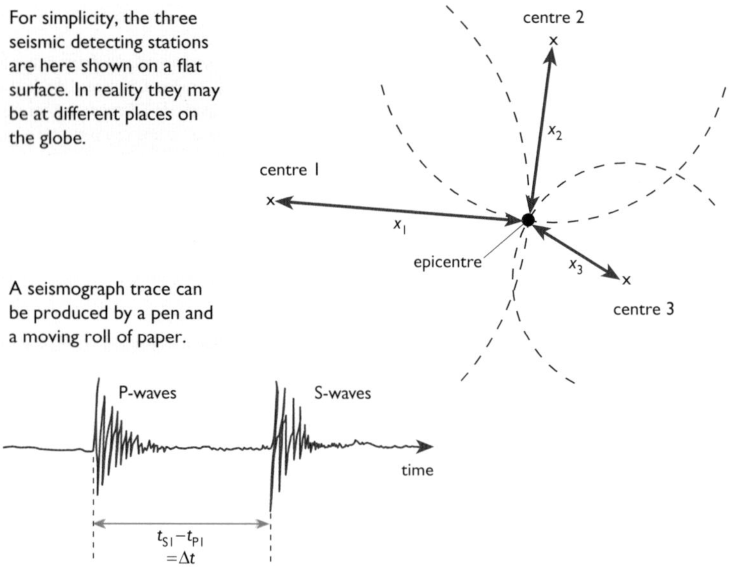

Figure 2.5
In an earthquake, S-waves and P-waves spread out from the epicentre at different speeds, so that scientists at a seismic detecting station can work out how far away they are from the epicentre.

For simplicity, the three seismic detecting stations are here shown on a flat surface. In reality they may be at different places on the globe.

centre 2

centre 1

epicentre

centre 3

A seismograph trace can be produced by a pen and a moving roll of paper.

P-waves   S-waves

time

$$t_{S1} - t_{P1} = \Delta t$$

For each seismic wave detecting centre, 1, 2 and 3, there is a time difference between the arrival of S-waves and P-waves from the earthquake epicentre. For example, for centre 1, the times taken for arrival of P-waves and S-waves are given by:

$$t_{P1} = \frac{x_1}{v_P}$$

and

$$t_{S1} = \frac{x_1}{v_S}$$

and the difference in arrival times is

$$t_{S1} - t_{P1} = \Delta t = \frac{x_1}{v_S} - \frac{x_1}{v_P}$$

$$\Delta t = x_1 \left(\frac{1}{v_S} - \frac{1}{v_P} \right)$$

Measurement of Δt and known values of v_S and v_P allow distance x_1 to be calculated. Data from the three centres can then be combined to find the epicentre location.

1 What factors limit the speed at which a wave travels along a rope?

2 Explain why an explosion in a gas results only in longitudinal waves, but an explosion underground results in longitudinal and transverse waves.

3 **a** P-waves travel between an epicentre and a detecting station at an average speed of $12.5\,\mathrm{km\,s^{-1}}$, while the average speed of S-waves for the same journey is $6.5\,\mathrm{km\,s^{-1}}$. What is the distance of the detecting station from the epicentre if the time lag for arrival of the two types of wave is $24\,\mathrm{s}$?
   b Sketch distance–time graphs, on the same axes, for a pulse of P-waves and a pulse of S-waves travelling from an earthquake epicentre at the speeds in **a**. Show how the distance calculated in **a** relates to the graph.

For the following three questions you might wish to refer to Chapter 2 in *Introduction to Advanced Physics*.

4 **a** List similarities and differences for travelling (progressive) waves, such as a pulse of waves on a rope, and stationary (standing) waves, such as the vibrations of a guitar string.
   b What causes the generation of stationary waves from travelling waves?

5 **a** In what circumstances might seismic waves generate stationary wave effects?
   b What might be the consequence of this for human and other populations?

6 Explain why the frequency of the sound produced by a vibrating air column in a pipe changes as one end of the pipe is opened and closed.

Free oscillation and natural frequency

If you flick a ruler on the edge of a table to produce the familiar twanging sound, then you set the ruler into **free oscillation**. You do not choose the frequency of the oscillation or of the sound. The ruler vibrates at a **natural frequency**, which is determined by the physical properties of the system.

You can, however, control the frequency indirectly by changing the physical properties of the system. The easiest way to do this would be to change the oscillating length of the ruler. Alternatively, you could use a ruler made of a different material. The ruler is acting as a simple musical instrument. Other physical systems, such as strings, columns of air and 'skins' under tension, freely oscillate at a natural frequency that is determined by their physical properties. Part of the skill of the musician is in altering these physical properties in order to create desired frequencies.

It is not only in musical instruments that systems show free oscillation. The frequencies of oscillations generated by an earthquake, for example, are determined by the properties of the system that oscillates – properties such as the density and stiffness of the rock.

Forced oscillation and resonance

A free oscillation may be induced by a single input of energy, and results in movement at the natural frequency. But oscillation can also be induced and maintained by a continuing source of energy. Then amplitude increases until the rate of input of energy is balanced by the rate of dissipation, when a state of equilibrium exists.

When you hold an operating hair-dryer, for example, the frequency of rotation of the hair-dryer motor has a tendency to force your arm to become an oscillating physical system. Your arm receives energy, from the hair-dryer, and also dissipates it. The rate of dissipation becomes equal to the rate of input when amplitude is still low. In a car, as another example, every physical system is forced to vibrate at the frequencies determined by the engine and the bumpiness of the road surface. Your arm and the car components and contents are subject to **forced oscillation**.

Figure 2.6
The amplitude of a system can be measured for a range of forcing frequencies. When forcing frequency equals the system's natural or resonant frequency, then amplitude shows a sharp peak. A graph that shows such a feature is called a **resonance curve**.

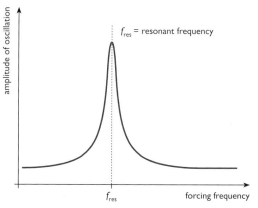

Most forced oscillation has low amplitude. But car components, for example, can begin to oscillate with high amplitude, perhaps causing audible rattling. This happens when the forcing frequency provided by the engine or the road is close to the natural frequency of the component. The effect is called **resonance** (Figure 2.6).

It is also possible that resonance of the arm holding the hair-dryer could take place. However, the frequency of vibration of the hair-dryer (the forcing frequency) is different from the natural frequency of the human arm, so resonance effects are avoided.

A lorry climbing a distant hill might be too far away to hear. The sound waves travelling through the air do not induce oscillation of our eardrums of sufficient amplitude for conscious sound. But a whole window may rattle with a frequency that is the same as that of the lorry engine. This is a resonance effect. The natural frequency or **resonant frequency** of the window is the same as the forcing frequency, which is the lorry engine frequency. Sound waves carry energy from the lorry to the window. Energy arrives at a low rate at the window, but for as long as it keeps on arriving faster than the window can lose it, the amplitude of vibration increases and we hear the window rattle.

Mechanical engineers usually take care to avoid resonance effects. If, to take one extreme example, a forcing frequency provided by a jet engine were to match the natural or resonant frequency of an aeroplane wing, the consequences would be dramatic and disastrous. The amplitude of oscillations of the wing would go on increasing for as long as the engine continued to provide energy faster than the wing could dissipate energy to the surroundings.

Civil engineers know that bridges are subject to resonance effects. The collapse of the Tacoma Narrows Bridge (Figure 2.7) in the 1940s was caused by resonant vibration initiated by the wind, and was captured on film. Other bridges, such as the Millennium Bridge (Figure 2.8) across the Thames in London, have had to be adapted, at great expense. For the Millennium Bridge, the frequency of human walking was close to a natural frequency of the structure. The problem was made worse because a small initial oscillation caused people to walk in step with each other, which caused still more oscillation, and so on.

Figure 2.7 (left)
The Tacoma Narrows Bridge collapse – resonance was created by wind.

Figure 2.8 (right)
The Millennium Bridge had to be adapted because of resonance caused by people walking.

Resonance is not always bad. It is not only the string of a violin that oscillates (Figure 2.9), but the body of the violin and the air inside. The shape means that there is not a single resonant frequency of these, but many. The string provides the forcing frequency, and oscillation takes place along different axes of the wood and air. The oscillations add together to give the sound its particular quality. An electrical circuit can also resonate (Figure 2.10) – see Chapter 6.

Figure 2.9
The strings oscillate, and so do the wood of the violin and the air inside. The shape of the violin therefore makes a big difference to the quality of the sound.

Figure 2.10
A radio aerial receives many frequencies of radio waves. The properties of the tuning circuit can be changed so that it 'resonates' at the frequency of the required radio station.

7 Does your voice put objects around you into free or forced vibration?
8 How is it possible for a voice of high frequency to shatter a drinking glass?
9 The frequencies of seismic waves are low – rarely more than a few hertz. Explain why, other than because of different construction techniques, earthquakes sometimes destroy buildings of certain sizes while leaving others in the same area relatively unharmed.

Oscillation of a mass on a spring

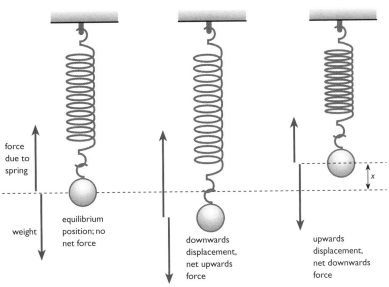

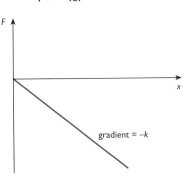

force
due to
spring

weight

equilibrium
position; no
net force

downwards
displacement,
net upwards
force

upwards
displacement,
net downwards
force

Figure 2.11 (above)
A mass on a spring is
subject to restoring forces
that act in the opposite
direction to the
displacement of the mass.

For laboratory investigation of oscillatory motion, it makes sense to use a simple system. A mass hanging from a vertical spring serves this purpose (Figure 2.11).

At its rest or equilibrium position the mass experiences a net force of zero. It is subject to its own weight, and to an equal upwards force that results from the extension of the spring. It will have an acceleration of zero whenever it is in that position.

If the mass is displaced a little downwards, then its weight stays the same but the upwards force increases due to the extra extension of the spring. There is a net upwards force. If the spring obeys Hooke's Law, which it will provided that it is not stretched too far, the size of this net force is proportional to the displacement of the mass.

Likewise if the mass is slightly displaced upwards, then the decrease in spring extension reduces the size of the upwards force. Net force is now downwards. And again, the size of the net force is proportional to the displacement.

Note also that the direction of the force is always in the opposite direction to the displacement. We can summarise the relationship between the net force, F, acting on the mass and the displacement, x, as follows:

$$F = -kx$$

Figure 2.12
For a mass on a spring, net
force is proportional to
displacement.

F

x

gradient = $-k$

This is a proportionality – a relatively simple relationship (Figure 2.12). The constant of proportionality, k, is the spring constant (or stiffness constant).

The motion of the oscillating spring is called **simple harmonic motion**. Any motion for which a force is proportional to displacement, and in the opposite direction, is simple harmonic motion.

Acceleration and force are related by $F = ma$, so for a mass on a spring acceleration is also proportional to displacement and in the opposite direction. We can say

$$ma = -kx$$

or

$$a = -\frac{k}{m}x$$

10 Explain why the ball-bearing on the runway shown (Figure 2.13) will not move in simple harmonic motion. (That is, show that it will not obey the relationship $F = -kx$.)

Figure 2.13

11 Upwards force acting on a body immersed (partly or completely) in a liquid is called upthrust. The upthrust acting on a weighted test tube floating upright in water is very nearly proportional to its depth of immersion. Will it move in simple harmonic motion when vertically displaced and then released? Explain your view fully by referring to the forces acting and the relationship of net force to displacement.

12 a Sketch graphs of acceleration, a, against displacement, x, for a body hanging from a vertical spring if its mass is i m, ii $2m$.
b Sketch graphs of acceleration, a, against force, F, for the same mass values.

Time dependence of the motion of a mass on a spring

The simple relationship between force and displacement, $F = -kx$, results in oscillation when the mass is released. Throughout the oscillation, force and displacement remain proportional, but they both vary with time. A relationship between displacement and time would provide useful new ways to model and analyse simple harmonic motion. To find such a relationship, we begin with the definition of acceleration expressed in mathematical form:

$$\text{acceleration} = \frac{dv}{dt} = \frac{d^2x}{dt^2}$$

This leads us to a relationship involving displacement and time and the constants k and m:

$$m\frac{d^2x}{dt^2} = -kx$$

The equation now contains displacement and its 'second-order differential with respect to time', d^2x/dt^2. It is a **differential equation**. (You will not be asked to deal with differential equations in your physics assessments. But as we have already seen, calculus is useful mathematics and you should be aware of its importance.)

On its own the differential equation provides little information about the relationship between displacement, x, and time, t. We need an equation that is not a differential equation. We can find this by **inspection** – by trying various formulae to see if they are consistent with the differential equation. There are trigonometric formulae that provide this match. They include:

$$x = x_0 \sin\sqrt{\frac{k}{m}}\,t$$

and

$$x = x_0 \cos\sqrt{\frac{k}{m}}\,t$$

where x_0 is a constant for any one example of simple harmonic motion. These relationships (Figure 2.14) between x and t are **solutions** of the differential equation (see *Comprehension and application*, page 35).

Note that sines and cosines have values that can only be between -1 and $+1$. So x varies between $-x_0$ and x_0; x_0 is the amplitude of the motion.

Figure 2.14 (below) Simple harmonic motion involves sinusoidal oscillation – displacement is related to time by trigonometric formulae involving sine and cosine.

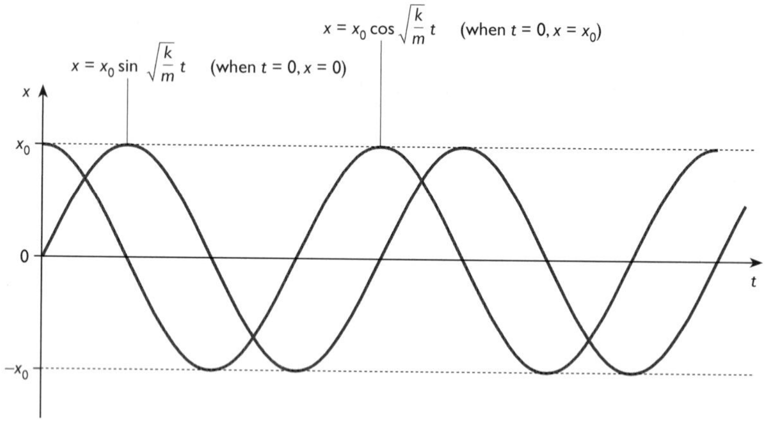

The relationships

$$x = x_0 \sin\sqrt{\frac{k}{m}}\,t \quad \text{and} \quad x = x_0 \cos\sqrt{\frac{k}{m}}\,t$$

show two possible forms of time dependence of the displacement of a body that is moving in simple harmonic motion. The only difference between the 'sin' and 'cos' formulae is the starting point. Both formulae and their graphs relate to the same amplitude and frequency.

13 The quantities $m\,d^2x/dt^2$ and $-kx$ are both forces. Explain why we are justified, above, in linking them by an equals sign.
14 a Write down the simple formula that defines simple harmonic motion.
  b Write down the same formula as a differential equation.
  c Write down one solution to the differential equation. (That is, write down a relationship between displacement and time for simple harmonic motion.)
15 For a mass that is displaced and then released, is

$$x = x_0 \sin\sqrt{\frac{k}{m}}\,t \quad \text{or} \quad x = x_0 \cos\sqrt{\frac{k}{m}}\,t$$

more appropriate?

Oscillations of atoms in pairs

A pair of masses on the ends of a spring (Figure 2.15) will both experience force due to extension and compression of the spring. The forces are equal. They are also proportional to displacement from the mean position, for each mass, and in the opposite direction to displacement. For each mass:

$$F = -kx$$

When x is zero, a mass is at its mean position and the force is also zero. So if the forces due to the spring are the only forces acting (such as will be the case in a weightless and resistance-free environment) the motion of each mass is simple harmonic motion.

Figure 2.15
The force–displacement graph for one of a pair of masses joined by a spring

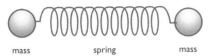

mass   spring   mass

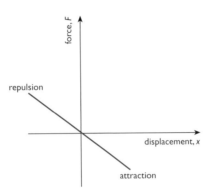

For a pair of atoms isolated from other atoms, the usual force–separation graph is as shown in Figure 2.16. The force acting on an atom is very nearly proportional to its displacement from its equilibrium position, provided that the displacement remains relatively small. Thus the oscillation of an atom in an atom pair approximates to simple harmonic motion.

Figure 2.16
The force–separation graph for one of a pair of atoms.

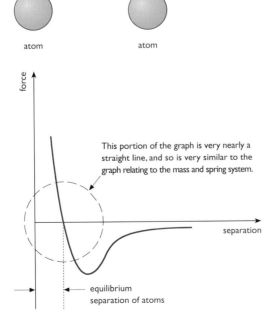

Simple harmonic motion and circular motion

Circular motion is periodic – with uniformly repeating patterns – as is simple harmonic motion. Trigonometric formulae allow us to create mathematical representations of periodicity for both types of motion. It is worth looking at ideas from circular motion to see if they can help us to describe simple harmonic motion.

Consider the motion of a body in a circle of radius x_0 with angular velocity ω, and consider just one component of the motion (Figure 2.17). This is like thinking about the shadow of the body projected on to a flat surface by a parallel beam of light. We could call this the motion of a 'linear projection' of the circular motion. It could be the vertical displacement of the body, for example, ignoring its horizontal displacement. We can represent such vertical motion and the corresponding circular motion together on the same diagram, with a rotating radius or **phasor**.

Figure 2.17
Circular motion and its linear projection.

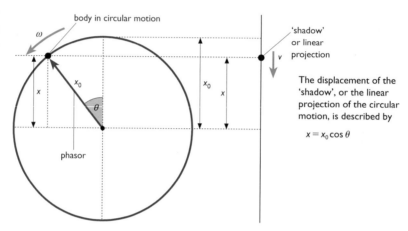

The displacement of the 'shadow', or the linear projection of the circular motion, is described by

$$x = x_0 \cos \theta$$

We can take a situation in which the initial vertical displacement is equal to the radius, x_0. We can say that the vertical displacement of the 'shadow' at some time later is x, and the angular displacement of the phasor at that time is θ; this angle θ can be called the **phase angle** of the motion. Then x and x_0 are related by:

$$\cos \theta = \frac{\text{adjacent}}{\text{hypotenuse}} = \frac{x}{x_0}$$

or

$$x = x_0 \cos \theta$$

If the body is moving in a circle at constant speed, then θ/t is its angular velocity, ω. So $\theta = \omega t$, and so for such circular motion we can say

$$x = x_0 \cos \omega t$$

We have seen that, for simple harmonic motion of a mass m,

$$x = x_0 \cos \sqrt{\frac{k}{m}}\, t$$

This is a graph of exactly the same form as $x = x_0 \cos \omega t$, and is identical with it if

$$\omega = \sqrt{\frac{k}{m}}$$

This equation, $\omega = \sqrt{(k/m)}$, relates the physical properties of an oscillating system to the angular velocity of the matching circular motion. But is it supported by observation of real systems? If it is, then simple harmonic motion can be described by the same mathematics as the motion of the linear projection of circular motion. We need to look at real systems to find out whether $\omega = \sqrt{(k/m)}$ is a reliable equation.

We know that, for circular motion, ω is related to period:

$$\omega = \frac{2\pi}{T} \quad \text{and} \quad T = \frac{2\pi}{\omega}$$

If $\omega = \sqrt{(k/m)}$ is a valid equation for simple harmonic motion, then it must also be true that, for an oscillating mass,

$$T = 2\pi\sqrt{\frac{m}{k}}$$

We can investigate the relationship $T = 2\pi\sqrt{(m/k)}$ by experiment (Figures 2.18 and 2.19).

Figure 2.18
An experiment with the same spring and different masses reveals that period, T, is proportional to the square root of mass, m. The gradient of the graph is $2\pi/\sqrt{k}$.

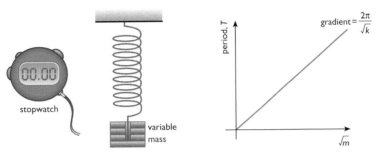

Figure 2.19
An experiment with the same mass and different springs of known spring constants reveals that period, T, is inversely proportional to the square root of the spring constant, k. The gradient of the graph is $2\pi \times \sqrt{m}$.

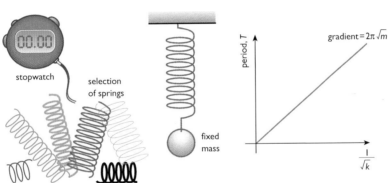

The graphs in Figures 2.18 and 2.19 confirm the relationship

$$T = 2\pi\sqrt{\frac{m}{k}} \quad \text{and therefore that} \quad \omega = \sqrt{\frac{k}{m}}$$

So for a body in simple harmonic motion whose initial displacement is maximum, x_0, we can now confidently use the same mathematical description as we would use for linear projection of circular motion. The formula

$$x = x_0 \cos \omega t$$

applies to simple harmonic motion as well as to circular motion (Figure 2.20).

Figure 2.20 (below)
An x–t (displacement–time) graph for simple harmonic motion, showing the relationship of ω to period, and also phasor motion.

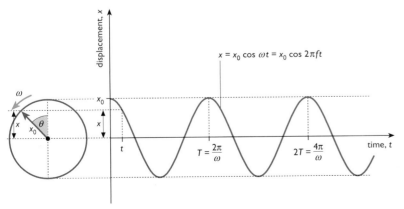

x_0 is the amplitude of the simple harmonic motion, and the radius of the corresponding circular motion. ω is the angular velocity of the matching circular motion. It is given by $\omega = \theta/t$. For one revolution, $\theta = 2\pi$ and $t = T$ (the period), so that

$$\omega = \frac{2\pi}{T}$$

and therefore

$$T = \frac{2\pi}{\omega}$$

and

$$2T = \frac{4\pi}{\omega}, \quad 3T = \frac{6\pi}{\omega}$$

and so on.

Frequency and period

We know that

$$T = \frac{2\pi}{\omega} \quad \text{and} \quad T = \frac{1}{f}$$

So

$$\omega = 2\pi f$$

Thus the equation

$$x = x_0 \cos \omega t$$

may also be written as

$$x = x_0 \cos 2\pi f t$$

Defining features and key formulae of simple harmonic motion revisited

We have seen that the defining formula for simple harmonic motion can be written as

$$F = -kx$$

or, in terms of acceleration,

$$a = -\frac{k}{m}x$$

Also, since,

$$\omega = \sqrt{\frac{k}{m}}$$

we can say that

$$a = -\omega^2 x$$

Again using the formula

$$\omega = 2\pi f$$

we obtain

$$a = -(2\pi f)^2 x$$

16 For simple harmonic motion with zero initial displacement, write down a relationship between displacement x and phase angle θ.

17 Explain, fully but using as few steps as possible, how $F = -kx$ leads to $a = -4\pi^2 f^2 x$.

18 If the motion of a mass of 0.5 kg on a spring has a period of 0.2 s, what is
  a its frequency
  b the spring constant, k
  c the value of the gradient of a graph of the acceleration of the mass (y-axis) plotted against displacement (x-axis)?

19 A mass of 0.3 kg on a spring, of spring constant 80 N m^{-1}, is displaced vertically by 0.05 m and released, neither losing nor gaining significant energy for several cycles of the motion.
  a What is the amplitude of the motion?
  b What is the frequency?

20 For a mass on a spring, what is the gradient of
  a a graph of period, T, against $1/\sqrt{k}$
  b a graph of ω against $\sqrt{k}$?

21 Suppose that measurements are made of period, T, and $1/\sqrt{k}$ for a spring and a graph is plotted, and that this is repeated several times with the same spring but different masses. How would the graphs vary? Sketch a family of such graphs.

Velocity and acceleration graphs for simple harmonic motion

Velocity–time graphs

The formula $x = x_0 \cos \omega t$ relates displacement to time, and the displacement–time graph is a sinusoidal curve. By differentiating a displacement–time relationship we obtain a velocity–time relationship (Figure 2.21). That is,

$$x = x_0 \cos \omega t$$

gives us

$$\frac{\mathrm{d}x}{\mathrm{d}t} = -x_0 \omega \sin \omega t = v$$

The most important point about differentiation here is that it is a way of taking a relationship for x and using it to find a relationship for its own gradient, $\mathrm{d}x/\mathrm{d}t$. At any instant, the value of velocity ($\mathrm{d}x/\mathrm{d}t$) is *always* given by the gradient of the graph of displacement (x) against time (t).

Figure 2.21
Displacement–time and velocity–time graphs for simple harmonic motion.

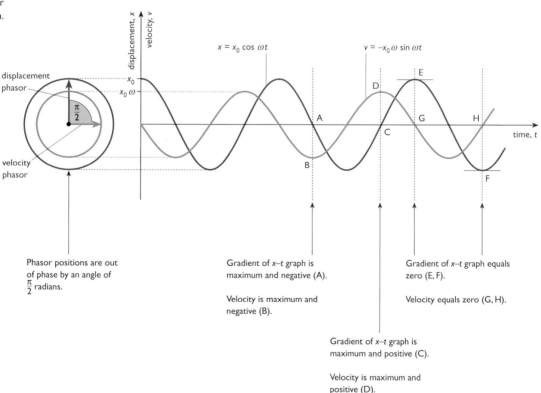

Phasor positions are out of phase by an angle of $\frac{\pi}{2}$ radians.

Gradient of x–t graph is maximum and negative (A).

Velocity is maximum and negative (B).

Gradient of x–t graph equals zero (E, F).

Velocity equals zero (G, H).

Gradient of x–t graph is maximum and positive (C).

Velocity is maximum and positive (D).

Note that ω is a constant for any particular case of simple harmonic motion, and so $x_0\omega$ is also a constant. The maximum displacement is x_0, and the maximum value of $\sin \omega t$ is 1, so the maximum value of the velocity is $x_0\omega$. Since $\omega = 2\pi f$, this maximum velocity can also be written as $2\pi f x_0$. Note also that displacement x and velocity v differ greatly at the start of the motion – displacement is zero when velocity is maximum, and so on. We can describe them as being $\pi/2$ out of phase, or say that there is a **phase difference** of $\pi/2$.

Acceleration–time graphs

We have seen how displacement–time and velocity–time relationships are linked. We can take this a step further, and consider acceleration–time relationships. We know that the value of an acceleration at any instant is equal to the gradient of the velocity–time (v–t) graph. By differentiating the velocity–time relationship, we obtain the relationship between acceleration and time.

Acceleration, by definition, is equal to $\dfrac{dv}{dt}$. Differentiating the expression for v on page 27, we have

$$\frac{dv}{dt} = -x_0\omega^2 \cos \omega t$$

or

$$a = -x_0\omega^2 \cos \omega t$$

Since the maximum value of $\cos \omega t$ is 1, the maximum value of acceleration is $x_0\omega^2$. Again, because $\omega = 2\pi f$, this maximum acceleration can also be written as $(2\pi f)^2 x_0$. Note that, since

$$a = -\omega^2 x$$

displacement is zero when acceleration is zero, but the negative sign shows that displacement and acceleration are in the opposite direction. We can describe them as π out of phase, as is shown by the graphs and the associated phasors (Figure 2.22).

Figure 2.22
Displacement–time and acceleration–time graphs for simple harmonic motion.

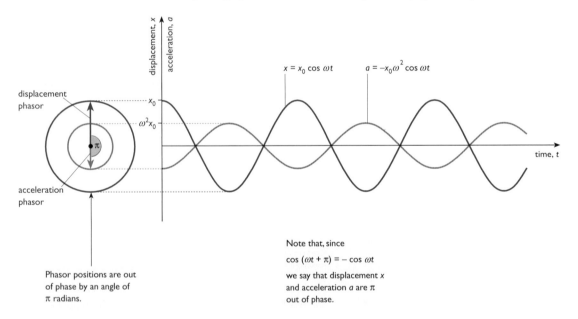

Phasor positions are out of phase by an angle of π radians.

Note that, since

$\cos (\omega t + \pi) = -\cos \omega t$

we say that displacement x and acceleration a are π out of phase.

22 For a mass on a spring, explain with the aid of a sketch at what stages in its motion it will have:
   a maximum downwards velocity
   b zero velocity
   c maximum downwards acceleration
   d zero acceleration.
23 At what point, relative to displacement, does a body in simple harmonic motion have
   a maximum velocity
   b zero velocity?
24 Sketch a displacement–time graph for a body in simple harmonic motion. On the same axes sketch a graph to show the variation of force acting on the body with time.
25 What is the value of time t after four cycles of motion,
   a in terms of the period T
   b in terms of ω?

Damping

Dissipation of energy from an oscillating physical system, unless matched by an energy input, causes the amplitude to decay. Oscillation that experiences such decay, or fails to experience an increase in amplitude when there is a net supply of energy to it, is said to be **damped motion**.

Figure 2.23
Exponential decay of amplitude.

For a mass on a spring, for example, air resistance and friction and heating of the spring all result in dissipation of energy from the oscillation. Addition of a vane to increase air resistance, for example, increases the level of damping. Heavier damping would be provided by immersing the mass in a more resistive medium such as water or oil.

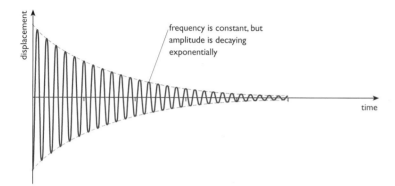

frequency is constant, but amplitude is decaying exponentially

In decaying damped simple harmonic motion, the amplitude is time dependent. A common form of time dependence, for a system that receives an initial input of energy and then oscillates freely, is exponential (Figure 2.23), just like the decay of a population of radioactive nuclei or the decay of the charge on a capacitor. In this case the dependence of amplitude on time is given by

$$x_0 = x_{01}\, e^{-ct}$$

where x_{01} is the initial amplitude and c is a constant that is related to the degree of damping.

With light damping (Figure 2.24a), the oscillation dies away slowly, and here amplitude may decay exponentially. Larger resistive force can prevent oscillation altogether. In the case of heavy damping (Figure 2.24b), the resistive force is so large that the oscillating body can return only slowly to its equilibrium position. For **critical damping** (Figure 2.24c), the resistive force is still quite large but allows the body to return to its equilibrium position in the minimum possible time. Suspension systems (Figure 2.25, overleaf) depend on damping for successful operation.

Figure 2.24
Graphs of displacement against time for lightly, heavily and critically damped oscillations.

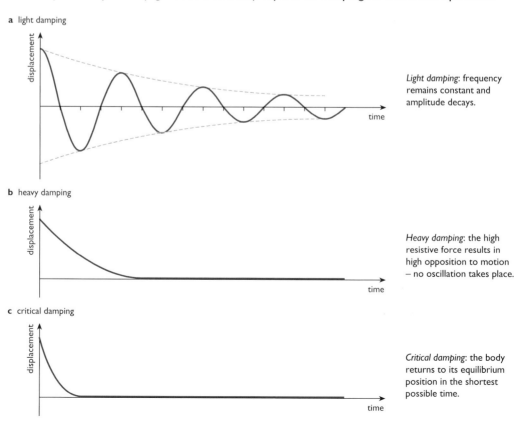

a light damping

Light damping: frequency remains constant and amplitude decays.

b heavy damping

Heavy damping: the high resistive force results in high opposition to motion – no oscillation takes place.

c critical damping

Critical damping: the body returns to its equilibrium position in the shortest possible time.

Figure 2.25
The suspension systems of ordinary vehicles have springs and shock absorbers. The shock absorbers contain a piston in a viscous liquid. The piston experiences quite large resistive force against its motion, and damps the oscillations of the spring.

The level of damping is important – it mustn't be so light that the spring (and the vehicle) continues to oscillate for some time after passing over each bump in the road. It mustn't be so heavy that the motion of the spring is too inhibited, or there will be very little benefit in terms of cushioning of the ride.

The effect of damping on resonance

Damping increases the rate of dissipation of energy by an oscillating system. So, as a forcing mechanism feeds energy into an oscillating system, the increased rate of energy dissipation means that a balance between rate of input and rate of dissipation of energy is reached at a relatively low amplitude. Damping thus reduces resonance effects.

To prevent troublesome rattling in a car, for example, screws may be used with energy-absorbing washers or set into energy-absorbing housings. The damping mechanism – whether a vane on a mass on a spring, or the liquid in a vehicle shock absorber, or a rubbery washer – is itself part of the oscillating system. Inevitably it changes the overall physical nature of the system. The effect is almost always to reduce resonant frequency (Figure 2.26).

Figure 2.26
Damping diminishes resonant effects. Heavy damping also alters the nature of the oscillating system enough to affect the value of the resonant frequency.

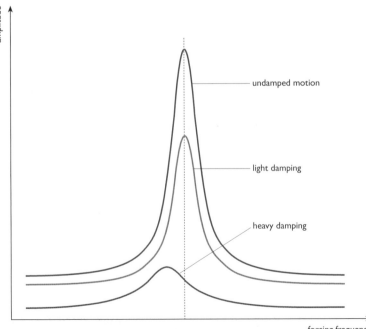

amplitude

undamped motion

light damping

heavy damping

forcing frequency

With light damping, the amplitude peak is smaller but the resonant frequency is unchanged.

With heavy damping, the resonance effect is greatly reduced, and the resonant frequency is changed.

The pendulum as a simple harmonic system

If we can show that the net force acting on a body is proportional to but in the opposite direction to displacement, then we can say that the body must be in simple harmonic motion. So can a pendulum's swing be thought of as simple harmonic motion?

The forces acting on a pendulum bob that is displaced from its rest or equilibrium position are its weight and the tension in the string (Figure 2.27). These are not in opposite directions – the tension has a horizontal as well as a vertical component.

Figure 2.27
At small angular displacement ϕ of a pendulum, the horizontal component of tension in the string provides a restoring force, while vertical forces can remain so very nearly in balance that we can assume vertical acceleration to be zero.

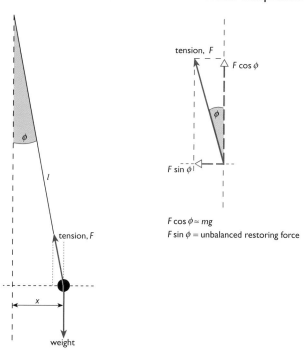

$F \cos \phi \approx mg$
$F \sin \phi$ = unbalanced restoring force

When the pendulum is at its equilibrium position, tension and weight are balanced:

tension, $F = -mg$

For displacements that are small relative to the length of the string, the vertical component of tension remains very closely in balance with weight:

tension, $F \sim F \cos \phi \approx -mg$

We can assume that $F \approx -mg$ can be taken to be $F = -mg$ provided that angle ϕ remains small. The result is that vertical acceleration of the pendulum can then be ignored.

The horizontal component of tension is unbalanced, and is given by $F \sin \phi$. This acts in the opposite direction to the displacement – it is a restoring force. Note also that

$$\sin \phi = \frac{\text{opposite}}{\text{hypotenuse}} = \frac{x}{l}$$

where x is the horizontal displacement and l is the length of the pendulum. So

$$F \sin \phi = \frac{Fx}{l} = -\frac{mgx}{l}$$

Thus, provided that the weight of the bob and the length of the pendulum are constant, and that the amplitude of the swings is small compared with the pendulum length,

restoring force, $F \sin \phi \propto -x$

Thus net force is proportional to and in the opposite direction to displacement. This means that a small-amplitude pendulum swing approximates to simple harmonic motion.

Use of the pendulum for measurement of *g*

Note that, since

$$\text{horizontal force} = F\sin\phi = -\frac{mgx}{l}$$

$$\text{horizontal acceleration} = \frac{F\sin\phi}{m} = -\frac{gx}{l}$$

But for a body in simple harmonic motion,

$$\text{horizontal acceleration} = -\omega^2 x$$

So,

$$-\frac{gx}{l} = -\omega^2 x$$

$$\omega = \sqrt{\frac{g}{l}}$$

Also, we know that

$$T = \frac{2\pi}{\omega}$$

where *T* is the period. So for a pendulum *T* is related to *g* and *l* by

$$T = 2\pi\sqrt{\frac{l}{g}}$$

We can use this as a starting point for measurement of gravitational field strength (and acceleration due to gravity), *g*, as outlined in Figure 2.28.

Figure 2.28
Measurement of *g* from the graph of T^2 against *l* for a pendulum.

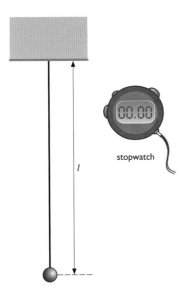

stopwatch

l

The period of swing of a pendulum is

$$T = 2\pi\sqrt{\frac{l}{g}}$$

and squaring both sides gives

$$T^2 = \frac{4\pi^2}{g}l$$

This formula predicts that a graph of T^2 (y-axis) against *l* (x-axis) will be a straight line through the origin, with gradient $4\pi^2/g$, from which *g* can be found.

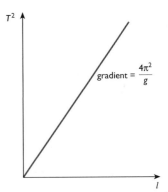

$$\text{gradient} = \frac{4\pi^2}{g}$$

26 For a pendulum in a laboratory, what could you add to increase the degree of damping? What effect would you expect this to have on the frequency?

27 You buy a grandfather (pendulum) clock that keeps perfect time at a place in the world where $g = 9.81\,\text{N kg}^{-1}$. You get home and discover that as a result of variation in the value of *g* the clock loses time – that is, its period has increased. What changes could you make to restore the clock's accuracy? Explain whether your home has a value of *g* bigger or smaller than $9.81\,\text{N kg}^{-1}$.

Energy of bodies in simple harmonic motion

Variation of potential energy and kinetic energy with time

Clearly a body in simple harmonic motion does not have constant kinetic energy, or constant potential energy. For the mass on a spring, the potential energy is partly gravitational and partly elastic due to deformation of the spring. For the pendulum, the potential energy is gravitational.

If a system is damped, then energy leaks away from it. The energy is dissipated, and the amplitude of the motion decreases. But for an undamped system, energy does not leak away. If energy does not flow into or out of the system then its total energy is constant, E.

Consider the kinetic energy of such a system,

$$E_k = \tfrac{1}{2}mv^2$$
$$= \tfrac{1}{2}mx_0^2\omega^2 \sin^2 \omega t$$

The maximum value of $\sin^2 \omega t$ is 1. So,

$$E_{k,\,max} = \tfrac{1}{2}mx_0^2\omega^2$$

At the times when the mass m is passing through its equilibrium position, potential energy is a minimum and kinetic energy is a maximum. At these times, *all* of the energy associated with the oscillation is kinetic. So maximum kinetic energy is equal to total energy, E:

$$E = \tfrac{1}{2}mx_0^2\omega^2$$

The potential energy of the oscillating system, E_p, is given by

$$E_p = E - E_k$$
$$= \tfrac{1}{2}mx_0^2\omega^2 - \tfrac{1}{2}mx_0^2\omega^2 \sin^2 \omega t$$
$$= \tfrac{1}{2}mx_0^2\omega^2(1 - \sin^2 \omega t)$$

We can show these relationships on a graph (Figure 2.29).

Figure 2.29
Total energy, kinetic energy and potential energy for simple harmonic motion, plotted against time.

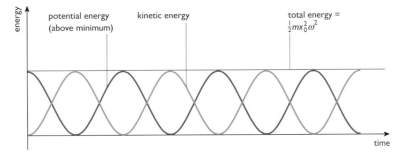

Variation of potential energy and kinetic energy with displacement

The potential energy stored by a system is equal to the area under a graph of force against distance. Now, since simple harmonic motion obeys the equation

$$F = -kx$$

such a graph for this motion will always be a straight line passing through the origin (Figure 2.30).

Figure 2.30
A graph of force against displacement is always a straight line through the origin for simple harmonic motion.

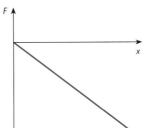

The potential energy stored is given by

$$E_p = \tfrac{1}{2}Fx$$
$$= \tfrac{1}{2}kx \cdot x \qquad \text{(ignoring the minus sign)}$$
$$= \tfrac{1}{2}kx^2$$

This is shown in Figure 2.31. The maximum potential energy of a body in simple harmonic motion occurs when displacement is maximum. That is, when displacement, x, is equal to amplitude, x_0. So

$$E_{p,\,max} = \tfrac{1}{2}kx_0^2$$

When potential energy is maximum, kinetic energy is zero, and the potential energy is equal to the total energy, E, of the oscillation:

$$E = \tfrac{1}{2}kx_0^2$$

Figure 2.31
The variation of potential energy and kinetic energy with displacement from the equilibrium position, for simple harmonic motion.

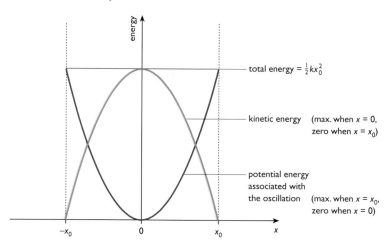

28 For a grandfather clock, amplitude is independent of time. What does the statement 'amplitude is independent of time' say about the energy of the oscillation?

The kinetic energy of the system (also shown in Figure 2.31) is given by

$$E_k = E - E_p$$
$$= \tfrac{1}{2}kx_0^2 - \tfrac{1}{2}kx^2$$
$$= \tfrac{1}{2}k(x_0^2 - x^2)$$

A formula for velocity in simple harmonic motion

We know that kinetic energy is given by $E_k = \tfrac{1}{2}mv^2$, so we can substitute this into the equation given above:

$$\tfrac{1}{2}mv^2 = \tfrac{1}{2}k(x_0^2 - x^2)$$

29 Use the formula $v = \omega\sqrt{(x_0^2 - x^2)}$ to
a show under what conditions $v = 0$
b show under what conditions v is maximum, stating an equation for this maximum
c calculate the maximum velocity for simple harmonic motion with frequency 20 Hz and amplitude 0.008 m.

We can then rearrange this to give a formula for velocity in terms of frequency, amplitude and displacement, as follows:

$$v^2 = \frac{k}{m}(x_0^2 - x^2)$$

$$v = \sqrt{\frac{k}{m}}\,\sqrt{(x_0^2 - x^2)}$$

$$= \omega\sqrt{(x_0^2 - x^2)}$$

$$= 2\pi f\sqrt{(x_0^2 - x^2)}$$

More mathematics and simple harmonic motion

We can show that

$$x = x_0 \cos \sqrt{\frac{k}{m}}\, t$$

is a solution of the differential equation

$$m\,\frac{d^2x}{dt^2} = -kx$$

by differentiating twice, as follows. That is, if

$$x = x_0 \cos \sqrt{\frac{k}{m}}\, t$$

then

$$\frac{dx}{dt} = -x_0 \sqrt{\frac{k}{m}} \sin \sqrt{\frac{k}{m}}\, t$$

and

$$\frac{d^2x}{dt^2} = -x_0\,\frac{k}{m} \cos \sqrt{\frac{k}{m}}\, t$$

(If you are not studying maths you will have to simply accept these differentiations.) But, if we substitute our new formula for d^2x/dt^2 into our initial differential equation we get:

$$m\left(-x_0\,\frac{k}{m}\cos\sqrt{\frac{k}{m}}\,t\right) = -kx$$

which simplifies to

$$-x_0\cos\sqrt{\frac{k}{m}}\,t = -x$$

or

$$x = x_0\cos\sqrt{\frac{k}{m}}\,t$$

So by making the guess that $x = x_0 \cos \sqrt{(k/m)}\, t$, unlikely as such a guess might seem, and feeding this guess into our starting equation, $m\,d^2x/dt^2 = -kx$, we end up with the conclusion that $x = x_0 \cos \sqrt{(k/m)}\, t$. This would not be enough to *prove* that $x = x_0 \cos \sqrt{(k/m)}\, t$ is the only correct formula relating x to t, but the formula is one member of a family of formulae that work when we perform the differentiations.

Note that the formula

$$x = x_0 \sin(\omega t + \phi)$$

would also be consistent with the initial differential equation,

$$m\,\frac{d^2x}{dt^2} = -kx$$

In this case ϕ represents an **initial phase angle**, which relates to the position of the mass when it is first released (Figure 2.32, overleaf).

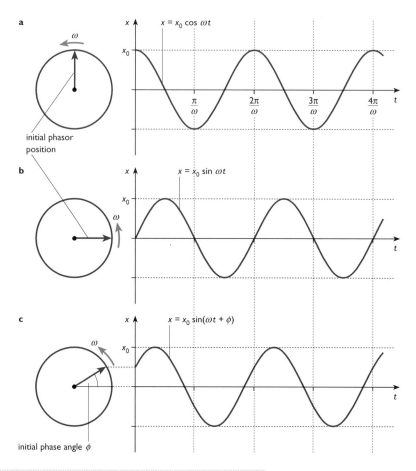

Figure 2.32
Simple harmonic motion with and without an initial phase angle.

a $x = x_0 \cos \omega t$

initial phasor position

b $x = x_0 \sin \omega t$

c $x = x_0 \sin(\omega t + \phi)$

initial phase angle ϕ

---

30 a i Express each of these times as multiples of period T:

$t = 6\pi/\omega$ $t = 7\pi/\omega$ $t = 20\pi/\omega$ $t = 21\pi/\omega$

ii Use Figure 2.32a to predict the value of x at each of the times in part **a i**.
b For Figure 2.32b, sketch a corresponding graph of velocity against time.
c For Figure 2.32c,
i predict the displacement phase angle at each of the times given in **a i**,
ii sketch a corresponding acceleration–time graph.

31 Given that $m \, d^2x/dt^2 = -kx$, if you are reasonably confident with differentiation, then try the following relationships between x and t to see if they provide a nonsense or a possible answer:

a $x = \dfrac{k}{m} t^2$

b $x = x_0 \cos \dfrac{k}{m} t$

● **Comprehension and application**

Loudspeaking

A loudspeaker cone oscillates in a sinusoidal manner, as in simple harmonic motion, when the current through its coil is varying sinusoidally. Of course, most speech and music is carried by waves that are more complex than a simple sinusoidal pattern. Nevertheless, it is useful to compare the behaviour of loudspeakers with simple harmonic motion.

The loudspeaker cone is subject to a restoring force, and if the cone is displaced slightly this force is approximately proportional to and in the opposite direction to the displacement. A freely moving loudspeaker cone oscillates in simple harmonic motion, with significant damping. However, the oscillations of the loudspeaker are not free oscillations but are forced by the interaction between the current in the coil and the surrounding permanent magnet.

A woofer is a loudspeaker with a large cone – able to oscillate at a range of low sound frequencies. A tweeter is a small loudspeaker that is better for producing high-frequency sounds. Some hi-fi speaker systems have a third loudspeaker to generate mid-range frequencies, between 500 Hz and 4 kHz. So that each loudspeaker only has to deal with a limited range of frequencies, a speaker system contains circuitry that transmits the suitable range of frequencies to each speaker. This has the further advantage of allowing independent control of the volumes of the three frequency ranges.

As a loudspeaker cone moves forwards it compresses the air in front of it and at the same time the air behind becomes rarefied. As it moves backwards the rarefaction is in front and the compression is behind. These pairs of compressions and rarefactions, originating so close together, can produce considerable cancellation of each other, according to the law of superposition of waves. Some high-quality speakers have a transmission line (or tunnel) that directs compressions and rarefactions that originate behind the oscillating cone through an extended pathway, so that they emerge at the front of the unit in phase with their partner sound waves that travel directly from the front of the cone.

32 Under what circumstances does a loudspeaker cone move sinusoidally?

33 By referring to a loudspeaker cone, distinguish between free and forced vibrations.

34 a On suitable axes, sketch displacement–time, velocity–time and acceleration–time graphs for a woofer emitting a single sinusoidal note of frequency 100 Hz.
b Explain, with further graphs if you wish, and by referring to suitable mathematical relationships, why the cone of a tweeter experiences higher accelerations than that of a woofer even at significantly lower amplitudes.

35 Explain why it is impossible for a loudspeaker cone to move in such a way as to produce perfect 'square wave' sound for which the displacement–time graph of a layer of air is as shown in Figure 2.33. Use graphs to aid your explanation.

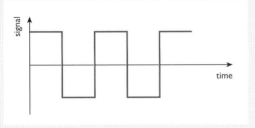

Figure 2.33 A square wave.

36 If a loudspeaker cone moves in simple harmonic motion at a frequency of 400 Hz at an amplitude of 0.5 mm,
a what is its maximum speed and maximum acceleration during the motion?
b what is the maximum force experienced if the oscillating mass is 0.020 kg?

37 Does the problem of superposition of waves originating from the back and front of a loudspeaker cone have any dependence on the frequency of the sound? Explain.

⬤ **Extra skills task** Communication

A pendulum can be used to measure the gravitational field strength, g.

1 Make notes, after research where needed, on how the value of g is relevant in *three* of the following fields:

- aircraft design
- design of large bridges or buildings
- water supply engineering
- one area of sports science
- animal size, shape and movement (you could mention giraffes, whales, birds, spiders)
- atmospheric science.

In your notes consider how
a our understanding of the world,
b our everyday lives,
would be different if we had no knowledge of the value of g.

2 Use your notes and thoughts to prepare a presentation for non-physics students which should persuade them of the importance of knowledge of the value of g to a wide range of human activities. Make extensive use of visual aids and include a demonstration of how the pendulum can be used to measure g.

Examination questions

1 **a** *Oscillation* of moving parts is essential in many everyday devices. In some cases the oscillations are *forced*, in some *resonance* occurs and in some the oscillations need to be *damped*.
  i What is meant by the four terms in italics?
  ii Give two different examples of oscillatory motion from everyday life. Discuss each example with reference to the terms above. (15)

 b Below is a graph of the variation with time of the displacement of an oscillating object.

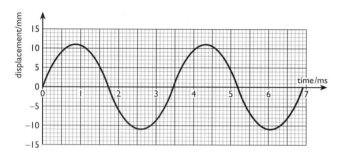

 Determine, for this oscillation,
  i the amplitude,
  ii the period,
  iii the frequency,
  iv the maximum speed of the oscillating object. (6)

 UCLES, AS/A level, 4831, June 1998

2 A spring is clamped at its upper end and a mass of 180 g is suspended from the lower end. The mass is set into vertical oscillation. The diagram shows the variation with time *t* of the velocity *v* of the mass.

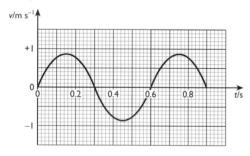

 a Mark with an X all the times at which the acceleration of the mass is zero. (2)
 b Show that the maximum kinetic energy of the mass is 0.065 J. (2)
 c On axes like the following, sketch graphs to show the variation with time *t* of
  i the kinetic energy of the mass (label this graph K),
  ii the potential energy of the mass (label this graph P).
  Assume that the potential energy is zero when the mass is in equilibrium. (5)
 d Explain why the shape of your graph P is determined by the shape of your graph K. (1)

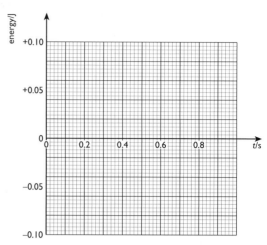

UCLES, A level, 4833, March 1998

3 **a** A simple pendulum has a period of 2.0 s and oscillates with an amplitude of 10 cm. What is the frequency of the oscillations? (1)
 b At what point of the swing is the speed of the pendulum bob a maximum? Calculate this maximum speed. (3)
 c At what points of the swing is the acceleration of the pendulum bob a maximum? Calculate this acceleration. (3)

London, A level, Module Test PH4, January 1996

4 **a** A body oscillates with simple harmonic motion. Sketch a graph to show how the acceleration of the body varies with its displacement. (2)
 b How could the graph be used to determine *T*, the period of oscillation of the body? (2)
 c A displacement–time graph for simple harmonic motion is drawn below.

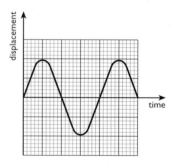

The movement of tides can be regarded as simple harmonic, with a period of approximately 12 hours.

On a uniformly sloping beach, the distance along the sand between the high water mark and the low water mark is 50 m. A family builds a sand castle 10 m below the high water mark while the tide is on its way out. Low tide is at 2.00 p.m.

On the graph

i label points L and H, showing the displacements at low tide and the next high tide,

ii draw a line parallel to the time axis showing the location of the sand castle,

iii add the times of low and high tide. (3)

d Calculate the time at which the rising tide reaches the sand castle. (3)

London, A level, Module Test PH4, June 1996

5 One simple model of the hydrogen molecule assumes that it is composed of two oscillating hydrogen atoms joined by two springs as shown in the diagram.

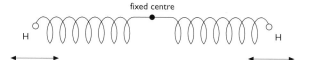

fixed centre

a If the spring constant of each spring is $1.13 \times 10^3 \, \text{N m}^{-1}$, and the mass of a hydrogen atom is $1.67 \times 10^{-27} \, \text{kg}$, show that the frequency of oscillation of a hydrogen atom is $1.31 \times 10^{14} \, \text{Hz}$. (2)

b Using this spring model, discuss why light of wavelength $2.29 \times 10^{-6} \, \text{m}$ would be strongly absorbed by the hydrogen molecule. (4)

London, A level, Module Test PH4, June 1997

6 A mass is suspended from a spring. The mass is then displaced and allowed to oscillate vertically. The amplitude of the oscillations is 6.0 mm. The period of the oscillations is 3.2 s.

a Calculate the maximum acceleration of the mass. (3)

b Sketch a graph showing how the acceleration of the mass varies with displacement. Add a scale to both axes. (4)

c State and explain *one* reason why the mass may not oscillate with simple harmonic motion. (2)

Edexcel, A level, Module Test PH4, June 2000

7 The graph shows the variation of acceleration a with displacement x for a body oscillating with simple harmonic motion.

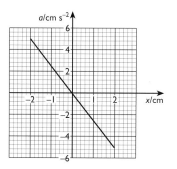

a Calculate the period of oscillation of the body. (2)

b At time $t = 0$ the body is momentarily at rest. On axes like those below, sketch a graph to show how acceleration of the body varies with time. Add a scale to the acceleration axis. (4)

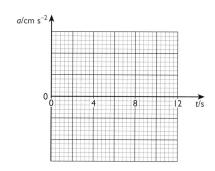

London, AS/A level, Module Test PH2, June 1999

8 The diagram shows a method for determining the mass of small animals orbiting the Earth in Skylab. The animal is securely strapped into a tray attached to the end of a spring. The tray will oscillate with simple harmonic motion when displaced as shown in the diagram and released.

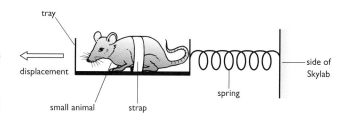

a Define *simple harmonic motion*. (2)

b The tray shown above has a mass of 0.400 kg. When it contains a mass of 1.00 kg, it oscillates with a period of 1.22 s. Calculate the spring constant k. (3)

c The 1.00 kg mass is removed and a small animal is now strapped into the tray. The new period of oscillation is 1.48 s. Calculate the mass of the animal. (2)

d The Skylab astronauts suggest that the calibration experiment with the 1.00 kg mass could have been carried out on Earth before take off. If a similar experiment were conducted on Earth would the time period be greater than, less than, or equal to 1.22 s? Explain your answer. (3)

London, AS/A level, Module Test PH2, June 1998

9 Diagram A shows a mass suspended by an elastic cord. The mass is pulled downwards by a small amount and then released so that it performs simple harmonic oscillations of period T. Diagrams B–F show the positions of the mass at various times during a single oscillation.

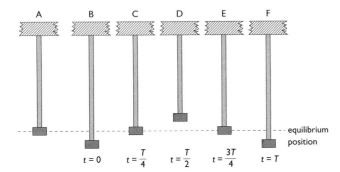

a Complete the table below to describe the displacement, acceleration and velocity of the mass at the stages B–F, selecting appropriate symbols from the following list:

maximum and positive   →   +
maximum and negative   →   −
zero   →   0

Use the convention that *downward* displacements, accelerations and velocities are positive.

	B	C	D	E	F
Displacement					
Acceleration					
Velocity					

(4)

In the sport of bungee jumping, one end of an elastic rope is attached to a bridge and the other end to a person. The person then jumps from the bridge and performs simple harmonic oscillations on the end of the rope.

People are bungee jumping from a bridge 50 m above a river. A jumper has a mass of 80 kg and is using an elastic rope of unstretched length 30 m. On the first fall the rope stretches so that at the bottom of the fall the jumper is just a few millimetres above the water.

b Calculate the decrease in gravitational potential energy of the bungee jumper on the first fall. (2)
c What has happened to this energy? (1)
d Calculate the force constant k, the force required to stretch the elastic rope by 1 m. (3)
e Hence calculate T, the period of oscillation of the bungee jumper. (2)

London, AS/A level, Module Test PH2, January 1998

10 This question is about the behaviour of various quantities during simple harmonic motion.

Suppose an object undergoes simple harmonic motion along a horizontal straight line between points X and Y as indicated:

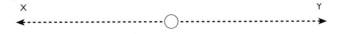

The diagram below represents a time sequence of the position of the body during one complete oscillation.

time/s
0.0
0.1
0.2
0.3
0.4
0.5
0.6
0.7
0.8
0.9
1.0
1.1
1.2

a Sketch the diagram and draw in vectors to represent (in direction and relative magnitude) the *velocity* of the body at each position. (2)
b Sketch the diagram again. Draw in vectors to represent (in direction and relative magnitude) the *resultant force* on the body at each position. (2)
c On your second diagram, label the positions which satisfy the conditions listed below:

Condition	Labels
Maximum elastic energy	A
Minimum elastic energy	B
Maximum kinetic energy	C
Zero kinetic energy	D

(There may be more than one position satisfying each condition.)
(2)

IB, Standard level 3, Paper 430, November 1998

3 Gravitational and electric fields

THE BIG QUESTIONS
- What are the similarities and differences between electrical and gravitational phenomena?
- How can seemingly abstract quantities be used to give us new ways to understand real situations?

KEY VOCABULARY

Coulomb's Law   dark matter   equipotential line   equipotential surface escape speed   Newton's Law of Gravitation   parabolic path permittivity of a vacuum   potential hills   potential wells   radial field uniform field   universal gravitational constant

BACKGROUND

Many of the ideas in this chapter may seem very abstract at first sight – seeming to have very modest practical importance. But the understanding of fields is fundamental to understanding all interactions between bodies, whether they are particles or galaxies.

Whether travelling in space, bouncing on a trampoline or just resting in bed, a human body experiences forces that have gravitational and electrical origins (Figure 3.1). At different scales, fields exist that govern every move you make, or don't make.

Figure 3.1
Particle–particle electric forces and planet–human gravitational forces are in balance until the diver jumps, and then the gravitational field becomes dominant.

Gravitational field strength and gravitational potential

Chapter 13 of *Introduction to Advanced Physics* dealt with force and motion in gravitational fields, and with the concept of field strength. This section provides a brief reminder of gravitational field strength and introduces the concept of gravitational potential (Figure 3.2).

In a region around a body with mass, another mass will experience force. This region is the body's gravitational field, and though in principle it stretches to infinity in all directions, it becomes weaker as distance from the body increases. We have a number of ways in which we can describe the field, beginning with the idea that every point in a field has two variable properties – field strength and potential.

Gravitational field strength of a point in a field is the force that will act on each unit of mass that might be placed at that point. It is often given the symbol g. The value of g at the Earth's surface, for example, is $9.8\,N\,kg^{-1}$.

gravitational field strength of a point = force per unit mass

$$g = \frac{F}{m}$$

Gravitational potential is, likewise, a property of a point in a field. But while field strength is defined in terms of force and mass, potential is defined in terms of energy (or work) and mass. The gravitational potential of a point is the work that we (or some other agency) would need to do to bring a unit of mass from an infinite distance (that is, from outside the field) to that point (Figure 3.2).

gravitational potential of a point = work done per unit mass

$$V = \frac{W}{m}$$

The unit of gravitational potential is the joule per kilogram, $J\,kg^{-1}$.

The value of V at the Earth's surface is $-6.26 \times 10^7 J\,kg^{-1}$. The negative sign is present because work does not have to be done on a mass by any third party (such as ourselves) to bring it to the Earth's surface from infinitely high in the sky (from beyond the Earth's gravitational field). In fact, in moving freely towards the Earth's surface a body gains kinetic energy from the field. The field itself does the work. So in the formula $V = W/m$, W is negative, and hence V is negative.

If, however, the force were repulsive then we (or some other agency) would have to do a positive amount of work against this repulsion. The potential at the Earth's surface would then have a positive value.

Figure 3.2
A negative amount of work must be done to move a body from outside the field (strictly from an infinite distance away) to a point P. The gravitational potential of P is the ratio of the work done to the mass transferred.

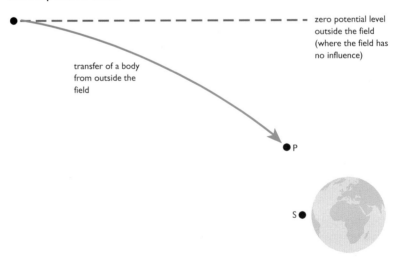

zero potential level outside the field (where the field has no influence)

transfer of a body from outside the field

P

S

Points P and S each possess different values of
• gravitational field strength
• gravitational potential.

Gravitational potential differences and changes in potential energy

Figure 3.3
Points P and S have different potentials. There is a potential difference ΔV between them. This is equal to the work that must be done per unit mass to move any body between the two points.

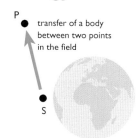

P

transfer of a body between two points in the field

S

Potential difference can exist between two points in a field (Figure 3.3). It is equal to the work that must be done to transfer unit mass, or unit charge in an electric field, between them. So, by definition,

$$\text{gravitational potential difference} = \frac{\text{work done}}{\text{mass}}$$

$$\Delta V_g = \frac{W}{m}$$

We can apply the same arguments (on a smaller scale) to the diver shown in Figure 3.1. This is done in Figures 3.4 and 3.5.

Figure 3.4
Points around a diving board and near the water surface both have gravitational potential. There is a potential difference. The diver must do work (or have work done, such as by a lift) to go to the higher level. Then, when the diver becomes a freely moving body, it is the field that does work – the diver loses potential energy and gains kinetic energy.

Figure 3.5
There is a gravitational potential difference between P and S. The diver has different potential energies when at P and at S.

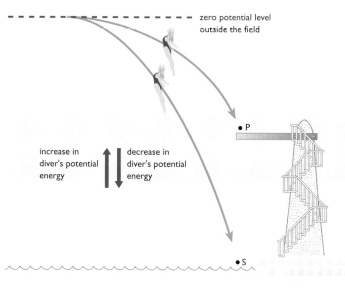

zero potential level outside the field

increase in diver's potential energy

decrease in diver's potential energy

• P

• S

Negative amounts of work must be done to transfer a diver from outside the field to either point P or point S. Points P and S have potentials that are below the zero level – they are negative potentials.

Note that the diver has potential energy. Points such as P and S have potential.

Electric field strength and electrical potential

For an electric field, also, we can define field strength and potential, again as properties of points. Electrical potential is covered in *Introduction to Advanced Physics*, Chapter 18 – so here we introduce electric field strength and provide a reminder of electrical potential (Figure 3.6). The definitions are very similar to those for gravitational field, with charge substituted for mass. So we have:

electric field strength of a point = force per unit charge

$$E = \frac{F}{q}$$

Its unit is the newton per coulomb, $N\,C^{-1}$.

Also, work done per unit charge in transferring a body from a point of zero potential (effectively outside the field) to a point P is the potential of the point P:

electrical potential of a point = work done per unit charge

$$V = \frac{W}{q}$$

The unit of electrical potential is the joule per coulomb, $J\,C^{-1}$. For the simple reason that electrical potential is an extremely useful quantity with which to think about electric fields, the joule per coulomb has been given its own name – the volt, V. Note that we use (italic) *V* for the quantity electrical potential, and (roman) V for its unit, the volt.

Figure 3.6
A positive amount of work must be done to transfer each unit of charge from outside the field (strictly from infinity) to a point P. The electrical potential at point P is the ratio of the work done to the charge transferred.

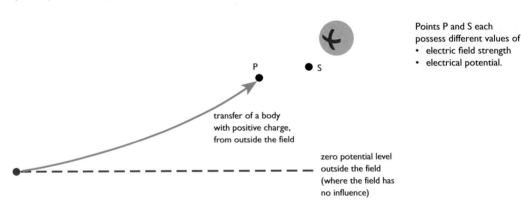

Points P and S each possess different values of
• electric field strength
• electrical potential.

transfer of a body with positive charge, from outside the field

zero potential level outside the field (where the field has no influence)

The situation is somewhat more complex for electric fields than for gravitational fields, in that charge may be positive or negative. An X-ray tube provides an illustration of some of the principles. The X-ray tube contains a cathode and an anode or target. It would take a different amount of energy to move a charged body from an infinite distance to the cathode than to the anode (Figure 3.7a, opposite). The cathode and the anode are at different electrical potentials.

Note that the X-ray tube works just as well whatever the potentials of the anode and cathode relative to the zero potential level – it is the size of the potential *difference* between them that matters (Figure 3.7b).

Electrical potential differences

The electrical potential difference between two points, such as points on the surfaces of the anode and cathode of an X-ray tube (Figure 3.7), is the work that must be done per unit charge to move charge between them:

electrical potential difference = $\dfrac{\text{work done}}{\text{charge}}$

$$\Delta V_e = \frac{W}{q}$$

a anode and cathode in a simple X-ray tube

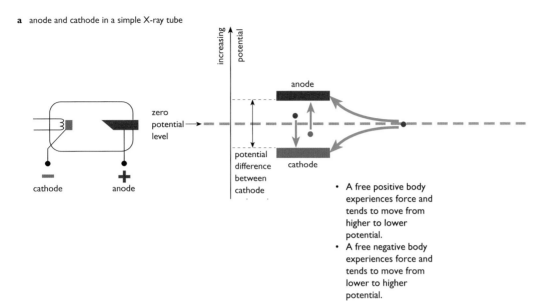

To move a positively charged body to the anode from a point of zero potential, an *increase* in both potential and potential energy is involved.

To move a positively charged body to the cathode from a point of zero potential, a *decrease* in both potential and potential energy is involved.

- A free positive body experiences force and tends to move from higher to lower potential.
- A free negative body experiences force and tends to move from lower to higher potential.

b anode–cathode systems with the same potential difference

Figure 3.7
Electrical potentials in an
X-ray tube

1 A charge of 1 μC experiences an electric force of 0.01 N at a point P in an electric field.
   a What is the electric field strength at point P?
   b What force would be experienced by a charge of 0.04 μC when at point P?
2 A gravitational field does 2×10^6 J of work in bringing a mass of 1 g from an infinite distance to a point A in the field.
   a What is the potential at point A?
   b How much work must be done by an external agency to remove the 1 g mass entirely from the field?
   c How much work must be done to remove a mass of 1 kg from point A to outside of the field?
3 Write down equivalent formal definitions of gravitational potential and electrical potential. (See also *Introduction to Advanced Physics*, page 154.) Highlight the differences and similarities in the definitions.
4 Compare the water surface/diving board system with the X-ray tube anode/cathode system. Use them to illustrate that gravitational potential must always be negative whereas electrical potential may be positive or negative.

Radial and uniform fields

Field strengths, both gravitational and electric, like force itself, are vector quantities. We can use vector visual representation to show their sizes and directions (Figure 3.8). From this we can develop a useful means of picturing the fields – with the use of field lines, which can also be called lines of force.

Figure 3.8
Field strengths can be represented by vector arrows at chosen points, while field line representations provide useful models of a field as a whole. They show the direction of the force that would act if a body were to be placed anywhere along their lengths. The fields shown are both radial fields.

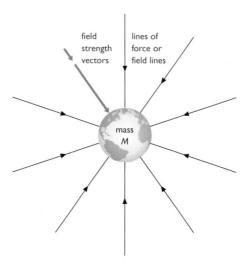

Figure 3.9
When two masses or two charges lie in the same region, then their fields combine. Each body contributes to the net field, and the net field strength at any particular point is found by addition of the two vectors that represent the individual field strengths.

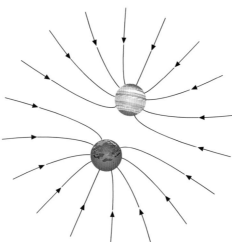

Note that for both gravitational and electric fields Figure 3.8 shows uniform spherical objects which behave like a point mass and a point charge. The forces acting are simply either directly towards or directly away from the point bodies. The field lines spread out in all directions, like the radii of spheres. The fields are therefore called **radial fields**. Fields around bodies with extended shapes, and around groups of two or more bodies, may be more complex, as in Figure 3.9.

In the case of gravitational fields, the point mass scenario is very important, because bodies with very significant gravitational fields (such as moons, planets and stars) have centres of mass that are almost exactly at the centres of their geometry. Apart from point bodies, only bodies like these, whose mass or charge can be considered to be at a simple point, have radial fields. There are more complex cases, such as galaxies of unusual shape, or clouds of matter that have quite low density but are nevertheless big enough for gravitational force to play a very major role in events (Figure 3.10).

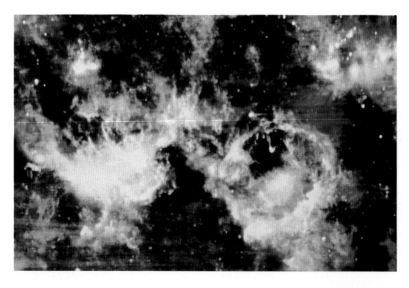

Figure 3.10
Stars are created by gravitational collapse of clouds of gas of relatively low density. The shapes of such clouds may be complex.

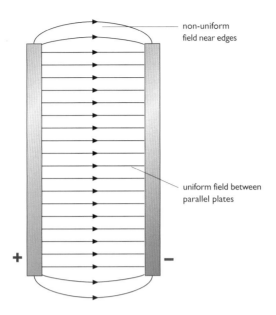

Figure 3.11
In a uniform field, field strength is the same everywhere. Field lines are parallel to each other, and evenly spaced.

non-uniform field near edges

uniform field between parallel plates

Systems of pairs of parallel plates maintained at different electrical potential are of considerable practical usefulness. In the region between a pair of parallel plates that are at different potentials, a charged particle experiences the same force at all points. The field lines are parallel to each other and evenly spaced. We describe the field as a **uniform field** (Figure 3.11). In a uniform field, field strength (and potential gradient) is the same at all points. Capacitors, the subject of Chapter 4, are usually parallel charged plates with a uniform field between them.

An oscilloscope uses an electric field between parallel plates to steer electron beams and produce a visual display of electrical information. The field is near-uniform at any one instant.

Note that in a region of a radial field that is small relative to the distance (radius) from the source of the field, the field lines are approximately parallel (Figure 3.12). We can and do often consider local gravitational field (in our immediate environment close to the Earth's surface) as uniform.

Figure 3.12
Locally (that is, over short ranges of distance relative to the distance from an effective point object), a radial field can approximate to a uniform field.

field lines

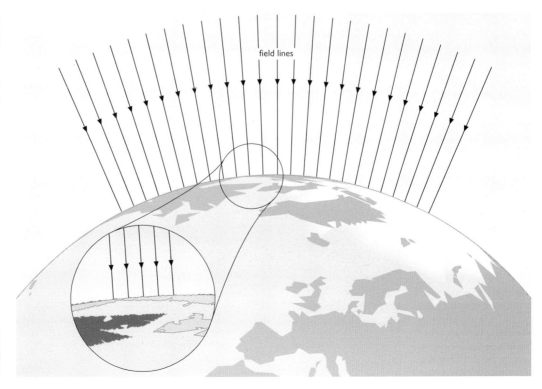

5 Why do field lines never cross each other?

Potential and equipotentials

Any pair of points that are the same distance from a point charge or point mass are at the same potential. No work needs to be done to transfer charge or mass between the pair of points. Since this is true for all points at the chosen distance, we can join them all together into an **equipotential surface**. An equipotential surface is simply one on which all points are at the same potential.

- For radial fields, equipotential surfaces are spherical shells (Figure 3.13).
- Equipotential surfaces in a uniform field, however, are plane parallel surfaces, evenly spaced (Figure 3.14).

Figure 3.13 (left) Equipotential surfaces for a radial field.

Figure 3.14 (right) Equipotential surfaces for a uniform field.

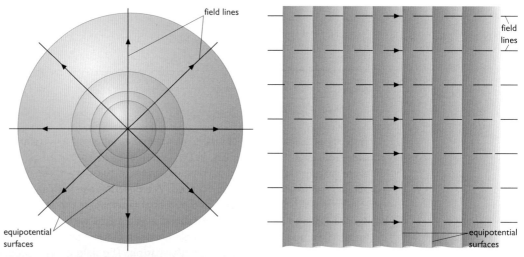

A satellite in circular orbit around the Earth (Figure 3.15) traces a pathway on an equipotential surface provided that its speed does not change. To move between equipotentials would involve loss or gain of potential energy, and since a satellite carries no fuel, the only way that it can do this is by corresponding gain or loss in kinetic energy. The surface of the ocean below the satellite is approximately an equipotential surface (Figure 3.16).

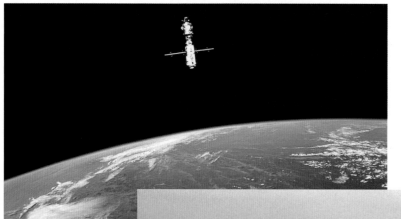

Figure 3.15 (above) The International Space Station remains on the same equipotential surface while in orbit.

Figure 3.16 (right) Ignoring the complications of waves, tides and other climatic effects, or averaging out their effects over a period of time, the ocean surface is an equipotential surface.

Figure 3.17
Equipotential lines in radial and uniform fields.

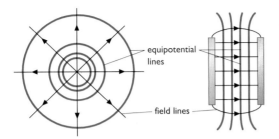

Equipotential surfaces are three-dimensional, but three-dimensional representations can be very complex. Also, we communicate using flat surfaces of paper and computer screens, and so two-dimensional aids to our thinking are often more useful. In a two-dimensional diagram, equipotential surfaces can be shown as **equipotential lines** (Figures 3.17 and 3.18). Note that in radial and uniform fields, and in any other kind of field, field lines are always normal to equipotential surfaces or lines.

Figure 3.18
A hill can be described in terms of gravitational potential.

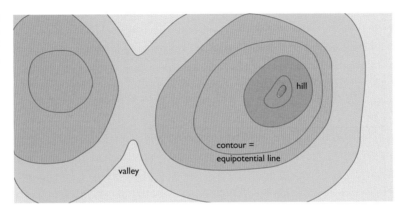

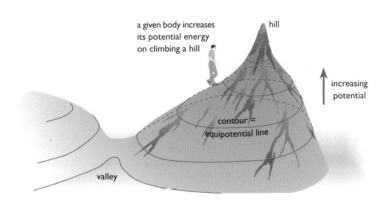

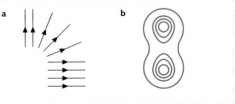

Contours on a map are lines joining points of equal elevation, though we could also think of them as lines joining points of equal gravitational potential. Certainly, to climb from one contour line to another, work must be done – there is a change in potential energy, and we can say that there is a potential difference between the two contours. There is no such potential difference between two points on the same contour. Note that the direction of force is always perpendicular to the contours.

6 a Sketch equipotential lines that correspond to the field lines in Figure 3.19a.
b Sketch possible field lines to match the equipotential lines in Figure 3.19b.

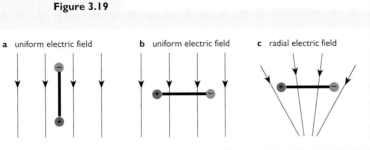

Figure 3.19

7 Two equal and opposite charges are linked by a rigid rod, as shown in Figure 3.20. Describe the forces acting on each charge in each case, and describe the overall consequences.

Figure 3.20

Newton's Law of Gravitation

Force and field strength due to a body of mass M, not surprisingly, depend on the value of this mass. For radial fields they also obey an inverse square law in the space outside the body. The relationships between field strength and the mass M or the distance r to the centre of the mass M can be shown graphically, as in Figure 3.21.

Figure 3.21
Dependence of field strength on M and r for a radial field due to mass M.

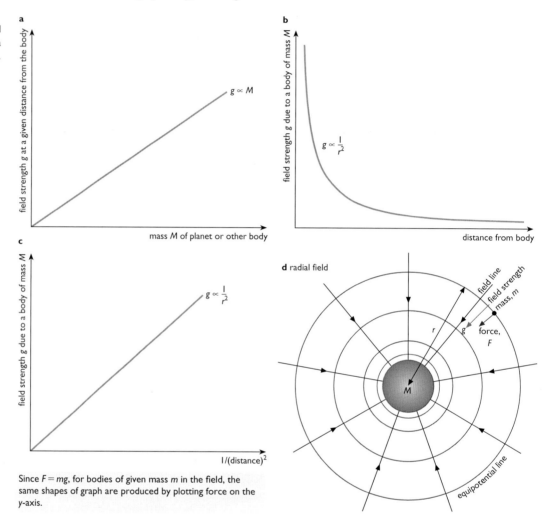

Since $F = mg$, for bodies of given mass m in the field, the same shapes of graph are produced by plotting force on the y-axis.

The graphs show how field strength g varies with mass M and with distance r. Force and field strength are simply related, by $F = mg$. So graphs of force against M and $1/r^2$ have the same shapes as graphs of field strength against M and $1/r^2$. **Newton's Law of Gravitation** provides an equation for force:

$$F = \frac{GMm}{r^2}$$

G is the **universal gravitational constant**. Since $G = Fr^2/Mm$, its units can be written as $N\,m^2\,kg^{-2}$. It has a small value, $6.67 \times 10^{-11}\,N\,m^2\,kg^{-2}$.

We also know that

$$g = \frac{F}{m}$$

So we can see that, for a radial field external to mass M,

$$g = \frac{GM}{r^2}$$

8 Express the unit of G in terms of SI base units (kilogram, metre and second).

9 For a planet of fixed mass, what graph should be plotted to obtain a straight line relating the force it exerts on a body of constant mass to the distance, r, of the body from the planet's centre?

10 Give expressions for the gradients of the graphs in Figures 3.21a and c.

Gravitational field strength inside spherical bodies

Figure 3.22
Effects of a shell of material.

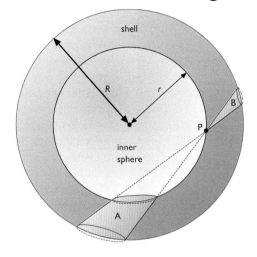

Portions A and B of the spherical shell have equal and opposite effects. A is bigger but is further away, resulting in this balance. The overall effect is that the shell of material (shaded) makes no contribution to the field strength at point P.

The centre of mass of a uniform sphere is at its geometric centre. For thinking about values of force and field strength *outside* it, we can treat a spherical mass as if it were a mass occupying just a single point of space, a 'point mass', at the centre. For irregular shapes, centre of mass is not so easily predicted. Fortunately, planets and stars are approximately spherical, or at least symmetrical.

A point P *inside* the sphere, however, at a distance r from the centre, has a sphere 'below' it, and also a shell of material that all lies further than distance r from the centre (Figure 3.22). The centre of mass of the shell is at the centre of the sphere, and analysis of its gravitational effect shows that it makes no net contribution to the field strength at point P. The inner sphere 'below' point P acts like any sphere, and its mass is related to its radius. If we suppose that the sphere has uniform density, then this relationship is given by:

$$\text{mass of sphere} = \text{density} \times \text{volume}$$
$$= \rho \times \tfrac{4}{3}\pi r^3$$

So field strength within a spherical mass of uniform density is given by:

$$g = \frac{GM}{r^2}$$
$$= \frac{G \times \rho \times \tfrac{4}{3}\pi r^3}{r^2}$$
$$= G \times \rho \times \tfrac{4}{3}\pi r$$
$$= kr$$

where $k = G \times \rho \times \tfrac{4}{3}\pi$ and is a constant.

Thus within the sphere, field strength is proportional to distance from the centre; while outside the sphere, field strength is inversely proportional to distance squared, as shown in Figure 3.23.

Figure 3.23
Variation of field strength with distance from the centre of a sphere.

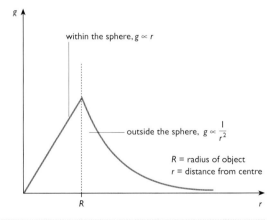

within the sphere, $g \propto r$

outside the sphere, $g \propto \dfrac{1}{r^2}$

R = radius of object
r = distance from centre

11 **a** What is the field strength at the centre of the Earth?
   b What would you experience if you were at the centre of the Earth?
12 The Earth's core is denser than surrounding regions. Sketch a graph to suggest the influence that this might have on the variation of field strength with distance from the centre of the Earth.

Measuring the universal gravitational constant, *G*

We make a claim that *G* is a *universal* gravitational constant. This is consistent with observations of moons, planets and galaxies. The behaviour of gravity seems to be the same everywhere. Such is the faith that people have had in this universality that it is assumed to be true when planning space missions. Spacecraft, whether 'routine' communications satellites or missions through the Solar System to distant planets (Figure 3.24), follow trajectories that are true to the predictions.

To measure *G*, we need to measure the force between two bodies, their masses and their separation. Unfortunately, *G* is extremely small and so the force between two human-sized masses is small. Measurement of such small forces is difficult, but not impossible. Henry Cavendish was the first to succeed in a laboratory measurement of *G*, in 1798.

Cavendish fixed two quite small metal spherical masses to a horizontal bar and hung the bar from a thread fastened to its centre (Figure 3.25). He then brought a pair of larger metal masses up to the original pair, and these had enough gravitational effect to cause the small masses to move and the thread to twist. It was easy enough to measure the masses, and their separations. It was possible to measure the force by making independent measurements of the couple required to twist the thread through different angles.

Figure 3.25
Cavendish's measurement of *G*.

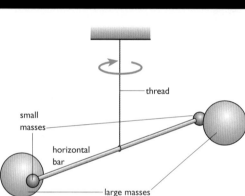

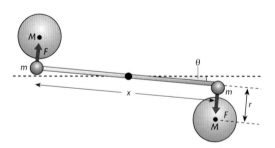

The couple acting on the horizontal bar is given by

$$\text{couple} = Fx = k\theta$$

where *k* is a constant, which can be measured by twisting the thread. *F* is the force between one pair of masses, *M* and *m*. So

$$F = \frac{k\theta}{x} = \frac{GMm}{r^2}$$

and hence

$$G = \frac{k\theta r^2}{xMm}$$

Measurement of *G* allows us to consider forces far beyond the Solar System. But at the time of Cavendish, one of the benefits of having a reliable value was that it allowed calculation of the mass of the Earth, using

$$g = \frac{GM}{R^2} \quad \text{where } R \text{ is the radius of the Earth}$$

so that the mass of the Earth is given by

$$M = \frac{gR^2}{G}$$

13 Write down an equation relating *G* to the force *F* acting between two masses *M* and *m* at separation *r*, with *G* as the subject.

The force between two point charges – Coulomb's Law

Whereas gravitational forces between small objects are tiny, electric forces between small objects are very significant, and can be more easily measured in the laboratory, as shown in Figure 3.26. Such forces of attraction between particles prevent your body from separating into a gas of its components, and such forces of repulsion prevent you and everything around you (the whole planet) from crashing inwards to become a tiny object of huge density.

Figure 3.26
Repulsion of two similar objects (metallised plastic balls) with the same charge can be detected and measured.

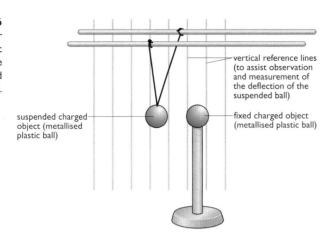

vertical reference lines (to assist observation and measurement of the deflection of the suspended ball)

suspended charged object (metallised plastic ball)

fixed charged object (metallised plastic ball)

Experimental investigation reveals that the force between two charged particles, whatever their polarities (positive or negative signs), obeys an inverse square law. The relationship between electric force and separation is the same as it is for gravitational force. The similarity does not end there, because the electric force depends on the magnitudes of the two charges. We can write a relationship between electric force and these variables as:

$$F = \text{constant} \times \frac{Qq}{r^2}$$

The general form of this relationship is the same as in the gravitational case, with charges in place of masses. One difference is that masses always have the same sign, whereas charges may have the same signs, resulting in repulsive force, or different signs, resulting in attractive force. There is also a difference in how we write the constant. In the gravitational case it is written as G. The constant in the electrical case is written as

$$\frac{1}{4\pi\epsilon_0}$$

A difference of very great importance is not how we write the constant, but its value. Whereas G is a small number, $1/(4\pi\epsilon_0)$ is a big number, with the value

$$\frac{1}{4\pi\epsilon_0} = 8.99 \times 10^9 \, \text{N} \, \text{m}^2 \, \text{C}^{-2}$$

We can write the electrical relationship as

$$F = \frac{1}{4\pi\epsilon_0} \frac{Qq}{r^2}$$

This relationship is known as **Coulomb's Law**. It is a historical accident that the constant is written in this way, in terms of the **permittivity of a vacuum** ϵ_0, which is another of nature's fundamental constants; ϵ_0 has a value of $8.85 \times 10^{-12} \, \text{C}^2 \, \text{N}^{-1} \, \text{m}^{-2}$ (see Chapter 4).

The origin of these numbers, and of other fundamental constants such as the speed of light in a vacuum or the charge on an electron, is one of the continuing great mysteries. If nature contained within it different values for these numbers, then it would not be the nature that we know. (And we would not be here to know it.)

Figure 3.27
This image, generated by interaction of electrons and atoms in a scanning tunnelling microscope (STM), shows an array of carbon atoms on the surface of graphite. The atoms line up in their arrays because of electric force, and its Coulomb Law behaviour.

As it is, gravitational force operates on the grand scale – over large distances, between large bodies. If the bodies are particle-sized, the gravitational force between them is tiny. Electric force dominates the world of smaller bodies (Figure 3.27), and the bodies do not need to have large amounts of charge in order to be much influenced by electric force. (Another reason that electric forces do not play a large part in interactions between very large bodies is that large accumulations of net charge do not gather – large bodies are very nearly perfect mixes of positive and negative charge.) Note also that electric forces do not dominate on the nuclear and sub-nuclear scale, where nuclear force rules.

The mathematics of both Newton's Law of Gravitation and Coulomb's Law assumes that the bodies between which the forces act are point bodies. The formulae can be applied to spherical bodies (such as planets) because they also have radial fields and mass or charge can be considered to act at their centres. More care, however, is required with extended bodies of other shapes.

Positive force and negative force

Electric force can be repulsive or attractive. We can show the difference in terms of the sign that we give to the force. It would be useful to have a consistent pattern for this – a convention. Since electric force is given by

$$F = \frac{1}{4\pi\epsilon_0} \frac{Qq}{r^2}$$

then if both charges are positive we obtain a positive value for force, and if both charges are negative we also have positive force. In both cases the force is repulsive, and this establishes our convention – repulsive electric force is thought of as positive. If the charges have opposite signs then their product is negative – attractive electric force is negative.

This does give us a small problem if we are trying to be completely consistent in considering electric fields and gravitational fields together. Masses are always positive, and yet the force between them is always attractive. If there were a need to be totally consistent we might want to say that G is a negative number, so that the attractive gravitational force would be negative. However, see page 55.

14 Given that electric field strength, E, is related to force by $E = F/q$, write down an equation that relates electric field strength to distance from a point charge Q.

15 Write down the units of G and $1/(4\pi\epsilon_0)$, highlighting the similarities and differences.

16 **a** A pair of deuterium nuclei each has charge of $+1.6 \times 10^{-19}$ C. Calculate the repulsive force between them when their separation is 2 fm. (1 femtometre = 1 fm = 10^{-15} m)
b Use your value to show why only those deuterium nuclei which have high initial speeds can get as close together as 2 fm centre-to-centre separation.

17 Two point charges X and Y are situated 1.0 m apart.
a What is the resultant field strength on a straight line joining X and Y at a point 0.4 m from X if
i X and Y both have charges of $+2 \times 10^{-12}$ C
ii X has a charge of $+2 \times 10^{-12}$ C and Y has a charge of -3×10^{-12} C?
b i Use vector addition to find the value of the resultant field strength at a point that is 1.0 m from X and 1.41 m from Y, for the charges in part **a i**.
ii Repeat for the charges in part **a ii**.

18 What are the ratios, with units, of
a proton charge and proton mass
b $1/(4\pi\epsilon_0)$ and G
c electric force and gravitational force on a pair of protons distance x apart?
d Imagine a Universe in which the ratio of part **c** is inverted, but in which gravitational force remained only attractive and electric force remained capable of being attractive or repulsive. Describe how matter would behave in that Universe.

19 **DISCUSS**
Does the existence of a Universe that has a particular set of fundamental constants and in which we live prove that
a the physical Universe must have been established by a God existing outside the physical Universe and setting the values of its physical constants
b our physics has only begun to scratch the surface of understanding the Universe
c both of these
d neither of these?

Variation of potential with distance for radial fields

We have formulae relating force with separation or distance for radial gravitational and electric fields. The two fully consistent formulae for force are

$$F = -\frac{GMm}{r^2} \quad \text{and} \quad F = \frac{1}{4\pi\epsilon_0}\frac{Qq}{r^2}$$

and those for field strength are

$$g = -\frac{GM}{r^2} \quad \text{and} \quad E = \frac{1}{4\pi\epsilon_0}\frac{Q}{r^2}$$

Here, for consistency, we include a minus sign in the formulae for gravitational field. The minus sign tells us that gravitational force between two masses is attractive, whereas electric force between two like charges is repulsive. Consistent use of signs is essential for consideration of potential and energy.

We know that work done is the product of average force F and distance x in the direction of the force. The problem that we now have is this: how can we determine the average force given the above relationships? The solution to our problem lies in the force–separation graphs (Figure 3.28). Work done is always equal to the area under the curve on such a graph. The mathematical process by which we can find the area under a curve is integration, a calculus process. Integration may be thought of as the reverse of differentiation. Differentiation of a relationship tells us about the gradient or gradients associated with that relationship. Integration of a relationship provides the total area of an infinite number of infinitesimally small slices of the graph that describes the relationship.

Uniform fields are in many ways simpler than radial fields. A body in a uniform field experiences a constant force (Figure 3.28a), which means that the work done in moving a given distance is simply the product of the displacement and the force parallel to the displacement. There is no need to use calculus, as we have to for radial fields (Figure 3.28b).

Figure 3.28
For all kinds of fields, work done is given by the area under the curve on a force–distance graph.

a

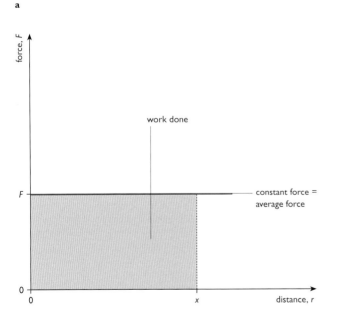

work done

constant force = average force

Work done in moving a body from distance 0 to distance x under the action of a constant force F

= area under curve between $r = 0$ and $r = x$ (shaded)
= Fx

Note that both x and r can be used for distance.

b

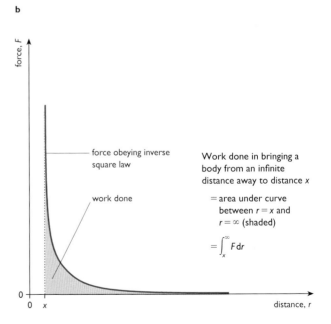

force obeying inverse square law

work done

Work done in bringing a body from an infinite distance away to distance x

= area under curve between $r = x$ and $r = \infty$ (shaded)

$$= \int_x^\infty F\,dr$$

Just as we say that the relationship $v = dx/dt$ is a more complete or more general formula than $v = x/t$, which tells us average velocity over a period of time and not instantaneous velocity, so also the definition $W = \int F\,dx$ is a formula that is always true. The simpler $W = Fx$ is only true where F is the average force. (In the case of a uniform force, the instantaneous force is equal to the average force at all times, so the formula $W = Fx$ is then valid.)

We do not need to worry here about how the integration works. What matters is that the process yields the following relationships. For a gravitational field, the work done in moving a body of mass m from infinity to a point in a field a distance r from a point mass M is given by

$$\text{work done} = -\frac{GMm}{r}$$

Note that, while force obeys the inverse square law, work obeys a simple inverse law. In fact that is the only difference in the formulae. Note also that gravitational potential is work done per unit mass:

$$V_g = \frac{\text{work}}{m}$$
$$= -\frac{GMm/r}{m}$$
$$= -\frac{GM}{r}$$

In a similar way as for a gravitational field, we can show that, for an electric field, the work done in moving a charge q from infinity to a point a distance r from a charge Q is given by

$$\text{work done} = \frac{1}{4\pi\epsilon_0}\frac{Qq}{r}$$

so the electrical potential (work done per unit charge) at a point in the field is given by:

$$V_e = \frac{\text{work}}{q}$$
$$= \frac{1}{4\pi\epsilon_0}\frac{Q}{r}$$

Graphs of g and V_g against r for a gravitational field (Figure 3.29) can be compared with those of E and V_e against r for an electric field (Figure 3.30).

Figure 3.29
The $g–r$ and $V–r$ graphs for a radial gravitational field.

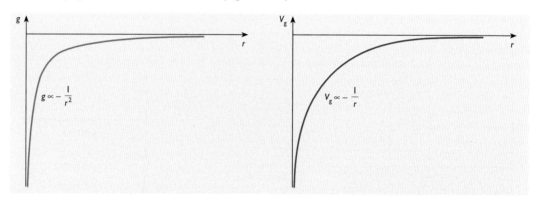

Figure 3.30
The $E–r$ and $V_e–r$ graphs for a radial electric field.

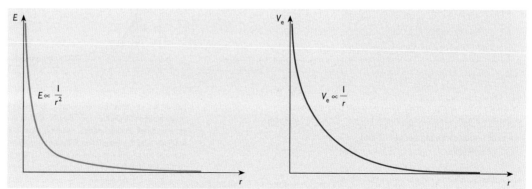

Potential gradient

As we have seen, the relationships for force and field strength versus separation obey the inverse square law, whereas the relationships for potential and work done versus separation are simpler inverse proportionalities.

A little exercise in calculus again proves useful. We can take the potential formulae, and differentiate them. Remember that differentiation is a way of finding the gradient of a graph. The process of differentiating a relatively simple relationship is itself not difficult. The mathematics of why this works is not something that we need to worry about here. So if, as we have found, the potential formulae are

$$V_g = -\frac{GM}{r} \qquad \text{and} \qquad V_e = \frac{1}{4\pi\epsilon_0}\frac{Q}{r}$$

then the formulae for the quantities $\dfrac{dV}{dr}$ are

$$\frac{dV_g}{dr} = \frac{GM}{r^2} \qquad \text{and} \qquad \frac{dV_e}{dr} = -\frac{1}{4\pi\epsilon_0}\frac{Q}{r^2}$$

In both cases dV/dr is the gradient of the potential–separation curve (Figure 3.31). It is therefore called potential gradient. Note that the formulae for potential gradient are the same as those for field strength, with opposite sign. Potential gradient and field strength are of the same size.

Figure 3.31
Graphs of V and dV/dr versus r.

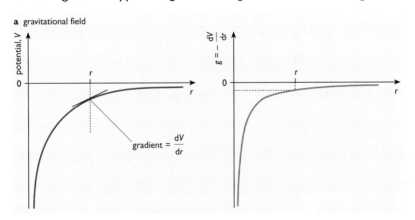

a gravitational field

The value of g at distance r is equal in size to the gradient of the V–r graph, with opposite sign.

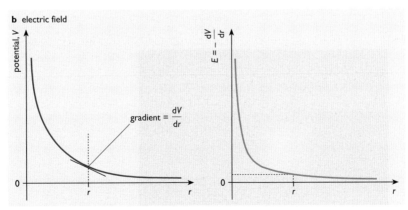

b electric field

The value of E at distance r is equal in size to the gradient of the V–r graph, with opposite sign.

20 Show that, for each type of field, the formulae for dV/dr are dimensionally consistent.

21 For a radial gravitational field, write down a relationship between field strength and potential.

22 Use Figure 3.31a to help you to sketch a corresponding pair of graphs for force acting on mass m against distance r, and potential energy of mass m against distance r.

- In a gravitational field:

$$\text{potential gradient,}\ \frac{dV}{dr} = -\text{field strength,}\ g$$

- In an electric field:

$$\text{potential gradient,}\ \frac{dV}{dr} = -\text{field strength,}\ E$$

Potential wells and hills

Graphs of potential energy against distance and potential against distance have the same shapes. The difference is that when considering potential energy we take the value of the mass or charge of a body into account, whereas for potential we do not need to do so. They are useful shapes because they predict direction of motion in a very easy way. The shapes behave like wells and hills (Figure 3.32), and motion of masses or positive charge is always 'down the slopes' of the graphs. The shapes can be called **potential wells** and **potential hills**. We can apply such concepts and models, say, for considering the electric field around one or more nuclei, or the gravitational field of two stars that make up a binary system (Figure 3.33).

Figure 3.32
Potential wells and potential hills.

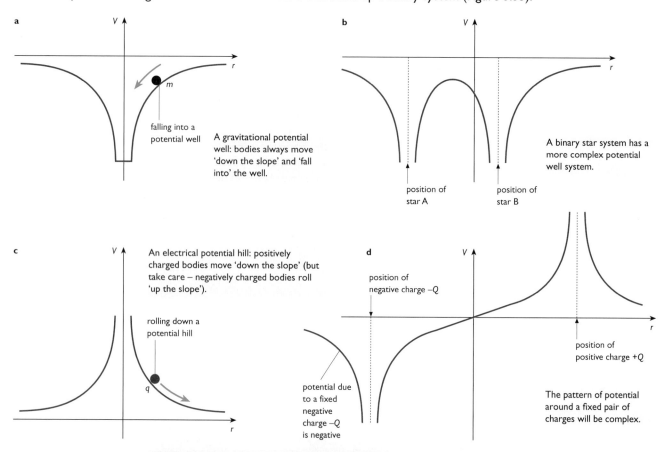

a

falling into a potential well

A gravitational potential well: bodies always move 'down the slope' and 'fall into' the well.

b

position of star A

position of star B

A binary star system has a more complex potential well system.

c

An electrical potential hill: positively charged bodies move 'down the slope' (but take care – negatively charged bodies roll 'up the slope').

rolling down a potential hill

d

position of negative charge −Q

potential due to a fixed negative charge −Q is negative

position of positive charge +Q

The pattern of potential around a fixed pair of charges will be complex.

Figure 3.33
A binary star system, Alpha Centauri A and B. Each object lies in the gravitational field of the other, and in fact their fields combine to produce a complex pattern of field strength and potential.

Gravitational potential and escape speed

The potential at a point on a planet's surface is given by the equation

$$V = -\frac{GM}{R}$$

where R is the radius of the planet and M is its mass. Thus the potential energy of a mass m resting on the planet's surface is given by

$$W = -\frac{GMm}{R}$$

This could be thought of as the binding energy of the mass to the planet. This is analogous to the binding energy of a nucleon to a nucleus. The amount of energy that must be given to the mass in order for it to escape is:

$$W = \frac{GMm}{R}$$

The potential energy of the body on the planet is negative, and the energy required for escape is positive.

One way in which this energy could be given to the mass is by firing it at high speed – that is, by giving it kinetic energy. Then as the mass moves away from the planet's surface, its kinetic energy will decrease and its potential energy will rise towards zero (Figure 3.34).

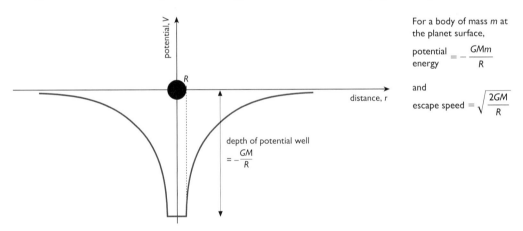

Figure 3.34
Potential well and escape speed.

potential, V

R

distance, r

depth of potential well
$$= -\frac{GM}{R}$$

For a body of mass m at the planet surface,

$$\text{potential energy} = -\frac{GMm}{R}$$

and

$$\text{escape speed} = \sqrt{\frac{2GM}{R}}$$

Only when outside the field will the body have zero potential energy. In order to achieve this escape, the minimum kinetic energy that must be given to the mass is

$$E_k = \frac{GMm}{R}$$

$$\tfrac{1}{2}mv^2 = \frac{GMm}{R}$$

which gives

$$v^2 = \frac{2GM}{R}$$

So the minimum speed v with which the mass must be fired from the surface is

$$v = \sqrt{\frac{2GM}{R}}$$

Note that the mass of the body, m, has disappeared from this expression. The minimum velocity required for a body to escape completely from a planet's gravitational field, starting at its surface, is independent of the mass of the body, but depends on the mass and radius of the planet. It is therefore a constant for the planet, and is called the planet's **escape speed**.

The escape speed of the Earth is $1.1 \times 10^4 \, \text{m s}^{-1}$. Interplanetary space missions effectively escape from the Earth's surface. But they are not projected into space from the Earth's surface with a speed of $11 \, \text{km s}^{-1}$. The acceleration involved would be too big for a spacecraft to withstand, and a speed of $11 \, \text{km s}^{-1}$ at the Earth's surface would result in both slowing of the spacecraft and heating due to collisions with air particles. So instead rockets (Figure 3.35) provide a relatively gentle (but still large) acceleration and spacecraft do not reach very high speed until above the atmosphere. Their speed may then be close to escape speed.

Of course, for spacecraft that remain in Earth orbit, as artificial satellites, complete escape from the Earth's gravitational field is not required. Satellites remain within the Earth's field.

Another very important point is that escape from the gravitational field of the Sun, or even movement further from the Sun, requires more energy than is provided by an escape speed of $11 \, \text{km s}^{-1}$ (see *Comprehension and application*, page 66).

Figure 3.35
The purpose of a rocket is to launch a payload. If the payload at the top of the rocket is to be launched into interplanetary space, then soon after leaving the Earth's atmosphere, when the main rocket fuel is all consumed, the payload will need to have the appropriate escape speed. (This will be slightly less than the escape speed from the Earth's surface because of the payload's height.)

The concept of escape speed is useful in understanding why the Earth has an atmosphere and the Moon does not. Since escape speed is given by

$$v = \sqrt{\frac{2GM}{R}}$$

and for the Moon, $M = 7.4 \times 10^{22} \, \text{kg}$ and $R = 1.74 \times 10^6 \, \text{m}$, we get the escape speed for the Moon as

$$v = \sqrt{\frac{2 \times 6.67 \times 10^{-11} \times 7.4 \times 10^{22}}{1.74 \times 10^6}}$$

$$= 2.38 \times 10^3 \, \text{m s}^{-1}$$

This is much smaller than that for the Earth, and at the range of temperatures found on the Moon sufficient gas molecules would have speeds at this level for a gradual leakage of gas into the surrounding space to take place. If the Moon ever had an atmosphere of gases, it leaked away. For the Earth, and other bodies of similar size, the rate of such leakage is very small.

Summary table for radial fields

The behaviour of a field due to a point mass M or a point charge Q can be summarised as shown in Table 3.1.

Table 3.1

	Gravitational field		Electric field	
	Formula	**Unit**	**Formula**	**Unit**
field strength at point P	$g = -\dfrac{GM}{r^2} = -\dfrac{dV}{dr}$	$N\,kg^{-1}$	$E = \dfrac{1}{4\pi\epsilon_0}\dfrac{Q}{r^2} = -\dfrac{dV}{dr}$	$N\,C^{-1}$
force on mass m, or charge q, at point P	$F = -\dfrac{GMm}{r^2}$	N	$F = \dfrac{1}{4\pi\epsilon_0}\dfrac{Qq}{r^2}$	N
potential at point P	$V = -\dfrac{GM}{r}$	$J\,kg^{-1}$	$V = \dfrac{1}{4\pi\epsilon_0}\dfrac{Q}{r}$	$J\,C^{-1}$ or V
work done in bringing mass m, or charge q, to point P	$W = -\dfrac{GMm}{r}$	J	$W = \dfrac{1}{4\pi\epsilon_0}\dfrac{Qq}{r}$	J
potential energy of a body at point P	$W = -\dfrac{GMm}{r}$	J	$W = \dfrac{1}{4\pi\epsilon_0}\dfrac{Qq}{r}$	J
change in potential energy for masses M and m, or charges Q and q, in moving from separation r_1 to r_2	$W = GMm\left(\dfrac{1}{r_1} - \dfrac{1}{r_2}\right)$	J	$W = -\dfrac{1}{4\pi\epsilon_0}Qq\left(\dfrac{1}{r_1} - \dfrac{1}{r_2}\right)$	J

23 A point charge of $+2 \times 10^{-11}$ C is brought from a point of zero potential to a point P, 0.02 m from a point charge of $+8 \times 10^{-11}$ C.
a What is the potential at point P?
b How much work must be done?
c How much extra work must be done to take the smaller charge from point P to a point 0.01 m from the larger charge?

Potential–distance and force–distance graphs for uniform electric fields

Figure 3.36
Representations of a uniform field between parallel plates.

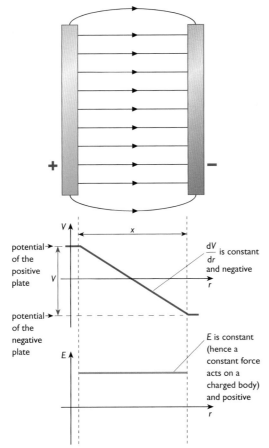

Field strength and potential gradient are constant in a uniform field between charged parallel plates (Figure 3.36). In such a relatively simple field, and in practical circumstances in which signs have little significance, $E = -dV/dr$ can be written simply as $E = V/x$, where x is the separation of the plates.

In the space between the plates, dV/dr is constant, and is equal to V/x. So we can say that

$$E = -\frac{dV}{dr} = \frac{V}{x}, \text{ ignoring the sign}$$

Motion in uniform fields

In a uniform field, because the force is constant, acceleration is constant. Pathways of bodies in uniform fields are generally easier to analyse than those in non-uniform fields. But not surprisingly, pathways depend on the initial state of motion of any such bodies.

If, for example, a body is initially at rest, then it accelerates uniformly in a straight line. The equations of uniformly accelerated motion may be used. This is also true for a body that has an initial velocity that is parallel to the field lines, with initial velocity, u, being non-zero.

For a body which has an initial velocity that is *not* parallel to the field lines, we can, following Galileo's analysis of cannonball motion (see *Introduction to Advanced Physics*, page 111), consider the motions perpendicular and parallel to the field lines separately (Figure 3.37). What we then find is that, if the perpendicular velocity is constant and the parallel velocity increases uniformly, as described by the equations of uniformly accelerated motion, the pathway followed is **parabolic**. Here we can consider a parabola as the pathway followed by a body whose velocity has two mutually perpendicular components, one of which is constant while the other is increasing uniformly with time.

Figure 3.37

Motion of a mass in a small region of the Earth's gravitational field and motion of a charge between a pair of parallel plates are both examples of motion in uniform fields. Pathways of cannonballs and electrons depend in the same way on particular combinations of acceleration and initial velocity. Note that acceleration is always parallel to the field, but velocity is not.

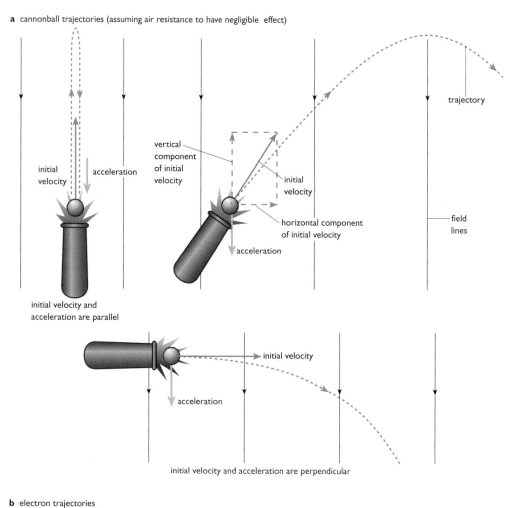

a cannonball trajectories (assuming air resistance to have negligible effect)

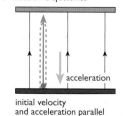
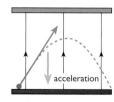
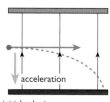

b electron trajectories

24 A pair of parallel plates separated by a distance of 0.02 m have a potential difference between them of 100 V.

    a What is

    i the potential gradient

    ii the field strength

    in the space between them?

    b What is the force acting on a charge of $+10^{-16}$ C when between them?

    c If the body has a mass of 2×10^{-20} kg, calculate its acceleration due to the field.

    d If the charged body were initially at rest on the plate at lower potential, how much work would have to be done to move it to the plate at higher potential?

25 A cannonball is fired at an angle of 45° to the horizontal.

    a Sketch its pathway if air resistance is negligible.

    b Draw arrows to show the horizontal and vertical components of the cannonball's velocity at the following three points in its motion:

    i when it is rising

    ii when it is at its maximum height

    iii when it is losing height.

    c Sketch velocity–time graphs for the horizontal motion and the vertical motion.

    d Sketch the pathway of the cannonball when air resistance is significant.

26 a The equations of uniformly accelerated motion (see *Introduction to Advanced Physics*, pages 345–6) can be written as:

$$v = u + at$$
$$v^2 = u^2 + 2as$$
$$s = ut + \tfrac{1}{2}at^2$$

Explain why they do not apply to behaviour in a radial field.

b Using one or more sketches, show that the inverse square law does not apply to uniform fields.

Potential inside isolated electrical conductors

Potential inside a solid conductor due to an external field

We have considered the gravitational field strength inside a body such as the Earth, but what about the field inside electrical conductors? In this section we consider 'isolated' conductors that cannot experience inwards or outwards current and so cannot gain or lose total charge. (Chapter 4 is all about capacitors, which are arrangements of conducting bodies that are usually *not* isolated.)

We know that in *all* places where there is net electric field strength and potential gradient, free charges experience force. They only experience no net force when field strength and potential gradient are zero. Thus within an isolated conductor (in which moving charge experiences resistive force) charge moves until these states of zero are achieved. That is, charge distributes itself to achieve such equilibrium. So if, for example, an isolated conductor of zero net charge is placed in an electric field, then free charges within the conductor move until potential gradients within it are eliminated. The result is that all points on and in the conductor are at the same potential. There are no potential differences (Figure 3.38, overleaf).

Potential inside a solid conductor due to its own net charge

Suppose that an isolated spherical conductor carries a net charge that is, say, positive. If this positive charge were evenly distributed within the conductor, then we could apply the same thinking as we used in considering the gravitational field within the Earth. A small positive charge at a point P within the sphere would experience no net force due to the charge in the shell of material that lies further out than point P from the centre of the sphere. But there would be a sphere of material below it, and our small positive charge would experience repulsion by the charge within this sphere. So the small positive charge would be initially repelled outwards, along with every other such small positive charge. Thus free charges tend to move outwards, and accumulate on or near the surface. This tendency ceases when the net force on each charge is zero, and the potential gradient is zero (Figure 3.39, overleaf).

Figure 3.38
In a region of potential gradient, charges will experience force. In an isolated conductor, any initial potential gradient results in force on free charges within it, so that the free charges redistribute themselves. Equilibrium is established when the charges have distributed themselves such as to eliminate potential gradient.

For a neutral isolated conductor there are no potential differences.

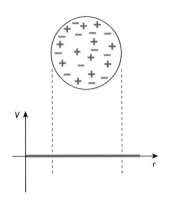

If an electric field exists around the conductor, the charge within it redistributes. This redistribution continues until there is no longer any potential difference between any two points within the conductor.

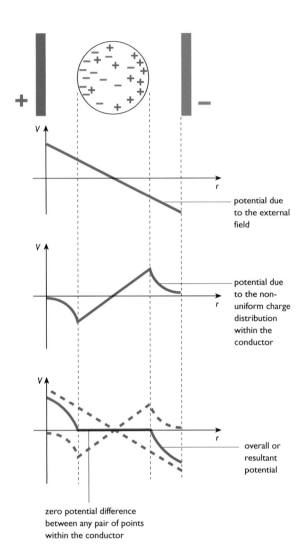

potential due to the external field

potential due to the non-uniform charge distribution within the conductor

overall or resultant potential

zero potential difference between any pair of points within the conductor

Figure 3.39
Free charge on an isolated conductor accumulates on the surface. Only in such a state does the conductor have no internal potential gradient. Free charge has distributed itself until no potential differences exist between any two points within the conductor.

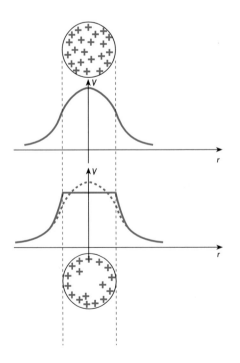

Potential–distance curve for a spherical body containing uniformly distributed net positive charge.

Potential–distance curve with positive charge having tended to move towards the surface of the sphere, resulting in no internal potential differences and no potential gradients.

Field inside a hollow conductor

Suppose that a conductor consists of a spherical shell that has a net positive charge, and that all that exists inside the shell is a single mobile positive charge (Figure 3.40). We have already used the idea that a shell of mass or charge exerts no net force on a body within it. The small mobile positive charge experiences no force. There is no potential difference between any two points within the shell. There is no electric field (Figure 3.41).

Figure 3.40
The small positively charged body inside the shell experiences no net force.

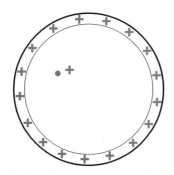

Figure 3.41
With no field, no potential gradient and no potential differences inside a metal shell, one of the safest places to be during a thunderstorm is inside a metal shell.

27 Imagine a large spherical hollow shell of charged material with uniformly distributed charge Q and radius R.
   a Write down formulae for the electric field strength and potential at a point that is outside the shell at a distance r from the centre.
   b A charge $-q$ at a point P inside the shell at a distance $x\,(=R/2)$ from the centre, as shown in Figure 3.42, experiences forces in opposite directions due to attraction by the charges of two small areas of the shell that are diametrically opposite. The small areas can be linked by 'cones' that pass through P. Show, using appropriate approximations, that the charge $-q$ experiences no net force as a result of the opposing attractions.

28 If a large cloud of interstellar dust were to have a non-uniform distribution of charged particles, what would happen?

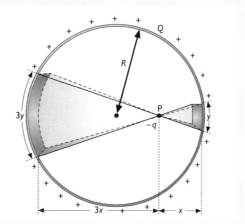

Figure 3.42

Exploring planets

Table 3.2 provides data about the Sun and the innermost six planets. Use the data in the table to answer the questions that follow.

Table 3.2

	Mass/kg	Mean radius/m	Mean orbital radius/m
Sun	1.989×10^{30}	6.960×10^8	
Mercury	3.347×10^{23}	2.44×10^6	5.800×10^{10}
Venus	4.900×10^{24}	6.05×10^6	1.082×10^{11}
Earth	5.976×10^{24}	6.371×10^6	1.496×10^{11}
Mars	6.574×10^{23}	3.390×10^6	2.228×10^{11}
Jupiter	1.900×10^{27}	7.14×10^7 (at equator)	7.78×10^{11}
Saturn	5.7×10^{26}	6.0×10^7 (at equator)	1.427×10^{12}

29 Create a table showing:
   a the gravitational field strength at the equator of each planet
   b the potential at the equator of each planet due to the gravitational field of the planet itself
   c the potential at each planet due to the gravitational field of the Sun
   d the total potential at the equator of each planet (by adding the previous two figures).

30 a Calculate the potential difference between the surfaces of Earth and Venus.
   b Sketch a potential–distance graph for the Sun's gravitational field. Mark the approximate positions of the six innermost planets on the distance axis.
   c A potential hill exists between any two planets. Sketch the shape of this hill for Jupiter and Saturn, taking account of the gravitational field of the Sun.
   d Explain whether a net amount of energy is needed to send a spacecraft to Venus.

31 The Earth's escape speed is $1.1 \times 10^4\,\mathrm{m\,s^{-1}}$, but this provides escape into local interplanetary space only. Calculate the speed with which an unpowered spacecraft would have to leave the top of the Earth's atmosphere in order to escape from the Solar System.

32 Explain why it is inevitable that a principal challenge in engineering a safe landing of a space capsule on the Moon or on a planet is to slow the capsule down.

33 The Cassini–Huygens space mission (Figure 3.43) is travelling from Earth to Saturn. The spacecraft has a mass of 5650 kg.
   a Calculate the net energy required for its journey, assuming that the whole mass lands on Saturn.
   b To gain energy, the spacecraft passed close to the Earth 22 months after its launch. It may seem therefore that it had travelled at high speed for some considerable time but had got nowhere. But during this fly-by the spacecraft was subject to the Earth's gravitational field, and because of the force between them the spacecraft was accelerated. A body that enters a potential well and then emerges again cannot experience a net gain in energy as a result *unless* the potential well is due to a *moving* mass. What is the source of the energy that the body then gains?
   c During most of the 22 months between launch and Earth fly-by, the spacecraft was free of the Earth and in independent elliptical solar orbit. Would it have made most sense for the spacecraft to have orbited the Sun in the same direction as the Earth or in the opposite direction?

Figure 3.43
The pathway of the Cassini–Huygens spacecraft. It was launched from Earth in October 1997, made a fly-by of Venus in April 1998, followed by an Earth fly-by and a further Venus fly-by, reached Jupiter in December 2000 and arrives at Saturn on 1st July 2004.

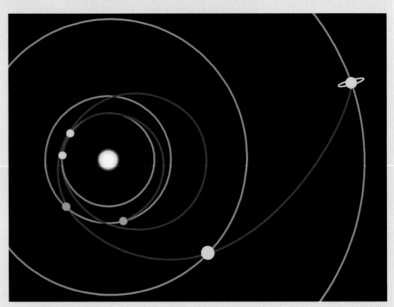

Comprehension and application

Millikan's oil drops

Robert Andrews Millikan was an American physicist who created a neat way to obtain quite precise measurements of the fundamental unit of charge, the charge on a single electron.

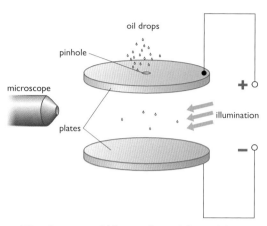

Figure 3.44
Millikan's apparatus.

At the heart of Millikan's apparatus was a pair of horizontal parallel metal plates, between which he could apply a varying potential difference (Figure 3.44). Between the plates he could establish a region of uniform field – constant potential gradient and field strength. He created a spray of oil drops, and a small number of these drops could enter the field through a hole in the upper plate. The drops were small and not easy to see, but Millikan found that he could control the motion of the oil drops. They did not just fall under gravitational attraction, but were subject to electric force. Clearly they carried charge.

The drops could be made to fall quickly or slowly, to rise upwards, and even with great care in controlling the applied potential difference, to stay still. For a motionless oil drop (ignoring the upthrust that is provided by the air), gravitational and electric forces were in balance:

gravitational force = electric force
$$mg = qE$$

Here E is the field strength, which is constant between the plates, and equal to the potential gradient, V/x, where V is the applied potential difference and x is the distance between the plates. So

$$mg = \frac{qV}{x}$$

and

$$q = \frac{mgx}{V}$$

By switching off the potential difference and allowing the drop to fall, Millikan could see that the drop acquired a new constant velocity. The drop quickly reached terminal velocity – a state of zero acceleration due to balance of the gravitational force and the resistive force. By measuring the velocity, Millikan was able to use standard formulae to obtain the mass m of an observed oil drop. He could then calculate q, the charge on the drop.

What he found was that the drops tended to have charges that were multiples of the same value, this value being 1.6×10^{-19} C. If we call this charge e, then in Millikan's experiment, $q = ne$, where n is a small whole number.

It seems that Millikan's oil drops were charged, perhaps in the process of being created at the spray nozzle, by gaining or losing just a small number, n, of electrons.

34 a Millikan assumed that both the electric and gravitational fields were uniform. Explain why this assumption is a reasonable approximation.
b Sketch electric and gravitational field lines between Millikan's plates.
c Sketch electric and gravitational equipotentials between the plates.
d Why was it reasonable for Millikan to use $E = V/x$ rather than the more generally true statement that $E = dV/dx$? (Signs here have little practical importance and have been ignored.)

35 a It is hard to imagine a human-sized object – something considerably larger than an oil drop – experiencing electric and gravitational forces that are equal and opposite. Given that the maximum potential gradient that dry air can sustain is $3 \times 10^6\,\text{V m}^{-1}$, calculate the minimum charge that would have to be given to a typical human being in order to suspend them in the air in the manner of the oil drops.
b If such a human body were considered to be a point charge, what would be the potential gradient and field strength at a distance of 3 m away?
c Would you want to get close to this person?

36 DISCUSS
Comment on the general statement that: "Electric fields have more applications than gravitational fields in human technologies because they can be manipulated."

37 DISCUSS
Speculate on how the Universe might be different if $e = -1.6 \times 10^{-16}$ C rather than -1.6×10^{-19} C.

● **Comprehension and application**

Circular motion in gravitational fields

Features of the Universe that we can look at with wonder are the spinning spiral galaxies – like our own. We can consider a star within such a galaxy, and we can use mathematics to do so.

Mathematics is a stunningly powerful way of modelling behaviour in and of the Universe. Not that it is possible to model the entire Universe – we have to accept that the only complete description (or model) of the Universe is the Universe itself. But mathematical models are still powerful predictors of events.

We can create mathematical models of circular motion, for example. It doesn't have to be of a particular motion of a particular body, but circular motion in general. We know that force and acceleration are linked by

$$F = ma$$

and for a body moving in a circle at constant speed (but not constant velocity, of course), then force and speed are linked by

$$F = \frac{mv^2}{r}$$

F is here a centripetal force. For a body whirling on a string, the tension in the string provides the centripetal force. For the Earth in orbit around the Sun, or the Sun as it moves around the galaxy centre, centripetal force is provided by gravity. We know the rules for gravity – we know that the force between two bodies is given by:

$$F = \frac{Gm_1m_2}{r^2}$$

where m_1 and m_2 are the masses of the two bodies, and r is the distance between them.

We can now combine the above two equations. For a body of mass m_2 in orbit around another of mass m_1, at radius r and speed v, we have

$$\frac{m_2v^2}{r} = \frac{Gm_1m_2}{r^2}$$

which simplifies to

$$v^2 = \frac{Gm_1}{r}$$

Note that the speed is independent of the mass of the body in orbit.

Io is a moon of Jupiter, in orbit at low radius compared with that of the other moons of Jupiter such as Europa, and therefore at relatively high speed. Our own Moon has a radius of orbit that is similar to that of Io, but is in orbit around a much less massive object. Data for these moons are given in Figure 3.45.

Figure 3.45
The orbits of Io, Europa and our Moon.

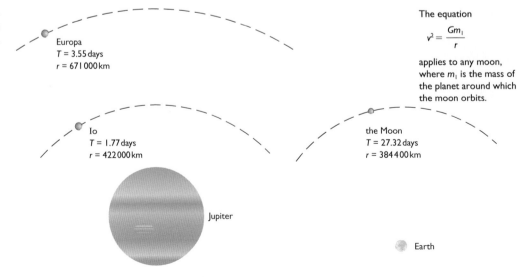

Europa
$T = 3.55$ days
$r = 671\,000$ km

Io
$T = 1.77$ days
$r = 422\,000$ km

the Moon
$T = 27.32$ days
$r = 384\,400$ km

Jupiter

Earth

The equation

$$v^2 = \frac{Gm_1}{r}$$

applies to any moon, where m_1 is the mass of the planet around which the moon orbits.

38 a In the relationship

$$v^2 = \frac{Gm_1}{r}$$

we have seen that the velocity of a star in a rotating spiral galaxy is independent of the star's mass. How would the observed motion of stars in galaxies be different (from that which we observe) if this were not the case?

b In what way does the above equation *not* agree with observation?

c Why is the mismatch between mathematical prediction and actual observation important?

39 Show that the data in Figure 3.45 is consistent with the equation

$$v^2 = \frac{Gm_1}{r}$$

40 a Sketch a graph of square of orbital speed, v^2, against the inverse of orbital radius, for the planets of the Solar System.

b Explain the shape of the graph.

c State the significance of the gradient.

Planets in orbit around the Sun follow the same mathematical rule. We would expect a star in its orbit around a galaxy centre to obey the same rule. Here again, the mathematics is reasonably straightforward. But for stars and galaxies it doesn't work. That is, the prediction that it makes is not what we *observe*. It is not too difficult to measure the radius, r, of the orbit of a star – its distance from a galaxy centre. It is also possible to make a good estimate of the number of stars in a galaxy and hence its mass, m_1. The universal gravitational constant G is known. But stars move faster in their orbits than this mathematics predicts.

Where are we going wrong? The only explanation that is taken seriously is that the mass of a galaxy is much bigger than the mass of its visible stars. Each galaxy must contain a lot of material that we cannot see. If we cannot see it that is because it emits no light. It is **dark matter**.

One of the quests that is motivating physicists, and mathematicians, of various descriptions (astronomers, cosmologists, particle physicists who study the fundamental nature of matter) is what *is* this dark matter? The answer may be relatively mundane – galaxies may be crowded with objects that are too small to burn like stars. Or it may be more exotic – black holes, antimatter, ...

● **Extra skills task**   Information Technology and Application of Number

1 Use the Internet or CD ROMs to find data, including velocity and radius, on the orbits of
  a all of the named moons (planetary satellites) in the Solar System
  b the Sun in the Milky Way.
  Also find the masses of the orbiting bodies.

2 For the moons, present the mass data as a distribution curve, and give values for the range, median and mean.

3 a For the moons of Saturn, plot a graph relating orbital speed to orbital radius, so as to obtain a straight line.
  b Show that this graph is consistent with the mathematics of centripetal force.
  c Predict how the graph would differ for the moons of Jupiter.

4 For one of the moons and for the Sun, use the data to calculate the centripetal forces that you would expect to be acting in each case. Use a computer to develop a visual presentation of this information, capable of showing the sizes of the forces but also emphasising their different scales.

Examination questions

1 a Define *gravitational field strength* at a point in a
gravitational field. (1)

b Tides vary in height with the relative positions of the
Earth, the Sun and the Moon which change as the
Earth and the Moon move in their orbits. Two
possible configurations are shown in the diagrams.

not to scale

Configuration A

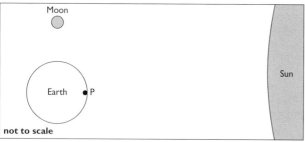

not to scale

Configuration B

Consider a 1 kg mass of sea water at position P. This
mass experiences forces F_E, F_M and F_S due to its
position in the gravitational fields of the Earth, the
Moon and the Sun respectively.

i Draw labelled arrows on *both* diagrams to indicate
the three forces experienced by the mass of sea
water at P. (3)

ii State and explain which configuration, A or B, of
the Sun, the Moon and the Earth will produce the
higher tide at position P. (2)

c Calculate the magnitude of the gravitational force
experienced by 1 kg of sea water on the Earth's
surface at P, due to the *Sun's* gravitational field. (3)

$$\text{radius of the Earth's orbit} = 1.5 \times 10^{11}\,\text{m}$$
$$\text{mass of the Sun} = 2.0 \times 10^{30}\,\text{kg}$$
$$\text{universal gravitational}$$
$$\text{constant, } G = 6.7 \times 10^{-11}\,\text{N m}^2\,\text{kg}^{-2}$$

AEB, A level, Module Paper 8, Summer 1999

2 a State what is meant by the *potential at a point in an
electric field*. (2)

b A Van de Graaff generator is a machine which is used
to produce very high potential differences. When in
use, charge builds up on a hollow spherical
conducting dome until the dome achieves the
required potential difference between it and the
Earth. The electric potential of the Earth is assumed
to be 0 V.

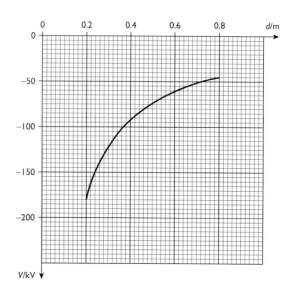

dome of
Van de Graaff
generator

The graph shows how the potential *V* varies with
distance *d* from the centre of the charged spherical
dome of a Van de Graaff generator, for values of *d*
greater than 0.2 m.

The spherical dome of radius 0.15 m is also
shown.

i Use the graph to determine the potential at a
distance of 0.30 m from the centre of the dome of
the Van de Graaff generator. (1)

ii Determine the potential at the surface of the
dome of the Van de Graaff generator. Show your
reasoning. (3)

iii Draw lines to represent the electric field near the
surface of the dome. (2)

iv Determine the electric field strength at a distance
of 0.30 m from the centre of the dome. (3)

v Calculate the force exerted on an electron when it
is 0.30 m from the centre of the dome. (2)

Charge on an electron, $e = -1.6 \times 10^{-19}\,\text{C}$

AEB, A level, Paper 2, Summer 1998

3 a State Newton's law of gravitation. (2)

b Explain why it is extremely difficult to obtain a
reliable value for the gravitational constant *G* in a
laboratory. (3)

c For the region outside the Earth's surface, the graph
opposite shows the variation of gravitational
potential *V* with distance *r* from the centre of the
Earth.

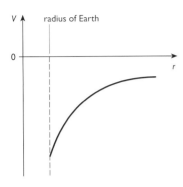

Explain why in this situation V approaches zero as r approaches infinity. Explain why in this situation V is always negative. (2)

London, A level, Module Test PH4, June 1998

4 a The diagram below shows two charged, parallel, conducting plates.

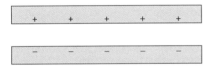

i Add to the diagram solid lines to show the electric field in the space between and *just beyond* the edges of the plates. (2)

ii Add to the diagram dotted lines to show three equipotentials in the same regions. (2)

b Define *electric potential* at a point. Is electric potential a vector or a scalar quantity? (3)

c An isolated charged conducting sphere has a radius a. The graph below shows the variation of electric potential V with distance r from the centre of the sphere.

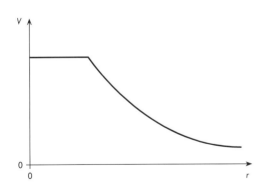

i Mark with an 'a' on the distance axis the point that represents the radius of the sphere. (1)

ii Add to the graph a line showing how electric field strength E varies with distance for the same range of values of r. (3)

London, A level, Module Test PH4, January 1996

5 The diagram (not to scale) shows a satellite of mass m_s in circular orbit at speed v_s around the Earth, mass M_E. The satellite is at a height h above the Earth's surface and the radius of the Earth is R_E.

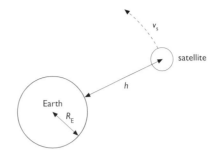

a Using the symbols above write down an expression for the centripetal force needed to maintain the satellite in this orbit. (2)

b Write down an expression for the gravitational field strength in the region of the satellite. State an appropriate unit for this quantity. (3)

c Use your two expressions to show that the greater the height of the satellite above the Earth, the smaller will be its orbital speed. (3)

d Explain why, if a satellite slows down in its orbit, it nevertheless gradually spirals in towards the Earth's surface. (2)

London, A level, Module Test PH4, June 1997

6 a What is meant by the term *gravitational potential V* at a point? In what unit is it measured? (3)

The gravitational potential at the surface of the Earth is given by the relationship

$$V = -\frac{GM}{r}$$

where
   G is the universal constant of gravitation
   M is the mass of the Earth
   r is the radius of the Earth

b Explain why V at the surface of the Earth is negative. (2)

c Show that $g = \dfrac{GM}{r^2}$ where g is the acceleration of free fall close to the Earth. (1)

d Use the relationships $V = -\dfrac{GM}{r}$ and $g = \dfrac{GM}{r^2}$ to help you to show that the escape speed for a projectile leaving the Earth is $11 \times 10^3\,\mathrm{m\,s^{-1}}$. Radius of Earth $= 6400\,\mathrm{km}$. (3)

e In calculating the escape speed for a space probe travelling outwards from the solar system what would the symbols M and r represent? (1)

London, A level, Module Test PH4, January 1998

7 a Using the usual symbols write down an equation for
  i Newton's law of gravitation
  ii Coulomb's law. (2)
  b State one difference and one similarity between gravitational and electric fields. (2)
  c A speck of dust has a mass of 1.0×10^{-18} kg and carries a charge equal to that of one electron. Near to the Earth's surface it experiences a uniform downward electric field of strength $100\,\mathrm{N\,C^{-1}}$ and a uniform gravitational field of strength $9.8\,\mathrm{N\,kg^{-1}}$.
  Draw a free-body force diagram for the speck of dust. Label the forces clearly. Calculate the magnitude and direction of the resultant force on the speck of dust. (6)

London, A level, Module Test PH4, January 1997

Question 8 also relates to Chapter 1

8 Io, the fifth of Jupiter's moons, was discovered by Galileo and Marius in 1610.

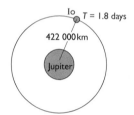

  a A spacecraft approaches Io; when it is 9.0×10^6 m away, the crew determines that Io subtends an angle of 22.6°. The spacecraft eventually lands on Io. To determine the mass of Io, a crew member drops a stone from a rocky ledge of height 19.0 m and finds that it takes 4.6 s to fall to the ground.
  i Show that the radius of Io is 1.8×10^6 m. (1)
  ii Determine the acceleration of the falling stone due to the gravitational attraction of Io near its surface. (2)
  iii State Newton's law of universal gravitation fully in words. (2)
  iv Calculate the mass of Io. (4)
  b A crew member has brought along a spring scale and a standard 1 kilogram mass from Earth. On Earth, when the mass was suspended from the scale, the scale reading was 10 N (within the precision of the scale), and when the mass was set in vertical oscillation on the end of the spring scale the period of oscillation was 0.20 s. The crew member tries the same experiments on Io.
  i On Io, what will be the scale reading? Explain. (1)
  ii On Io, what will be the period of oscillation? Explain your answer. (1)

c As shown in the diagram, Io orbits Jupiter in an approximately circular orbit.
  i Draw a free-body force diagram for Io, with vectors representing the force or forces acting on it while in orbit. For any force you include, state the type of force and name the object which exerts it. (2)
  ii While Io is in orbit, is it accelerating or not? If you say no, explain why not. If you say yes, explain why and state in what direction the moon accelerates. (2)
  iii The crew member notes that he was able to calculate the mass of Io from the motion of a stone in its gravitational field, and realises that they can likewise calculate the mass of Jupiter from knowledge of the motion of Io in Jupiter's gravitational field. Io orbits at a distance of 422 000 km from the centre of Jupiter, with a period of 1.8 earth days. Determine the mass of Jupiter. (5)

IB, Subsidiary level 3, Paper 430, Specimen Paper

9 a Define
  i electric potential; (2)
  ii electric field strength. (2)
  b Consider the following arrangement of two charged metal spheres of the same radius.

  i Sketch a graph of the electric potential along the line segment PQ. (3)
  ii Is there any location along PQ where the electric potential is zero? If not, explain why not, if so, locate the point. (3)

An initially neutral conducting rod is placed between the charges, without touching them, as shown.

  iii Does the conducting rod now have a net charge? Explain. (3)
  iv Sketch another graph of the electric potential along the line segment PQ. (3)
  v Determine the approximate electric field, in magnitude and direction, due to this collection of three objects, at a distance 2.0 m away from the centre of the rod, that is at a distance which is great compared to the distances between the objects. (4)

IB, Higher level 2, Paper 430, November 1997

4 Capacitors

THE BIG QUESTIONS	● What are capacitors and their principles?
	● What can capacitors do for us?

KEY VOCABULARY absolute permittivity   capacitance   capacitor   dielectric   farad
polarisation (electrical)   relative permittivity (or dielectric constant)   time constant

BACKGROUND We know about resistors, and about the simple relationship between current and potential difference applied to them. Capacitors are different, not least in their relationship between current and potential difference, as we shall see.

Tiny capacitors are found in integrated circuits and larger ones in systems of radio tuning and power adapters. The human body – a very electrical entity – contains a vast network of capacitors. Clouds and the Earth can behave as capacitor systems, with dramatic events when discharge – loss of stored charge – takes place (Figure 4.1).

Figure 4.1
Capacitor discharge can be dramatic.

Electron movement and storage of charge

Both a metal body and an insulating body may have a net charge. Such a charged body can be said to be storing charge. A metal body that is not isolated – that is, one that is connected by conducting pathways to other bodies – can more easily gain and lose charge. A household bath and, on a larger scale, a reservoir provide useful analogies. They can store water, and easily gain or lose it through inwards and outwards flow.

The charge stored on a metal body can be either positive or negative. Our knowledge of the structure of materials tells us that a positive charge is due to an excess of protons over electrons, and a negative charge is an excess of electrons. While the protons have fixed, immovable positions within the metal structure, a deficit or excess of electrons is created by the flow of electrons on to and off the body.

I DISCUSS

Relate the 'reservoir' or 'household bath' analogy to the processes illustrated in Figure 4.2.

Potential difference and charge

Free electrons flow whenever the space in which they exist is subject to a potential gradient. A tendency of their flow is to cause potential difference to decrease. In a circuit, however, the action of the power supply is to maintain the potential difference, which may then exist across a resistor or a capacitor (Figure 4.2). A practical **capacitor** is simply a pair of conducting plates that are not touching. Charge cannot flow directly from plate to plate. The usual circuit symbol for a capacitor is very simple and sensible – a pair of parallel lines with a space between.

Figure 4.2
Potential differences across
a a resistor and
b a capacitor.

a potential difference across resistor

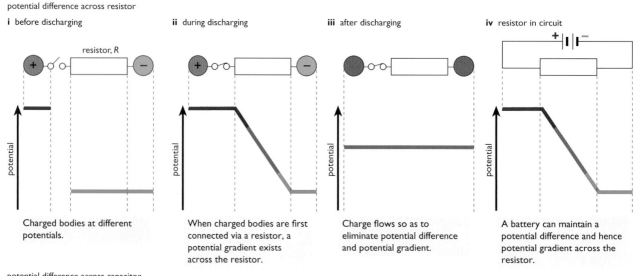

b potential difference across capacitor

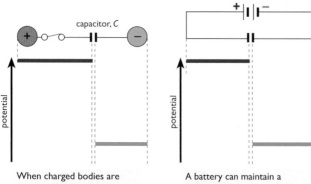

When connected to the battery, the capacitor stores charge (Figure 4.3). A capacitor can hold, or store, charge even if it is disconnected from the battery. Discharge will then take place only if flow of charge between the plates becomes possible (Figure 4.4).

Figure 4.3
A capacitor when connected to a battery becomes fully charged once the potential difference between its plates reaches the same value as the battery potential difference. The system of capacitor and battery will remain in this state until the circuit is changed.

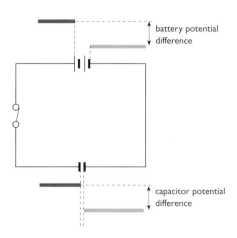

battery potential difference

capacitor potential difference

Figure 4.4
If the capacitor is disconnected from the battery and instead switched to a loop that joins the two plates together, then discharge takes place. The loop normally has some resistance.

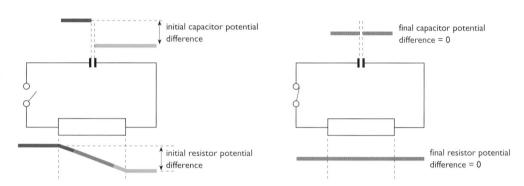

initial capacitor potential difference

final capacitor potential difference = 0

initial resistor potential difference

final resistor potential difference = 0

By storing charge, a capacitor maintains a potential difference between its plates. This is useful in a digital circuit (Figure 4.5), where the existence of a significant potential difference across a capacitor can represent the binary value of 1 and the existence of little or no potential difference across an uncharged capacitor represents 0.

Figure 4.5
Capacitors in a computer circuit.

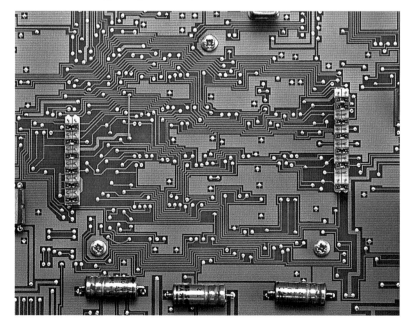

Capacitance

We measure capacitors in terms of the ratio of charge stored to the applied potential difference. This ratio is called **capacitance**:

2 What is the capacitance of a capacitor that stores 200 μC of charge when a potential difference of 6 V is applied to it?

3 What voltage is required across a 20 nF capacitor for it to store 2×10^{-10} C of charge?

capacitance = charge stored per unit of applied potential difference

$$C = \frac{Q}{V}$$

Capacitance is measured in coulomb per volt, CV^{-1}, this unit also being known as the **farad**, F. (Be careful not to confuse C for capacitance and C for coulomb.) One farad (1 F) is a relatively large capacitance, and capacitances of practical capacitors are normally given in μF (microfarad, 10^{-6} F), nF (nanofarad, 10^{-9} F) or pF (picofarad, 10^{-12} F).

Comparison and measurement of capacitance

Capacitance is a ratio of charge and potential difference. The potential difference may be known because it is equal to the e.m.f. of the source used to charge the capacitor. The capacitor acquires a potential difference that is equal to this e.m.f. Or it may be measured using a high-resistance voltmeter (Figure 4.6a). The high resistance is necessary so that little discharge takes place through the voltmeter.

Charge is more difficult to measure directly. For a constant current, charge that moves is simply a product of current and time. But when a capacitor gains or loses charge, the current is *not* constant (see pages 81–84). However, where a capacitor loses all of its charge rapidly, when it is discharged through a low resistance, the coil of a free-moving meter (a ballistic meter) experiences a rapid and brief deflection (Figure 4.6b). This 'throw' of the coil is an indication of the total charge transferred, and can be used to compare capacitors. If this is done with a capacitor whose value is known, at a given potential difference, then other capacitors can be compared with this and values ascribed to them.

Figure 4.6
Measurement of the potential difference across and the charge on a capacitor.

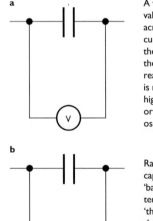

a A voltmeter can provide a value of potential difference across a capacitor, *but* current will pass through the meter coil, discharging the capacitor so that the reading falls. The problem is minimised by use of a high-resistance voltmeter or a cathode ray oscilloscope.

b Rapid discharge of a capacitor through a 'ballistic' meter produces a temporary deflection or 'throw' that depends on the total charge transferred.

ballistic meter

Dependence of capacitance on physical dimensions

There is a simple relationship between the capacitance of a parallel plate system and the separation of the plates. Measurements that reveal this relationship are most easily done with air between the capacitor plates. For such a capacitor, of fixed area of plate overlap, capacitance is inversely proportional to plate separation, r (Figure 4.7):

$$C \propto \frac{1}{r}$$

Figure 4.7
Experimental investigation
of the effects of separation
of the plates and area of
overlap of the plates on
capacitance yield graphs
like these.

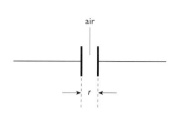

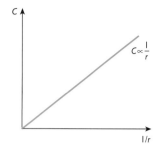

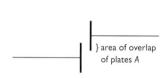

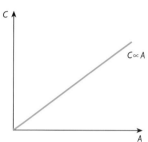

Further experimenting, keeping plate separation r constant, shows that the area of overlap of the two plates also plays a part in determining capacitance value. Capacitance is proportional to area of overlap, A (Figure 4.7):

$$C \propto A$$

Capacitance can be varied by varying the area of overlap of plates. A variable capacitor such as that illustrated in Figure 4.8 is a key part of a ratio tuning system that selects the frequency of a radio station (see Chapter 6).

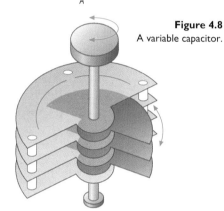

Figure 4.8
A variable capacitor.

Dependence of capacitance on the material between the plates

The material between the capacitor plates is called the **dielectric**. Capacitors with identical physical dimensions but different dielectric materials are found to have different capacitances. For any one chosen material we can plot capacitance C (as the output variable) against A/r, and (provided that the area of the plates is large and the distance between them is small) we find that the graph is a straight line passing through the origin, showing that

$$C \propto \frac{A}{r}$$

The gradient of the line is a constant, which depends on the dielectric between the plates. With a vacuum between the plates, the gradient is found to be $8.85 \times 10^{-12} \, \text{C}^2 \text{N}^{-1} \text{m}^{-2}$. This is the permittivity of a vacuum, ϵ_0 (which has already been met in Chapter 3). So we can say, for a vacuum,

$$C = \frac{\epsilon_0 A}{r}$$

For different dielectric materials we have to replace ϵ_0 with the required empirical value. 'Empirical' here means determined from experiment (from the gradients of graphs of capacitance against A/r, for example). Such an empirical value is called the **absolute permittivity** of the dielectric material. For a capacitor whose plates are separated by a dielectric material of absolute permittivity ϵ we could say that

$$C = \frac{\epsilon A}{r}$$

However, it is normal practice to compare the absolute permittivity of a material to that of a vacuum. The ratio of these is called the **relative permittivity** of the dielectric material:

$$\text{relative permittivity of a material} = \frac{\text{absolute permittivity of the material}}{\text{permittivity of a vacuum}}$$

$$\epsilon_r = \frac{\epsilon}{\epsilon_0}$$

which means that

$$\epsilon = \epsilon_0 \epsilon_r$$

Relative permittivities (sometimes called dielectric constants) of a range of dielectric materials are shown in Table 4.1. For example, for nylon,

$$\epsilon = 3.00 \times 10^{-11}\,C^2\,N^{-1}\,m^{-2}$$

$$\epsilon_r = \frac{\epsilon}{\epsilon_0} = \frac{3.00 \times 10^{-11}}{8.85 \times 10^{-12}} = 3.4$$

Table 4.1
Relative permittivities. To obtain a value of absolute permittivity, multiply the permittivity of a vacuum $(8.85 \times 10^{-12}\,C^2\,N^{-1}\,m^{-2})$ by the relative permittivity.

Material	Relative permittivity (dielectric constant)
vacuum	1.00
dry air	1.00*
bakelite	4.9
Pyrex glass	5.6
neoprene rubber	6.7
nylon	3.4
paper	3.7
water	80
tantalum pentoxide	20

*A higher degree of precision of figures reveals that the permittivity of air is not quite the same as that of a vacuum. To six significant figures, the relative permittivity of air is 1.00059 while that of a vacuum is by definition 1.00000

Note that relative permittivity is a ratio of quantities that have the same units and dimensions. Relative permittivity therefore is dimensionless.

So finally, for any material (anything other than a vacuum), we can write

$$C = \epsilon_0 \epsilon_r \frac{A}{r}$$

Plots of C against $\epsilon_0 A/r$ yield straight line graphs whose gradients give the values of ϵ_r (Figure 4.9).

4 If a capacitor is made of two flat strips of aluminium foil, 1 cm by 10 cm, separated by a layer of nylon 0.05 mm thick, what is its capacitance?
5 What will be the area of a 5 pF capacitor made of two layers of metal separated by a 0.05 mm thickness of tantalum pentoxide?
6 Charge Q on a isolated sphere can be considered to be at a single point at the centre of the sphere, and the sphere has a potential difference V between itself and regions outside the field, that is, at infinity. We know from Chapter 3 that the potential at the surface of the sphere is given by

$$V = \frac{1}{4\pi\epsilon_0}\frac{Q}{R}$$

where R is the distance of the surface from the point of action of the charge (the radius of the sphere) and V is the potential difference between the sphere and a point at infinity.
a What is the capacitance of such a sphere of radius 0.1 m?
b What size of sphere would have a capacitance of
i 1 pF   ii 1 µF?

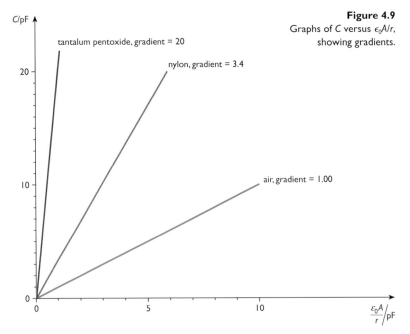

Figure 4.9
Graphs of C versus $\epsilon_0 A/r$, showing gradients.

Capacitors, potential difference and dielectric action

A single charged metal plate produces an electric field in the space around it, and it is possible to plot a graph of potential against distance from the plate along a central normal. We can do the same for an identical plate with an equal but opposite charge. If the plates are placed in close proximity, they then lie in each other's fields, and the potential at all points along a central normal, or central axis, of the system is a sum of the individual potentials (Figure 4.10).

Figure 4.10
Potentials due to individual plates and a pair of plates.

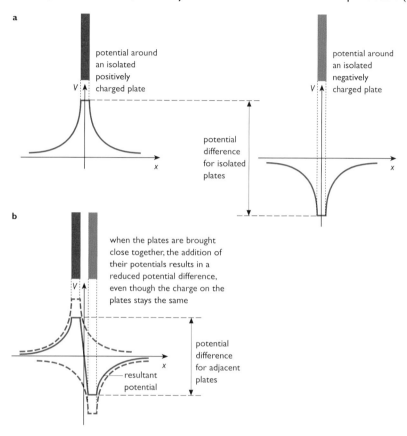

As a result of this addition of potentials, the potential difference between the plates is reduced, without any change in the charge on each. And the closer together the plates become, the more the potential difference is reduced. A decreasing potential difference and a constant total charge means an increasing capacitance – so this is consistent with the empirical observation that capacitance increases as plate separation decreases.

The material between the plates, the dielectric, is not a passive part of the system. That is, it changes, making a contribution to how the system behaves. Molecules within a material may already have some asymmetric charge distribution, or polarity, and some level of **polarisation** of the whole material is then induced by an external electric field. Molecules of a material rotate and align themselves with the field lines (Figure 4.11).

Figure 4.11
Effect of a field on polar molecules – polarisation.

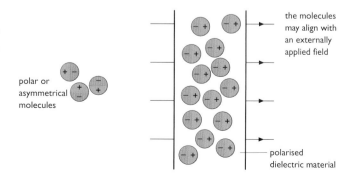

Polarisation of the dielectric means that there is an uneven charge distribution within it. This uneven charge distribution, if isolated, would result in a potential–distance graph, along a central axis, as shown in Figure 4.12b. The polarised dielectric thus makes a contribution to the overall potential–distance pattern of a capacitor, Figure 4.12c. The effect is to *reduce* the potential difference between the plates (for a given amount of charge), and so to increase capacitance.

Figure 4.12
Polarisation of the dielectric has a consequence for the potential–distance graph for a capacitor.

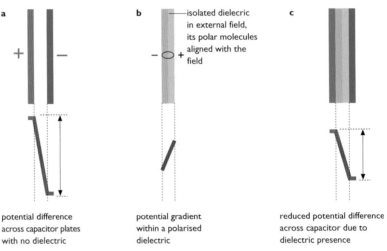

a

b — isolated dielectric in external field, its polar molecules aligned with the field

c

potential difference across capacitor plates with no dielectric

potential gradient within a polarised dielectric

reduced potential difference across capacitor due to dielectric presence

Practical capacitors

Figure 4.13
Examples of capacitor construction.

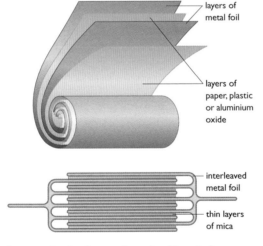

layers of metal foil

layers of paper, plastic or aluminium oxide

interleaved metal foil

thin layers of mica

To maximise capacitance, the plates must be as close together as possible – but not so close that current can pass between them. The dielectric should have a large relative permittivity (also called dielectric constant), and the area of overlap of the plates should be as large as possible. The effective area of overlap can be maximised – either by using multiple plates or by making a roll out of the plates and the dielectric (Figure 4.13). Some practical examples are shown in Figure 4.14.

A dielectric has a dielectric strength, which is the value of potential gradient (or field strength) that causes it to cease to insulate. In the case of air, such breakdown of insulating ability can be seen in the form of sparks. Dry air becomes conducting, or 'breaks down', as a result of the creation of mobile charged particles by ionisation of molecules of the gases, at potential gradients in the region of $3 \times 10^6 \, \mathrm{V \, m^{-1}}$. Commercial capacitors are often labelled with the maximum potential difference that should be applied to them to avoid breakdown occurring and charge passing directly from plate to plate.

Figure 4.14 (below right)
Mica, paper and electrolytic capacitors.

7 Explain in general terms what will happen to the capacitance of the arrangement described in question 4 if the external surfaces of the foil are also coated with a layer of nylon and the foil and nylon are then rolled into a cylindrical shape.

8 Two conductors have a potential difference of 3 kV. What is the minimum distance by which they must be separated in dry air in order to prevent conduction between them by sparks?

9 In a simple electric motor, sparks are visible at the moving contacts as the motor turns, driven by a 6 V supply. What is the maximum separation at which sparks are produced in dry air?

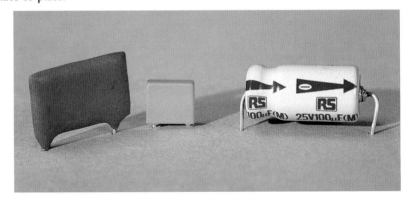

Discharging a capacitor

Suppose that a capacitor has charge Q_0 when a potential difference of V_0 is applied to it. We know that its capacitance is given by

$$C = \frac{Q_0}{V_0}$$

If the capacitor is discharged through a resistance R (Figure 4.15), then at some time later the charge on the capacitor has fallen to Q and the potential difference has fallen to V. The capacitance is a constant for the capacitor, and it has not changed. At this later time we can say that

$$C = \frac{Q}{V}$$

Figure 4.15

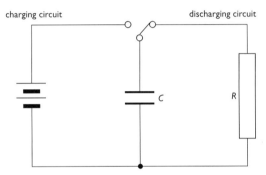

charging circuit   discharging circuit

What pattern does the changing value of the charge on the capacitor follow? The rate of change of the charge is identical to the value of the current that carries the charge away, through the resistor. Current and charge are related by:

$$I = -\frac{dQ}{dt} \qquad (1)$$

The minus sign is necessary since current is positive, but the change in charge on the capacitor is negative – it is losing charge.

As always (from the definition of resistance) we know that

$$I = \frac{V}{R}$$

and so using $C = Q/V$ to substitute for V, we get

$$I = \frac{Q}{CR} \qquad (2)$$

Combining equations (1) and (2),

$$-\frac{dQ}{dt} = \frac{Q}{CR}$$

or

$$\frac{dQ}{dt} = -\frac{Q}{CR}$$

A graph of dQ/dt versus Q for this equation is shown in Figure 4.16a, overleaf. We have come across a similar situation before, in which the rate of change of a value is proportional to the value itself. In radioactive decay, rate of change of a population of undecayed nuclei is proportional to the population itself (Figure 4.16b). Such a proportional relationship between rate of change of value and value itself always leads to an exponential relationship. In the case of capacitor discharge, rate of change of charge is proportional to the charge itself.

Figure 4.16
a The rate of change of charge is proportional to the charge itself. The graph has a gradient $-1/CR$.
b We see the same type of behaviour in radioactive decay, where the rate of change of a population of nuclei, dN/dt, is proportional to the population, N. In that case the gradient is $-\lambda$.

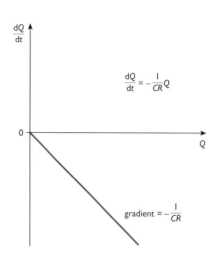

a

$$\frac{dQ}{dt} = -\frac{1}{CR}Q$$

gradient $= -\dfrac{1}{CR}$

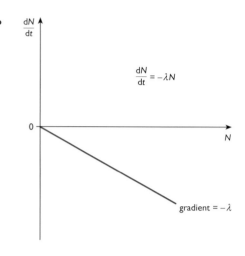

b

$$\frac{dN}{dt} = -\lambda N$$

gradient $= -\lambda$

- For radioactive decay:

$$\frac{dN}{dt} = -\lambda N \qquad\qquad \text{gives us} \qquad N = N_0\,e^{-\lambda t}$$

- For capacitor discharge:

$$\frac{dQ}{dt} = -\frac{Q}{CR} \qquad\qquad \text{gives us} \qquad Q = Q_0\,e^{-t/CR}$$

The step from

Figure 4.17 (below)
A decay curve for a discharging capacitor.

$$\frac{dQ}{dt} = -\frac{Q}{CR} \qquad \text{to} \qquad Q = Q_0\,e^{-t/CR} \qquad \text{is an integration.}$$

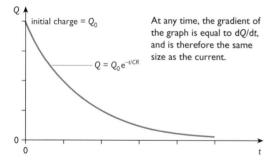

initial charge $= Q_0$

$Q = Q_0\,e^{-t/CR}$

At any time, the gradient of the graph is equal to dQ/dt, and is therefore the same size as the current.

You will not be asked to carry out mathematical integration in an A level physics exam. You can take the step on trust, but you should be aware of the importance of differentiation and integration in mathematical modelling and in physics.

The exponential decay of charge on a capacitor (Figure 4.17) produces the same shape of graph as we have seen for radioactive decay. It would be possible to determine a 'half-life' for the capacitor discharge (Figure 4.18a). However, rather than do this, we relate the decay curve to the actual circuit components and their values, C and R (Figure 4.18b).

Figure 4.18 (below)
Decay curves showing
a halving time and
b time constant.

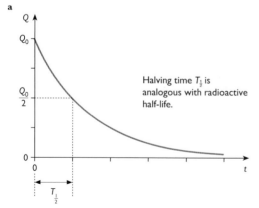

a

Q_0

$\dfrac{Q_0}{2}$

Halving time $T_{\frac{1}{2}}$ is analogous with radioactive half-life.

$T_{\frac{1}{2}}$

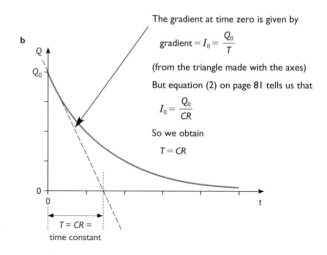

b

Q_0

$T = CR =$ time constant

The gradient at time zero is given by

$$\text{gradient} = I_0 = \frac{Q_0}{T}$$

(from the triangle made with the axes)

But equation (2) on page 81 tells us that

$$I_0 = \frac{Q_0}{CR}$$

So we obtain

$$T = CR$$

At time zero the rate of change of charge (dQ/dt), as well as the charge (Q), is maximum.

rate of change of charge at time zero = gradient of the decay curve at time zero

If we draw a tangent to the curve at time zero, to show the gradient, we see that it makes a triangle with the axes with height Q_0 and base CR (Figure 4.18b). This means that we can measure the product CR as a time, and indeed it can be shown by dimensional analysis that the units of CR can be reduced to those of time. CR is called the **time constant** of the circuit.

Variation of potential difference with time for a discharging capacitor

Note that since $Q = CV$, the equation

$$Q = Q_0 e^{-t/CR}$$

can be written as

$$CV = CV_0 e^{-t/CR}$$

where V is the potential difference across the capacitor at time t and V_0 is the initial potential difference. So,

$$V = V_0 e^{-t/CR}$$

The graph of this equation is shown in Figure 4.19a below.

Variation of current with time for a discharging capacitor

We know that the current in the circuit is the current in the resistor, so

$$I = \frac{V}{R} \quad \text{and initial current} \quad I_0 = \frac{V_0}{R}$$

Substituting for potential difference into the equation of the previous subsection,

$$IR = I_0 R e^{-t/CR}$$

which becomes

$$I = I_0 e^{-t/CR}$$

The graph of this equation is shown in Figure 4.19b.

Figure 4.19
P.d.–time and current–time graphs for a discharging capacitor.

For a discharging capacitor, the potential difference and the current, as well as the charge, decay exponentially. Note that the curves on the graphs have identical time constants, CR.

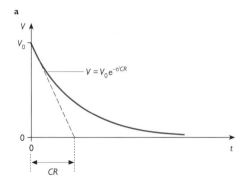

a

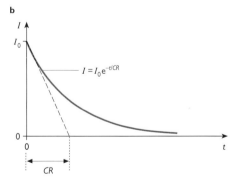

b

10 By dimensional analysis, show that the quantity CR has dimensions of time.

11 Sketch graphs of charge against time, on the same axes, for capacitor discharge for which:
   a $C = 1\,\mu F$, $R = 1\,k\Omega$
   b $C = 2\,\mu F$, $R = 1\,k\Omega$
   c $C = 2\,\mu F$, $R = 2\,k\Omega$.
   In each case, label and give the value for the time constant.

12 For a discharging capacitor with initial charge of $1\,\mu C$, what is the value of charge Q when:
   a $t = CR$
   b $t = 2CR$
   c $t = 3CR$?

13 In what ways is time constant CR similar to and different from
   a radioactive decay constant, λ
   b radioactive half-life?

Charging a capacitor

We have seen how charge stored on a capacitor decreases when it is disconnected from the source of potential difference and connected to a resistor. We can also ask how the charge increases when the potential difference is first applied to a capacitor.

A capacitor is charged by connecting it to a potential difference. The charging process would be instantaneous were it not for resistance of the connecting wires, which may be small but will not be zero. Suppose that the resistance in series with the capacitor is R (Figure 4.20). The potential difference supplied by the source is equal to the sum of the potential difference across R and the potential difference across the capacitor, C. This follows from Kirchoff's Second Law (see *Introduction to Advanced Physics*, page 423) and is always true. So

$$V_{source} = V_R + V_C$$

$$= IR + \frac{Q}{C}$$

Figure 4.20

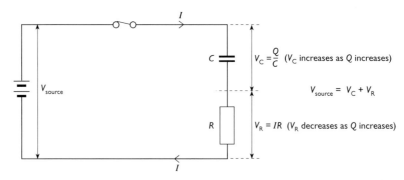

During the charging process, Q is initially zero but increases. We can consider the source potential difference to be constant, and so, as potential difference across the capacitor increases, that across the resistance decreases. At the end of the charging process, current ceases to flow and the charge on the capacitor is the maximum allowed by the source potential difference. This maximum charge, Q_m, is related to V_{source} by

$$V_{source} = \frac{Q_m}{C}$$

Note also that $IR = \frac{dQ}{dt}R$, and so

$$\frac{Q_m}{C} = R\frac{dQ}{dt} + \frac{Q}{C}$$

Rearranging this,

$$\frac{dQ}{dt} = \frac{Q_m}{CR} - \frac{Q}{CR}$$

Note that we now have a relationship between rate of change of a quantity and the quantity itself that is not a simple proportionality but is of the form $y = mx + c$ (Figure 4.21). Again, the process of integration leads to an exponential relationship (Figure 4.22):

$$Q = Q_m(1 - e^{-t/CR})$$

Figure 4.21
The graph of dQ/dt against Q is a straight line.

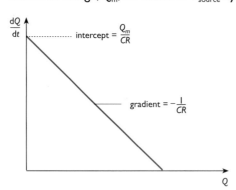

Figure 4.22
The graph of charge against time for a charging capacitor. The quantity CR is again called the time constant, and can be found from the initial gradient of the graph.

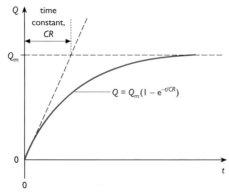

Some uses of capacitor charging and discharging

A capacitor in parallel with a resistor charges during times of high supply potential difference, and discharges through the resistor when supply potential difference falls. It is possible that the capacitor potential difference falls more slowly than the supply potential difference, maintaining a more constant voltage across a load. The potential difference applied to the load is thus smoothed (Figure 4.23).

Figure 4.23
Circuits and p.d.–time graphs for **a** a breaking circuit and **b** a smoothing circuit.

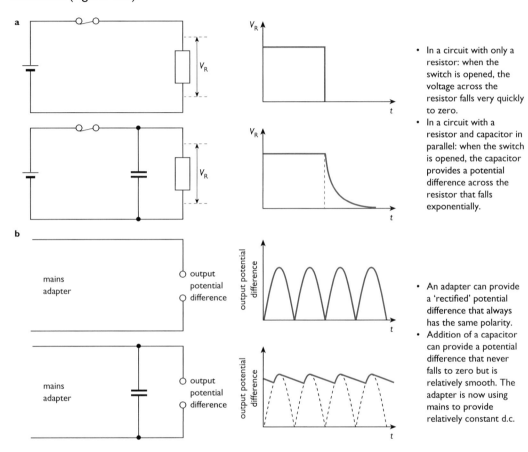

- In a circuit with only a resistor: when the switch is opened, the voltage across the resistor falls very quickly to zero.
- In a circuit with a resistor and capacitor in parallel: when the switch is opened, the capacitor provides a potential difference across the resistor that falls exponentially.

- An adapter can provide a 'rectified' potential difference that always has the same polarity.
- Addition of a capacitor can provide a potential difference that never falls to zero but is relatively smooth. The adapter is now using mains to provide relatively constant d.c.

If a capacitor is either charged or discharged through a resistor, the potential difference across it passes through some value at which switching takes place in a logic circuit to provide binary output 1 rather than 0. The values of capacitance C and resistance R can be chosen so that this switching occurs after a required time (Figure 4.24).

14 A capacitor is connected in series with a resistor and charged.
   a Explain why the potential difference across the resistor decreases with time during the charging.
   b Sketch a graph to show the variation of this resistor potential difference with time.
15 Sketch graphs of potential difference against time for
   a a discharging capacitor
   b a charging capacitor.
16 Sketch graphs of current against time for
   a a discharging capacitor
   b a charging capacitor.
17 **a** A 10 μF capacitor is connected in series with a 1 kΩ resistor, to a potential difference of 6 V. Use the expression $\log_e(V/V_0) = -t/CR$ to calculate the time taken for the potential difference to fall to 1 V. (A scientific calculator will provide the means to do this without too much involvement with the mathematical principles of logarithms.)
   b Show this on a sketch graph.

The discharge of a capacitor can be used for controlled timing of the switching of digital signals. At a time t_s after discharge begins, the system switches from reading '1' to reading '0'. Time t_s can be set by adjusting the values of C and R.

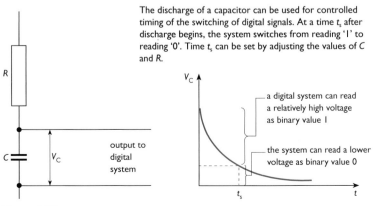

- a digital system can read a relatively high voltage as binary value 1
- the system can read a lower voltage as binary value 0

Figure 4.24
R and C in series, connected to a digital system.

● Combinations of capacitors

Several capacitors, like resistors, can be connected either in series or in parallel. And just as for resistors and their combined or total resistance, total capacitance can be calculated from their individual values. There are, however, some differences.

Capacitors in series

Suppose that there are three capacitors in series (Figure 4.25), with capacitances C_1, C_2 and C_3. The potential differences applied to each one add together to equal the supply voltage – capacitors obey Kirchoff's Second Law (see *Introduction to Advanced Physics*, page 423) just as resistors do. This allows us to write:

$$V = V_1 + V_2 + V_3 \qquad\qquad (3)$$

Figure 4.25
Capacitors in series.

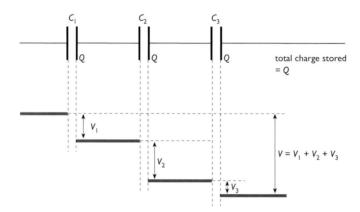

Capacitors in series must all carry the same charge, but the size of the voltage step depends on the capacitance.

Now since the capacitors are in series they must each carry the same charge at all times. To understand this we can think about the capacitors uncharged, and then during the charging process. Only the end two plates are actually connected to the supply. Everything in between becomes charged only because the end two plates are charged. Charge cannot be created here during charging, but can only be redistributed. Effectively the whole system only holds charge Q; this is the charge that must be lost by the system during discharge. So

$$V_1 = \frac{Q}{C_1} \qquad V_2 = \frac{Q}{C_2} \qquad V_3 = \frac{Q}{C_3}$$

and total applied difference is related to total capacitance by

$$V = \frac{Q}{C}$$

So for capacitors in series equation (3) becomes:

$$\frac{Q}{C} = \frac{Q}{C_1} + \frac{Q}{C_2} + \frac{Q}{C_3}$$

or

$$\frac{1}{C} = \frac{1}{C_1} + \frac{1}{C_2} + \frac{1}{C_3}$$

This formula does not follow the same pattern as the formula for resistors in series.

Capacitors in parallel

Figure 4.26
Capacitors in parallel.

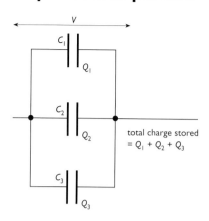

Capacitors in parallel must all have the same potential difference, but the charge stored on each depends on capacitance.

total charge stored
$= Q_1 + Q_2 + Q_3$

Capacitors in parallel, like resistors in parallel, all have the same potential difference. But in parallel there is no reason why they all have to have the same charge. Suppose that there are three capacitors in parallel (Figure 4.26), with capacitances C_1, C_2 and C_3. We can say that

$$V = \frac{Q_1}{C_1} \qquad V = \frac{Q_2}{C_2} \qquad V = \frac{Q_3}{C_3}$$

and for the system as a whole we can say that

$$V = \frac{Q}{C}$$

where Q is the total charge stored and C is the effective combined capacitance. Each capacitor stores charge independently, and the total charge stored is the sum of the individual charges. So for capacitors in parallel:

$$Q = Q_1 + Q_2 + Q_3$$

$$CV = C_1V + C_2V + C_3V$$

or

$$C = C_1 + C_2 + C_3$$

18 Calculate the total capacitance of capacitors of 2 nF, 5 nF and 7 nF connected
   a in series
   b in parallel.
19 Compare the formulae for capacitors in series and parallel with those for resistors in series and parallel (see *Introduction to Advanced Physics*, pages 423–5). Explain why the patterns are different.
20 A sheet of dielectric material of relative permittivity ϵ_r can be inserted between two plates, replacing air. If the sheet is just partially inserted then the arrangement can be considered as two capacitors in parallel, one with air as the dielectric and the other with the sheet as the dielectric. Sketch a graph of total capacitance against area of inserted sheet, and give algebraic expressions for the intercept (on the y-axis) and the gradient.

Energy storage by capacitors

Charge flows on to a capacitor, and can stay there until the capacitor is discharged some time later. During the discharge, the current can cause heating of a resistor or can even drive a motor. So the charged capacitor is acting as an energy store.

To consider whether the amounts of energy involved might be of practical value, we can first plot a graph of the potential difference across the capacitor against the charge stored by it (Figure 4.27a). The graph is a straight line, with gradient being a constant for the capacitor, and equal to 1/C. What is particularly interesting here is the area under the graph. This is a measure of the energy stored, in a way that is analogous to the area under a force–distance graph providing a measure of the work done. We can continue the analogy to the particular case of a spring, which has a simple proportionality relating force and extension (Figure 4.27b).

Figure 4.27
For capacitors and springs, energy stored is equal to the area under the appropriate graph.

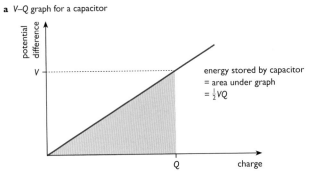

a V–Q graph for a capacitor

energy stored by capacitor
= area under graph
= $\frac{1}{2}VQ$

In general, potential difference is work done per unit charge transferred. For a capacitor, the potential difference is changing. The *average* potential difference is the work done per unit charge transferred, and since the graph is a straight line the average potential difference is half the final value.

work done = energy stored

$= \dfrac{\text{average potential}}{\text{difference}} \times \text{charge}$

$= \frac{1}{2}VQ$

b force–extension graph for a spring

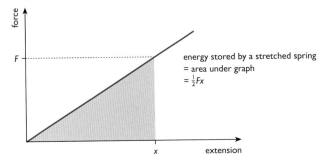

energy stored by a stretched spring
= area under graph
= $\frac{1}{2}Fx$

For a capacitor, we obtain

energy stored = area under V–Q graph
$= \frac{1}{2} \times$ final voltage $\times$ final charge
$= \frac{1}{2}VQ$

Note that since $C = Q/V$, we can also say

energy stored $= \frac{1}{2}CV^2$

or

energy stored $= \frac{1}{2}\dfrac{Q^2}{C}$

21 Which stores more energy, a 1.0 μF capacitor at a potential difference of 1 kV, or a spring of spring constant $5 \times 10^3\,\text{N m}^{-1}$ when at an extension of 15 mm?

22 Explain why capacitors are of little use for storage of energy for normal domestic purposes of lighting, heating and so on.

The amounts of energy that can be stored by individual capacitors in practical circuits are small. However, computer circuitry has relatively low energy consumption, and here capacitors can provide back-up energy sources, if only for short periods.

● **Comprehension and application**

Capacitance and the human nervous system

A strand, or axon, of a resting human nerve cell has a potential difference between the inside and outside of about 65 mV, and yet is only about 5 nm thick (Figure 4.28). There is no current through the membrane, except when channels open to allow ions to flow through, and this happens when an impulse passes along the length of the axon (Figures 4.29 and 4.30). In the meantime, the axon remains at rest, with a 'soup' of charged particles on either side. It is because these two liquids hold different concentrations of ions that there is a potential difference across the membrane. Thus the axon has capacitance and the membrane acts as the dielectric between two charged layers. The relative permittivity of the membrane is 8.0.

Figure 4.28
An axon membrane acts as a capacitor.

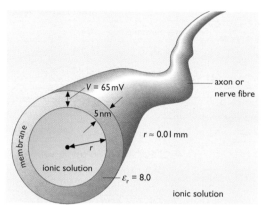

Figure 4.29 (left)
A human brain contains about ten billion (10^{10}) neurons or nerve cells. The long threads of nerve cells are called axons, here coloured purple. The axons make contact with other nerve cells (yellow) at junctions called synapses.

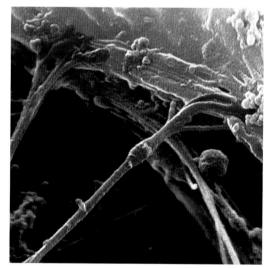

Figure 4.30 (right)
The human nervous system at work!

23 **a** What is the potential gradient across the resting axon membrane?
  b What is the minimum value of dielectric strength required by the membrane material? Compare this with the dielectric strength of dry air.
24 **a** Calculate the capacitance of an axon fibre 1 m long. (Surface area of a cylinder is given by $2\pi rl$ where r is the radius and l is the length.)
  b Calculate the capacitance per square metre of axon surface.
25 At a potential difference of 65 mV, what is the total energy stored by the axon fibre in question 24?

26 Channels may open and close through the axon membrane to allow ions to pass through. Simplifying a complex subject somewhat, a small area, say 1×10^{-9} m², of an axon becomes approximately discharged in this way when an impulse passes along the nerve, and then reverts to its normal 'resting' state.
  a Explain why energy is required to revert to the natural state.
  b Calculate the amount of energy required for this area.
  c Hence explain why avoidance of mental activity is regarded by some as a valuable survival technique!

● **Extra skills task**

Communication

Use a wide range of resources, including CD ROMs, books and popular and specialist periodicals, to carry out research into the uses of capacitors in computer circuitry. Gather and edit information to create a short (10 to 20 minutes) illustrated presentation suitable for other physics students. It should contain technical information at an appropriate level, with an explanation of the technical illustrations and the vocabulary that you use.

Examination questions

1 a i Sketch a graph to show the variation with charge Q [x-axis] of the potential difference V [y-axis] across the plates of a capacitor.
ii Use your graph to show that the energy E stored in a capacitor of capacitance C with a p.d. V across its plates is given by

$$E = \tfrac{1}{2}CV^2 \qquad (4)$$

b In the diagram, each capacitor has a capacitance of 2200 µF. The p.d. across each capacitor must not exceed 25 V.

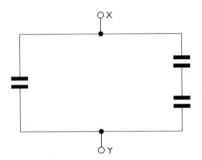

For the arrangement shown, calculate
i the total capacitance [µF] between X and Y,
ii the maximum energy that can be stored. (4)

UCLES, A Level, 4833, March 1998

2 a A 100 µF capacitor is connected to a 12 V supply. Calculate the charge stored. Show on the diagram the arrangement and magnitude of charge on the capacitor. (3)

b This 100 µF charged capacitor is disconnected from the battery and is then connected across a 300 µF uncharged capacitor. What happens to the charge initially stored on the 100 µF capacitor?
Calculate the new voltage across the pair of capacitors. (4)

London, AS/A level, Module Test PH1, June 1997

3 The equations describing the decay of a sample of a radioactive isotope and the discharge of a capacitor, capacitance C, through a resistor, resistance R, show many similarities. The number of radioactive atoms N in the sample and the charge Q on the capacitor both decay exponentially.

Complete the table below to obtain the equivalent formulae for the capacitor discharge.

	Radioactive decay	Capacitor discharge
Time constant/s	$1/\lambda$	
Half life/s	$\ln 2/\lambda$	
Rate of decay	$dN/dt = -\lambda N$	$dQ/dt =$
Decay law	$N = N_0 e^{-\lambda t}$	

(4)

London, A level, Module Test PH4, June 1997

4 The circuit below is used to investigate the discharge of a capacitor.

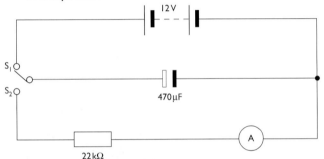

With the switch in position S_1 the capacitor is charged. The switch is then moved to S_2 and readings of current and time are taken as the capacitor discharges through the resistor. The results are plotted on the graph below.

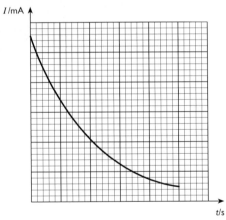

a Calculate the maximum charge stored in the capacitor. (1)
b Make suitable calculations to enable you to add scales to both axes of the graph. (4)

A second 470 µF capacitor is connected in series with the original capacitor. The switch is moved back to S_1 to recharge the capacitors.

c State the new charge stored. (1)

The switch is moved to S_2 and another set of discharge readings is taken.

d Draw a second line on the graph to show how the current varies with time during this discharge. (2)

e How could the charge stored in the capacitors be estimated *from your graph*? (1)

London, A level, Module Test PH4, June 1996

5 a A capacitor is constructed from two sheets of aluminium foil separated by a dielectric film of relative permittivity 2.70. The sheets are square, of side 870 mm, and are 0.02 mm apart. Show that the value of the capacitance is about 1 μF. (2)

This capacitor is charged to 9 V.

b Calculate the charge stored. (2)

The charged capacitor is then connected across a 2.2 MΩ resistor.

c Calculate the initial current in the circuit. Calculate the time constant of the circuit. (3)

d Draw a graph of how the current would vary over the first 10 s. Label the axes and include appropriate scales. (4)

Edexcel, A level, Module Test PH4, June 2000

6

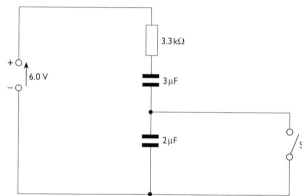

Calculate the maximum energy stored in the 3 μF capacitor in the circuit above

a with the switch S closed, (2)

b with the switch S open. (4)

London, AS/A level, Module Test PH1, January 1998

7 The switch S in the circuit below can be used to charge and discharge a capacitor of capacitance C.

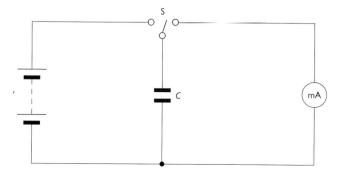

When this charging and discharging is carried out at a frequency *f*, the average current registered by the milliammeter is given by

$$I_{av} = fCV$$

where *V* is the e.m.f. of the battery.

a i The effective capacitance *C* of two capacitors of capacitance C_1 and C_2 connected in series is given by

$$\frac{1}{C} = \frac{1}{C_1} + \frac{1}{C_2}$$

Explain how you could use the circuit to verify this relationship. (4)

ii Show that the expression $I_{av} = fCV$ is homogenous with respect to units. (2)

b The graph shows the variation of current *I* in the milliammeter with time *t* for one discharge.

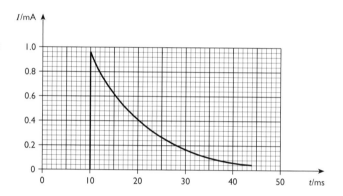

i The resistance of the milliammeter is 1500 Ω. Use the graph to calculate the capacitance of the capacitor. (4)

ii Sketch a graph of the charge *Q* on the capacitor against time *t* for one discharge.

State two features of the above graph and the graph you have drawn which are related quantitatively. (4)

iii Name one other branch of physics in which a graph of this shape occurs. What quantity is then plotted on the y-axis? (2)

London, A level, Synoptic Paper PH6, January 1997

8 A capacitor of capacitance 220 μF is charged so that the potential difference across its plates is 9.0 V.

a Calculate, for the capacitor,

i the charge stored,

ii the energy stored. (2)

b A resistor is connected across the terminals of the capacitor. You may assume that the capacitor discharges completely through the resistor in 2.0 ms. Calculate

i the mean power dissipated in the resistor,

ii the mean current flowing in the resistor during discharge. (2)

c The capacitor is now charged so that the potential difference across its plates is 18 V. A resistor is connected across its terminals so that it may be assumed that the capacitor discharges completely in the same time, 2.0 ms.

What is the mean power dissipated in the resistor? (2)

NEAB, AS/A level, Module Test PH01, June 1998

9 The plates of an air-filled parallel plate capacitor are each of area $1.5 \times 10^{-2}\,m^2$ and are initially separated by 2.0 mm.

a Calculate
i the capacitance of the capacitor,
ii the charge on the capacitor when connected across a 2500 V d.c. supply. (3)

b The plates of the capacitor are brought closer together while it is still connected to the 2500 V supply. State and explain what happens to the size of the charge on the plates. (3)

c The maximum electric field strength that the air can support before an electrical breakdown occurs is $3.0 \times 10^6\,V\,m^{-1}$. The plates of the capacitor are brought together, with the 2500 V supply connected, so that they are as close as possible without causing breakdown.

Calculate the separation of the plates when this condition is reached. (2)

NEAB, AS/A level, Module Test PH03, June 1998

10 The figure shows the diagram of a laboratory power supply which has an e.m.f. of 3200 V and internal resistance *r*.

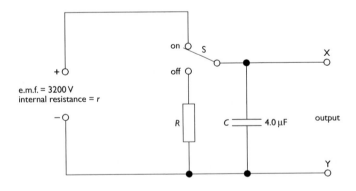

The capacitor *C* ensures a steady output voltage between X and Y when the switch S is in the *on* position.

The resistor *R* allows the capacitor to discharge when S is switched to the *off* position.

a For safety reasons, the power supply is designed with an internal resistance, *r*, so that if X and Y are short-circuited, the current that flows does not exceed 2.0 mA.

Calculate the value of *r*. (1)

b For safety reasons, when the supply is switched off, the capacitor must discharge from 3200 V to 50 V in 5.0 minutes.

i Calculate the time constant of the circuit when the capacitor discharges from 3200 V to 50 V in 5.0 minutes. (3)

ii Calculate the value of the discharge resistor *R*. (2)

iii Explain briefly whether or not increasing the value of *R* would make the supply safer. (1)

c **i** Calculate the charge on the capacitor when the potential difference across it is 3200 V. (1)

ii Sketch a graph to show the variation of the charge on the capacitor with time, as the capacitor discharges through the resistor. Include appropriate scales on the axes of your graph. (2)

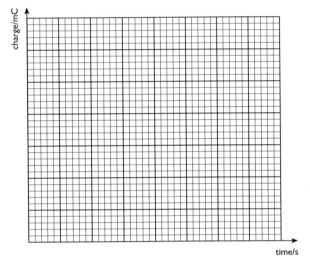

iii Sketch a graph to show the variation of the charge on the capacitor with voltage across the capacitor as it discharges. Include appropriate scales on the axes of your graph. (2)

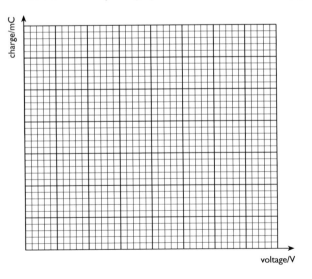

AEB, AS/A level, Module Paper 6, January 1998

5 Magnets and magnetic forces

THE BIG QUESTIONS
- What are the origins of magnetic force?
- How can we measure magnetic fields?
- How can we apply magnetic fields and the resulting forces?

KEY VOCABULARY
diamagnetic (behaviour)   dipole   domain   ferromagnetic (behaviour)   Hall effect
Hall probe   Hall voltage   magnetic moment   monopole   paramagnetic (behaviour)
permeability of a vacuum   relative permeability   solar wind   solenoid

BACKGROUND
Magnetic fields are at work in both the magnificent and the mundane.

Figure 5.1
Knowledge of magnetism and awareness of nature's patterns of magnetic behaviour enhances our experience of the wondrous sight of the polar aurorae.

Figure 5.2
This poster from 1906 shows that knowledge of magnetic effects was giving rise to life-changing technologies.

The aurora (Figure 5.1) are huge displays of shimmering and shifting colours that result from the natural interactions of a magnetic field, the field of the Earth, and the charged particles that stream from the Sun.

During the 19th and early 20th century, people learned about interactions of magnetic field and flowing charge, resulting in technologies such as the electric motor that liberated people from much physical drudgery both in industry and at home (Figure 5.2).

Later in the 20th century, magnetic force on particle beams provided the basis for entertainment technologies and for tools of further inquiry into the nature of the world.

Fields, field lines and the Earth's magnetism

Figure 5.3
Combinations of the magnetic field of a bar magnet and that of the Earth.

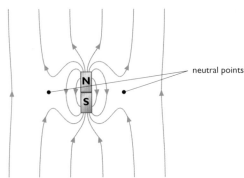

neutral points

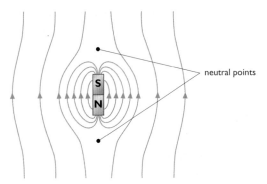

neutral points

Figure 5.4
The north poles of compasses are attracted by the south poles of other magnets, including that of the Earth. The Earth's magnetic south pole is in the north.

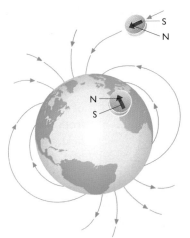

We use field lines to show the directions of the forces that would act on a small isolated magnetic north pole. This is a somewhat imaginary situation since an isolated magnetic pole, called a **monopole**, is thought not to exist. Magnetic poles come in pairs, called **dipoles**. The Earth behaves as a magnet, with two poles, just like all other magnets. The field lines of other magnets combine with those of the Earth, showing the directions of the net forces (Figure 5.3). This can create points where there are no net forces, or neutral points.

Compass needles point (approximately) along the lengths of field lines, so the field lines of the Earth can be traced with compasses (Figure 5.4). We know that the north pole of a compass points north, and was once more fully called a 'north-seeking' pole. But we also know that like poles repel and unlike poles attract. So if we are to continue to call the north-seeking pole of a compass needle simply its 'north pole', then we have to accept that towards the north of our planet there is a magnetic south pole.

We know about past reversals of the Earth's magnetic field from the study of rock along the middle of the oceans, where new soft rock oozes up to the ocean floor. There are bands of rock, some magnetised in one direction and some in the opposite direction (Figure 5.5), telling us about the polarity of the Earth's field at the time of solidification. These bands of alternately magnetised rock are the chief evidence we have for tectonic plate theory and 'continental drift'.

Figure 5.5
Mid-ocean patterns of magnetisation of rock.

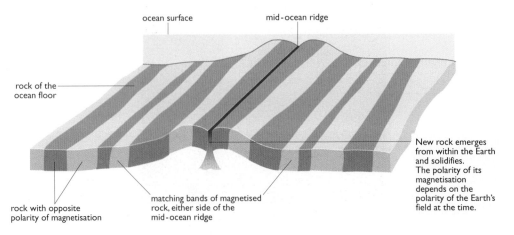

ocean surface

mid-ocean ridge

rock of the ocean floor

New rock emerges from within the Earth and solidifies. The polarity of its magnetisation depends on the polarity of the Earth's field at the time.

rock with opposite polarity of magnetisation

matching bands of magnetised rock, either side of the mid-ocean ridge

Magnetic force and flux density

In dealing with gravitational field we use field strength as a useful ratio of force and mass. For electric field, likewise, we use a ratio of force and charge. Mass and charge are the properties of bodies that experience the forces. Magnetic force acts between *moving* charges, and so here we use a quantity that depends on force and on both charge and velocity, and also on the relative orientation of the field and the velocity. The quantity, though analogous to gravitational and electric field strengths, is not called field strength but *flux density* (see *Introduction to Advanced Physics*, page 149).

Consider a charge q moving at velocity v perpendicular to a field of flux density B (Figure 5.6a). The flux density is related to the force acting on the particle by:

$$B = \frac{F}{qv}$$

The unit of flux density is the tesla (T).

Where velocity is *not* perpendicular to the field lines, we must consider its component that is (Figure 5.6b). This perpendicular component can be written as $v\sin\theta$, where θ is the angle between the field and the velocity. In that case,

$$B = \frac{F'}{qv\sin\theta}$$

Flux density provided by a field, the velocity of a charged particle through it, and the resulting force, are all vector quantities. Magnetic field lines show the direction of the flux density. For a given field and charge velocity, the size of the force that acts on the charge is maximum when flux density and velocity are mutually perpendicular. The force is zero when flux density and velocity are parallel.

Figure 5.6
The force on a moving charge in a magnetic field of flux density B.

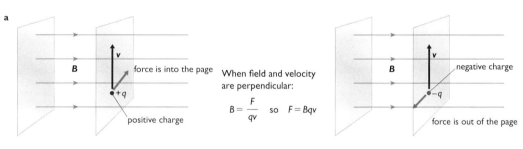

a

force is into the page

positive charge

When field and velocity are perpendicular:

$$B = \frac{F}{qv} \quad \text{so} \quad F = Bqv$$

negative charge

force is out of the page

b

$v\sin\theta$

θ = angle between B and v

for positive charge, force is into the page
for negative charge, force is out of the page

When field and velocity are *not* perpendicular:

$$B = \frac{F'}{qv\sin\theta} \quad \text{so} \quad F' = Bqv\sin\theta$$

If (as here) the flux density B is the same in both cases, then since $qv\sin\theta < qv$,

$$F' < F$$

Figure 5.7 (below)
For a positive charge carrier, the relative directions of perpendicular field, velocity and the resulting force are given by the 'left-hand motor rule'.

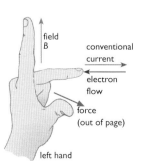

field
B

conventional current

electron flow

force (out of page)

left hand

As you can demonstrate to yourself with a pair of pens or pencils, any two intersecting straight lines lie in a single plane. For our moving charged particle, B and v, and indeed $B\sin\theta$ and $v\sin\theta$, all lie in the same plane. The force that acts is perpendicular to the plane, and its direction depends on the sign of the charge.

For a positive charge carrier and conventional current, that direction is given by the 'left-hand motor rule' (Figure 5.7), also known as Fleming's left-hand rule. The second finger represents the direction of the velocity of the positive charge, the first finger represents the direction of the field lines. The thumb will then indicate the direction of the force, as shown.

For a negative charge carrier, an electron, we can use the left-hand rule but imagine the charge carrier travelling in the opposite direction to the pointing finger.

The electrical origins of magnetism

We know that there is a magnetic field around any current-carrying conductor, including a loop of wire (Figure 5.8a). We could use a simple 'plotting compass' exercise to trace some of the field lines, and we would see a pattern that is quite familiar. Analogously, a single electron in a simple atom, say a ground-state hydrogen atom, moves in a pathway that can be thought of as a loop (Figure 5.8b). An atom can indeed develop a magnetic field due to orbital motion of electrons, provided that the effects of all of its electrons do not balance out. But such movement of the electron around the nucleus does not provide its only contribution to the observed magnetic field. The magnetic field is more complicated than that, and this is explained by supposing that the electron also has spin around its own axis (Figure 5.8c). Indeed, it seems that the protons (and neutrons) of the nucleus are also capable of spin, though the magnetic contribution that this makes to the magnetic behaviour of the atom as a whole is much smaller than that of the electron.

Figure 5.8
a A coil of wire has a magnetic field around it. In a hydrogen atom, **b** orbital motion and **c** spin motion of the electron both contribute to the atom's magnetic behaviour.

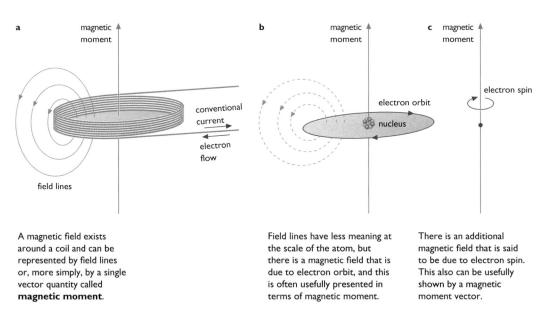

A magnetic field exists around a coil and can be represented by field lines or, more simply, by a single vector quantity called **magnetic moment**.

Field lines have less meaning at the scale of the atom, but there is a magnetic field that is due to electron orbit, and this is often usefully presented in terms of magnetic moment.

There is an additional magnetic field that is said to be due to electron spin. This also can be usefully shown by a magnetic moment vector.

A hydrogen atom has a single electron and significant permanent magnetic field. Helium has an extra electron, and no permanent magnetic field at all. The magnetic effects of the two electrons, it seems, act to cancel each other out.

All materials have the potential to behave in a **diamagnetic** way. This behaviour involves disturbance of the orbital motion of electrons by an external field, resulting in some net (but always weak) magnetism of the atoms of material, which *opposes* the direction of the applied field.

Some materials show **paramagnetic** as well as diamagnetic behaviour, and for these the paramagnetic behaviour dominates. The spins of electrons in the atoms of the material align with an external field in a coherent way. This *reinforces* the applied field. However, thermal agitation of atoms tends to cause their electron spins to fall out of alignment, so paramagnetic reinforcement of applied field is temperature dependent.

Hydrogen, for example, is described as a paramagnetic material. For helium, on the other hand, the effects of its two electrons cancel each other, and it does not show paramagnetic behaviour, but behaves only in a diamagnetic way. That is, it does not show net magnetic behaviour as a result of spin of its electrons.

A few materials – including iron, nickel and cobalt – show a particularly strong type of behaviour, called **ferromagnetic** behaviour (Figure 5.9) or ferromagnetism. As in paramagnetic materials, each atom acts as a tiny magnet and these magnets can be made to align with an external magnetic field. The difference in the ferromagnetic materials is that the atoms align strongly, even in a relatively weak field, and remain aligned to some extent after the field is removed. Within the structure of these elements, atoms seem to be grouped together in **domains**, each domain holding about 10^{17} to 10^{20} atoms. The atoms in each domain are permanently aligned with each other, whether or not an external field is applied. The alignment of one domain, however, does not always match that of its neighbours.

Figure 5.9
Ferromagnetic materials, when placed in a field, 'concentrate' the field lines and strongly intensify flux density in the space they occupy. Paramagnetic materials do so weakly, while materials that behave only diamagnetically 'dilute' the field lines and the flux density.

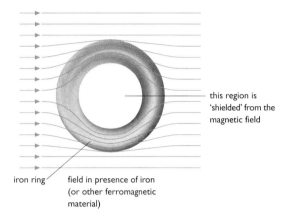

field in vacuum
(or air)

iron ring   field in presence of iron
(or other ferromagnetic
material)

this region is
'shielded' from the
magnetic field

When iron is newly solidified, the domains within the material are randomly aligned (Figure 5.10a). An external field can then make the domains line up together (Figure 5.10b), creating a strong magnet. The alignment of one domain with another seems to be self-sustaining – so that it is semi-permanent. Domain alignment can be destroyed, by heating, by intense physical vibration, or by rapidly changing magnetic field.

Figure 5.10
Domains.

a   randomly orientated domains

b   aligned domains

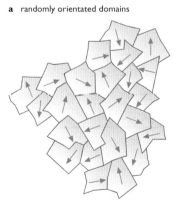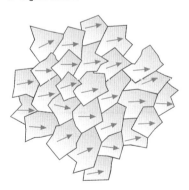

We know that the Earth contains iron, but we also know that this iron is too hot to retain much coherent magnetic alignment of domains. An alternative explanation for the Earth's magnetic field is that there is coherent flow of charged material, as if around a coil. If that were the case, we might expect any such flow to be related to the Earth's rotation. From observing the other planets of the Solar System, it does seem that the faster the rate of rotation, the stronger the planetary magnetic field. However, the Earth's magnetic field does not just wander from year to year (and even from day to day) but it also completely reverses in polarity from time to time, whereas the direction of spin of the Earth does not change. So though people have used compasses for many hundreds of years, we still don't know what are the origins of the Earth's magnetic field.

Magnetic permeability

The magnetic behaviour of a material – its tendency to reinforce or oppose an external field – is quantified in terms of its permeability, μ. Permeability of a material is very dependent upon conditions, such as temperature. Nevertheless, it is a useful quantity.

Permeability of a material can be compared to that of a vacuum, μ_0. This **permeability of a vacuum** is one of nature's fundamental constants. It has a value $4\pi \times 10^{-7}\,\text{H}\,\text{m}^{-1}$, where H is the abbreviation for the henry, the unit of inductance, which is dealt with further in Chapter 6.

Suppose that a point in space has a flux density B_0 in a vacuum (that is, without the presence of a material), and a flux density B_m when the material is present. The permeability of the material is then given by the ratio

$$\frac{\mu}{\mu_0} = \frac{B_m}{B_0}$$

This ratio of permeability of a material to that of a vacuum can be called **relative permeability**, μ_r:

$$\mu_r = \frac{B_m}{B_0}$$

Table 5.1
Some relative permeability values.

Material	μ_r	Magnetic behaviour
vacuum	1.000 000	
copper	0.999 990	diamagnetic
water	0.999 991	diamagnetic
silicon	0.999 986	diamagnetic
tungsten	1.000 079	paramagnetic
platinum	1.000 300	paramagnetic
iron	up to 2000	ferromagnetic
nickel	up to 2000	ferromagnetic
nickel/iron/molybdenum alloy (supermalloy)	100 000	ferromagnetic

Both μ and μ_0 have units henry per metre, $\text{H}\,\text{m}^{-1}$. Since μ_r is a ratio of quantities with the same units and dimensions, it has no units or dimensions. Some values of μ_r for a variety of materials are given in Table 5.1. Paramagnetic materials increase flux density, relative to that in a vacuum under the same conditions, and so have $\mu_r > 1$. Diamagnetic materials decrease flux density, and so have $\mu_r < 1$. For ferromagnetic materials, $\mu_r \gg 1$.

Materials change magnetic field patterns. Diamagnetic effects, due to disturbance of electron orbits, reduce flux density. In paramagnetic materials, unbalanced effects of orbit and spin of electrons tend to reinforce the field. The effects are still quite weak, however, except in ferromagnetic materials.

Some metal alloys are ferromagnetic, but compounds such as oxides of iron also provide significant intensification of magnetic fields. Iron oxides are compounds and not metals, and they are electrical insulators. This reminds us that magnetic behaviour is not dependent on the presence of free electrons. In fact, in iron itself the free electrons, which originate in the outermost atomic shell, play little part in magnetic behaviour; the magnetic effects are due to electrons that remain bound to individual atoms.

1 What is the relative permeability of a material that has permeability of $4.1\pi \times 10^{-7}\,\text{H}\,\text{m}^{-1}$?

2 Hydrogen atoms are paramagnetic but hydrogen molecules are diamagnetic. What does this suggest about the relative orientation of the orbits and spins of electrons in the pair of atoms in a hydrogen molecule?

3 a Interatomic spacing in an iron crystal is in the region of 0.25 nm. Estimate the dimensions of a domain containing 10^{18} atoms.
   b Would such a domain be visible using
   i an optical microscope
   ii an electron microscope?

4 Compare the effects of diamagnetic and paramagnetic materials on magnetic field with the electrical polarisation that occurs in dielectrics in the presence of an electric field, as described in Chapter 4.

5 Why are compass needles designed for use in Sweden (Northern Hemisphere) weighted differently about the pivot to needles for use in South Africa (Southern Hemisphere)?

6 Predict the motion of the magnet in each of the cases shown in Figure 5.11, assuming that there are no resistive forces acting.

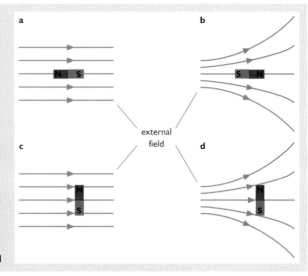

Figure 5.11

Some field patterns and flux densities

The flux density at distance r from a straight conductor carrying current I is known to obey a standard formula:

$$B = \frac{\mu_0 I}{2\pi r}$$

(Strictly, the formula applies to a vacuum, but behaviour in air is a close approximation to that in a vacuum.)

Figure 5.12
The direction of the field around a straight conductor is given by the 'right-hand grip rule'.

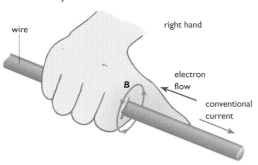

wire · right hand · electron flow · B · conventional current

The field lines are concentric circles, becoming more widely spaced with distance from the conductor, as flux density decreases with increasing distance from the conductor. The direction of the field lines, and hence of the force on a north pole, also behaves in a standard way, which is best remembered in terms of a hand. If the conductor is imagined held in the right hand as in Figure 5.12, then if the thumb points along the direction of conventional current, the four fingers are curved in the same direction as the field lines.

(Those who prefer to think in terms of electron flow will have to use the thumb of the left hand to indicate this, with the fingers again showing the direction of the field.)

The easiest kind of field to consider is usually a uniform one, where flux density B is the same everywhere, and the field lines are parallel and equally spaced. But we know that the field lines around coils and bar magnets are curved. Some arrangements, though, provide a good approximation to a uniform field (Figure 5.13). One of these is inside a tightly wound long coil, or **solenoid**. Another is between the closely spaced opposite poles of two bar magnets. And a third is within a Helmholtz coil arrangement, which has a pair of coils whose separation is equal to their radii.

Figure 5.13
Methods for creating uniform fields:
a solenoid,
b bar magnets,
c Helmholtz coils.

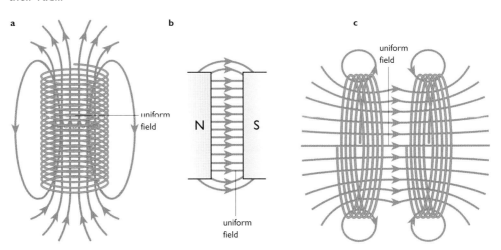

a · b · c · uniform field · N · S · uniform field · uniform field

Inside a solenoid, the flux density is found to be given by:

$$B = \mu_0 n I$$

where μ_0 is again the permeability of a vacuum, I is current, and n is the number of turns of the solenoid per unit length, and therefore has units m^{-1}. Note that distance from the central axis of the coil, r, does not appear in the formula. Flux density can be said to be independent of radial position, or in other words the same at all distances from the centre. So the field is uniform.

If a solenoid is given an iron core, then the field strength within it is increased, as given by

$$B = \mu_0 \mu_r n I$$

7 What is the flux density at a point 0.1 m from a wire carrying a current of 1.0 A?

8 Sketch graphs of flux density, B, against r, for:
 a a long straight conductor, where r is distance perpendicularly from the conductor
 b the inside of a solenoid, where r is distance perpendicularly from its central axis.

9 Compare the equations relating magnetic flux density and the permeability of a vacuum, μ_0 for a straight conductor and a solenoid, to the equations relating electric field strength and the permittivity of a vacuum, ϵ_0 (see Chapters 3 and 4), for a field around a point charge and the uniform field between two charged plates. List differences and similarities in the forms of the equations. Include a comparison of the values of μ_0 and ϵ_0.

10 The graph in Figure 5.14 shows the relationship between relative permeability of the iron in the coil of a solenoid against the current the coil carries.
 a Does relative permittivity of a dielectric show any such variation that depends on conditions or is it more nearly a constant for the material?
 b What is the relative permeability of the iron for a current of 0.2 A?
 c If, at this current, the flux density at the centre of the solenoid is 0.5 mT when it has a non-ferromagnetic core, what will it be when an iron core is put in place?

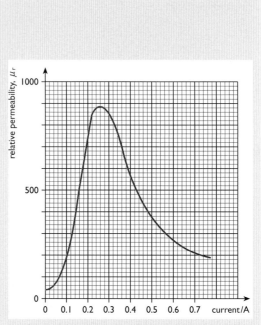

Figure 5.14

Particle pathways in uniform and non-uniform fields

A particle may enter a magnetic field with an initial velocity. Magnetic force is always perpendicular to the velocity. The effect of the force is to change the direction but not the magnitude of the velocity. If the field is uniform then force is constant, and in that case it is behaving exactly as a centripetal force. Magnetic force tends to produce circular motion.

Such circular motion occurs in the case of J. J. Thomson's famous 'crossed fields' experiment (which demonstrated the particle nature of electrons) when magnetic field only is applied to an electron beam. It also happens in the case of the mass spectrometer. (See *Introduction to Advanced Physics*, pages 58 and 77 respectively.) In both of these the initial velocity is perpendicular to the field (Figure 5.15a). The result is that velocity, field and the force experienced by the particles are all continuously perpendicular.

However, it is possible that initial velocity is not perpendicular to the field (Figure 5.15b). In the absence of resistive forces, the component of the initial velocity that is parallel to the field remains constant. The component that is perpendicular to the field tends to produce circular motion. The combination of these two components of motion is a helical motion.

There is another interesting scenario, in which the particle experiences another force in addition to the magnetic force. This additional force may cause velocity to change in magnitude as well as direction (Figure 5.15c). Then the pathway will be tend to spiral. This happens where there is resistive force, reducing velocity and causing inwards spiralling motion, such as for particle tracks (see Figure 5.16, page 102). In a cyclotron, where there is an electric force acting intermittently on particles to accelerate them, the particle pathways spiral outwards. (For a description of a cyclotron see pages 228–30.)

Figure 5.15
Trajectories of particles
entering a magnetic field
and of particles accelerated
within a field.

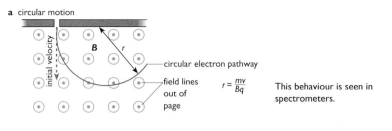

a circular motion

initial velocity

B

r

circular electron pathway

field lines
out of
page

$r = \dfrac{mv}{Bq}$

This behaviour is seen in
spectrometers.

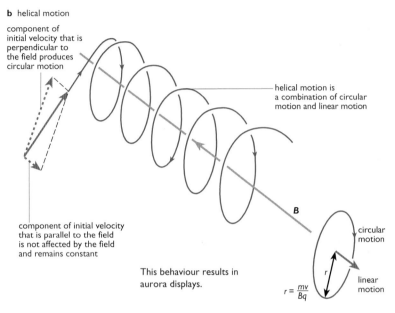

b helical motion

component of
initial velocity that is
perpendicular to
the field produces
circular motion

helical motion is
a combination of circular
motion and linear motion

component of initial velocity
that is parallel to the field
is not affected by the field
and remains constant

B

circular
motion

r

linear
motion

This behaviour results in
aurora displays.

$r = \dfrac{mv}{Bq}$

c spiral motion

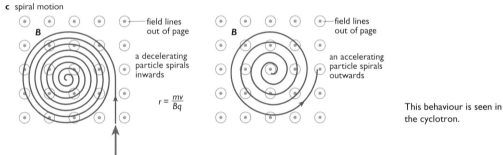

B

field lines
out of page

a decelerating
particle spirals
inwards

$r = \dfrac{mv}{Bq}$

B

field lines
out of page

an accelerating
particle spirals
outwards

This behaviour is seen in
the cyclotron.

The motions of charged particles can be summarised as follows:

a when entering a uniform magnetic field with an initial velocity perpendicular to the field and no resistance to motion, the result is circular motion

b when entering a uniform magnetic field with an initial velocity that is not perpendicular to the field, the result is helical motion

c when accelerated within the field by additional force that has, for some or all of the time, a component parallel to velocity, the result is a tendency to spiral.

We can say, in the simple case where velocity and flux density are mutually perpendicular (as in Figure 5.15a), that

$$Bqv = \frac{mv^2}{r}$$

so that the radius of the pathway is given by

$$r = \frac{mv}{Bq}$$

Figure 5.16
Particle detectors may have a magnetic field applied to them, so that charged particles within them follow pathways that are arcs of circles, obeying $r = mv/Bq$. Radius of the pathway can be measured from the images of the pathway, and the size of the flux density is known. Together with other data, this can be used to determine information about particle mass, speed or charge.

In TVs and electron microscopes the priority is to deflect beams of electrons to targets or screens, rather than to produce motions that must be circular. In many cases non-uniform fields are used to achieve the required effects. This kind of deflection is sometimes called magnetic lensing.

The Earth's magnetic field steers the **solar wind** of charged particles, mostly electrons and protons, that come our way from the Sun (Figure 5.17). The force acting on such a particle is always perpendicular to the field lines, but the particles have very large initial velocity relative to the Earth. This, together with the non-uniformity of the Earth's field, prevents circular motion, and instead the particles tend to travel in elaborate near-helical pathways. The result is that the energetic particles are directed towards the Earth's poles, where collisions with particles of the air cause excitation and ionisation. The air particles then emit light, providing the huge displays of the northern lights (aurora borealis) and the southern lights (aurora australis) – see Figure 5.1.

Figure 5.17
Solar wind and particle pathways.

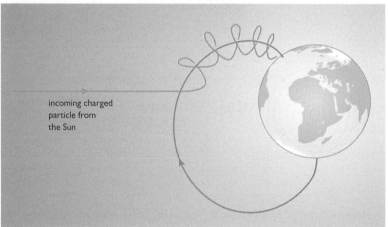

incoming charged particle from the Sun

Suitably shaped magnetic fields can hold charged particles within them indefinitely. This occurs naturally in the Van Allen belts high above the Earth, where charged particles, mostly from the solar wind, become trapped in the Earth's magnetic field. The particles are energetic, and therefore behave as ionising radiation. The Van Allen belts are dangerous to astronauts. (See Chapter 7, page 163 for a further example of magnetic containment of moving charged particles – plasma containment for controlled nuclear fusion.)

11 Calculate the radius of an electron pathway in a field of flux density $10 \, \mu T$ if it has a perpendicular speed of $10^5 \, m \, s^{-1}$. (electron mass $= 9.1 \times 10^{-31} \, kg$, electron charge $= -1.6 \times 10^{-19} \, C$)

12 **a** Explain why the pathway of a particle of given initial velocity entering a uniform electric field has a different shape to that of the same particle entering a uniform magnetic field.
b Explain why the pathways of charged particles have different shapes in cyclotrons and mass spectrometers.

13 If electrons travel through the vacuum of TV sets at speeds of $3 \times 10^7 \, m \, s^{-1}$, and the Earth's magnetic field strength in equatorial regions is $5 \times 10^{-5} \, T$ parallel to the ground, do designers of TV sets need to take account of whether sets will be used in northern Canada or in equatorial Africa?

Magnetic force on a straight conductor

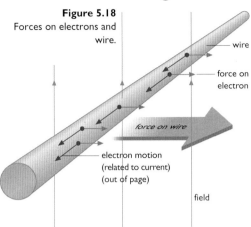

Figure 5.18
Forces on electrons and wire.

wire

force on electron

force on wire

electron motion
(related to current)
(out of page)

field

A magnetic field exists around a current-carrying conductor. Not surprisingly, it interacts with other magnetic fields, and as a result the conductor exerts and experiences force.

A wire carrying a current acts as a beam of electrons, except that the beam is now relatively rigid – it cannot bend without taking the mass of the whole wire with it. Each individual free and moving electron experiences force, and the sum of these forces provides a force on the wire (Figure 5.18) that can result in its acceleration. If it lies entirely within a uniform magnetic field, then each unit length of wire experiences the same force and hence (in the absence of force that opposes its motion) the same acceleration.

Consider length l of wire, carrying current I. In the simpler case where field and current – and hence electron velocity – are mutually perpendicular (Figure 5.19a):

force on each electron $= Bev$ 　　(using e for charge)
force on the wire $= Bev \times$ number of electrons in length l of wire

Here v is the average electron velocity, which we know from *Introduction to Advanced Physics*, page 415, is the drift velocity, and therefore related to current by $I = nAve$. In this equation, n is the number of electrons per unit volume, and A is the cross-sectional area of the conductor. So for a wire of cross-sectional area A,

number of electrons in length $l =$ number of electrons per unit volume $\times$ volume of length l
$= nAl$

So,

force on the wire $= Bev \times nAl = B \times nAve \times l$
$= BIl$

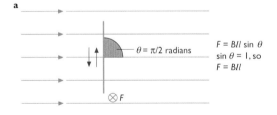

Figure 5.19
The force on a wire depends on its orientation relative to the field.

a

$\theta = \pi/2$ radians

$\otimes F$

$F = BIl \sin \theta$
$\sin \theta = 1$, so
$F = BIl$

This formula applies when field and wire are mutually perpendicular. When they are not, but there is an angle θ between them (Figure 5.19b), then we must consider the component of one of the quantities that is perpendicular to the other. This gives us

force on the wire $= BIl \sin \theta$

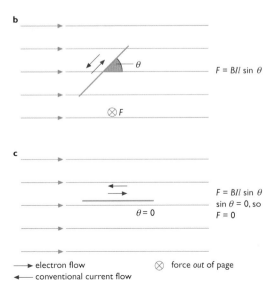

b

θ

$\otimes F$

$F = BIl \sin \theta$

c

$\theta = 0$

$F = BIl \sin \theta$
$\sin \theta = 0$, so
$F = 0$

→ electron flow
← conventional current flow

⊗ force *out of page*

14 Explain why thermal motion of electrons in a wire produces no net magnetic force on the wire.

15 A current-carrying wire of length 0.10 m lies in a magnetic field of flux density 12 mT and is subject to a force of maximum value 5 mN that resists its motion. What is the minimum current that is required to produce acceleration?

Couple acting in moving coil meters and in electric motors

In a moving coil meter or a d.c. electric motor the coil rotates but the field lines and the length of the side of the coil are arranged to be nearly mutually perpendicular at all times. So, considering a rectangular coil, the force acting on *one* side of the coil is given by

$$F = BIlN$$

where N is the number of turns in the coil. The same force acts on the opposite side of the coil, so that the two forces form a couple (Figure 5.20). There are also forces acting on the ends of the coil and these forces are also in opposite directions. However, unlike the 'side' forces, these 'end' forces are co-linear, and they are in balance, producing no net acceleration of the coil.

Figure 5.20
In a uniform magnetic field, the forces on the side of a coil act as a couple.

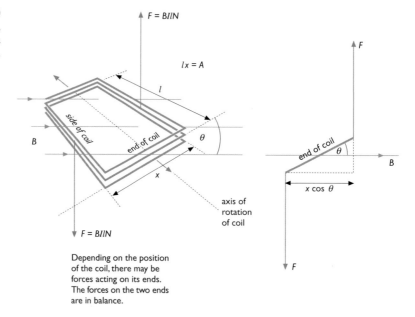

Depending on the position of the coil, there may be forces acting on its ends. The forces on the two ends are in balance.

The forces on the sides of the coil form a couple which can produce rotational acceleration (until the couple is balanced by an opposing couple, as is supplied by a spring in a meter, or by other opposing forces in a motor once a given rate of rotation is reached). But in a uniform field the effective value of the couple does not remain constant as the coil rotates. In general the size of the couple is given by:

$$\text{couple} = \text{force} \times \text{perpendicular component of separation}$$
$$= BIlN \times x\cos\theta$$

where x is the width of the coil and θ is the angle between the field lines and the plane of the coil. But lx is the area of the coil, which we could write as A, so that

$$\text{couple} = BANI\cos\theta$$

This formula applies to circular and other coils, as well as to rectangular coils.

Meter deflection

For a meter with a uniform field (Figure 5.21a, opposite), once it is settled in its equilibrium position, so that a reading can be made,

$$BANI\cos\theta = \text{restoring couple} = k\theta$$

assuming that at zero reading the plane of the coil is parallel to the field lines. Here k is the torsional spring constant of the meter's restoring spring and θ is the angular deflection of the coil.

However, many meters use magnetic fields that are not uniform but are provided by permanent magnets with cylindrical pole faces (Figure 5.21b). Close to these faces, the field lines are not parallel but (very nearly) radial, so that the force on a side of the coil is constant whatever the rotational position of the coil. The couple is therefore independent of angle θ. For a meter with such a radial field,

$$BANI = k\theta$$

B, A, N and k are all constant for a particular meter. So for a meter with a radial field, the angle of deflection of the coil, θ, is simply proportional to the current. Such a meter can use a 'linear' scale.

Figure 5.21
A radial field allows use of a linear scale.

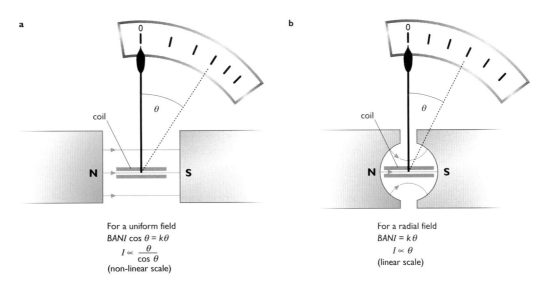

For a uniform field
$BANI \cos \theta = k\theta$
$I \propto \dfrac{\theta}{\cos \theta}$
(non-linear scale)

For a radial field
$BANI = k\theta$
$I \propto \theta$
(linear scale)

Motor coil rotation

For the meter in use, θ settles to a constant value, but for the motor there is no spring providing a restoring couple. For the motor, θ is time dependent.

The force on one side of a coil in a uniform field will always be in the same direction if it always carries current in the same direction. If, say, the force is upwards then it will always be upwards. The force on the other side of the coil is then always downwards. This means that the coil settles in the position with its plane perpendicular to the field lines, at which the forces on the two sides are co-linear, and the value of the couple is zero (see Figure 5.22 overleaf, inset). Rotation does not continue. So in a d.c. motor, in order to achieve further rotation, it is necessary to reverse the current in the coil, and therefore in each of its sides, as the plane passes through the position at which it is perpendicular to the field lines. This is achieved by connecting the coil to a battery by way of a split ring commutator arrangement and carbon brushes (Figure 5.22). There is relative rotation of these (usually the brushes are stationary and the commutators turn with the coil) but they remain in contact. The effect is to reverse the current in each side of the coil at the required times. The use of commutators and brushes solves a further problem – if a simple coil were connected to a power supply and allowed to rotate, the connecting wires would twist and twist together.

Figure 5.22
A split ring commutator and carbon brushes reverse the current when the plane of the coil is vertical.

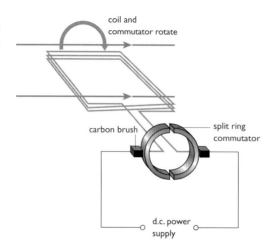

coil and commutator rotate

carbon brush

split ring commutator

d.c. power supply

Without reversal of direction of current, the force on each side of the coil is always in the same direction. The coil may vibrate, but it will not rotate continuously.

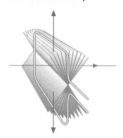

Parallel conductors and current measurement

Although current is defined in terms of charge, it is impossible to detect charge itself inside wires. All that we can do is to detect the effects of the charge in the outside world, and even this is seldom observable unless the charge is moving. We detect moving charge in a wire by its magnetic and heating effects.

To establish a system of practical measurement we are forced to start with what we can observe. For current, we can start with magnetic effect. We can place two wires close to each other, so that their fields interact, and the wires exert forces on each other. Because it is possible to set up such systems of straight conductors and to measure the force between them, potentially anywhere in the world, we can use this as the basis for defining a practical international unit of current. The amp, or ampere, is defined as the current flowing in each of two infinitely long conductors one metre apart when the force acting on one metre of them is equal to 2×10^{-7} N.

Once established by practical measurement, we can use the amp as the basis for defining other electrical units, starting with the coulomb. (The coulomb is the amount of charge passing a point when a current of one amp flows for one second.)

16 Draw a sketch to show why there is no couple acting on the ends (which are perpendicular to the coil axis) of a rectangular coil, regardless of the rotational position of the coil.

17 a What is the maximum couple that can act on a coil with 100 turns of area 5.0×10^{-3} m² carrying a current of 40 mA in a field of flux density 10 mT?
   b What is the minimum couple that acts on this coil?
   c Sketch a graph of the size of the couple against time for a coil that rotates with constant angular velocity (that is, such that $d\theta/dt$ is constant).

18 A meter coil has 50 turns and area 2×10^{-3} m², and lies in a radial magnetic field of 5.5 mT. When there is a current of 10 mA in the coil, the angular deflection, θ, is 0.1 radians.
   a Express 0.1 radians in degrees.
   b What is the value of the meter spring's torsional spring constant, k?

19 Ignoring effects of distortion of the magnetic field due to curvature of the ends of wires, the force between two wires each 1 m long and 1 m apart is 2×10^{-7} N when a current of 1 A flows in each.
   a When is the force repulsive and when is it attractive?
   b Give an equation for calculating the field due to a long straight wire.
   c What would you expect to happen if
     i the current in one wire is doubled
     ii the current in both wires is doubled
     iii the distance between the wires is halved?

The Hall probe for measuring flux density

The **Hall probe** is a small piece of semiconductor with connections to a circuit, which is used to measure flux density, B. It is based on an effect called the **Hall effect** – which is the tendency for flowing charged particles to be deflected towards one edge of a conductor due to magnetic force. ('Conductor' here is used to mean any material that carries a current, and so includes a semiconductor.) This sets up a charge difference, and so a potential difference, between opposite edges. The potential difference is called the **Hall voltage**, V_H.

Figure 5.23
A Hall probe is used to measure magnetic flux density, B.

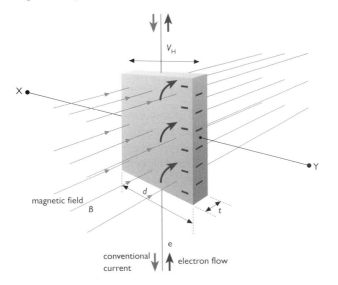

There is a tendency for electrons to accumulate at one side of the Hall probe (as shown), setting up a 'Hall voltage', V_H. The Hall voltage can be detected and measured using contacts X and Y.

An electron moving in the conductor experiences a magnetic force, given by

$$F_B = Bev$$

Where it is electrons that are the charge-carrying particles, they accumulate on one side of the conductor (Figure 5.23). This sets up the Hall voltage, and hence introduces an electric force on each of the particles, given by

$$F_E = eE$$

where E is the electric field strength. For a constant potential gradient the electrical field strength is related to the potential difference by

$$E = \frac{V_H}{d}$$

where V_H is the Hall voltage and d is the width of conductor across which the voltage exists (see Figure 5.23).

Particles cease to drift under the action of the magnetic force, and the Hall voltage ceases to grow, when the magnetic force is balanced by the electric force, that is, when

$$F_B = F_E$$

$$Bev = \frac{eV_H}{d}$$

We take v to be the average velocity of charge carriers in the conductor, i.e. the drift velocity,

$$v = \frac{I}{nAe}$$

Then we get

$$\frac{BI}{nAe} = \frac{V_H}{d}$$

and

$$B = \frac{V_H nAe}{Id}$$

$A = t \times d$, where t is the thickness of the probe in a direction that is perpendicular to the orientation of the Hall voltage and to the electron flow or current, and so

$$B = \frac{V_H nte}{I}$$

For a given piece of material, n is constant, as is e. V_H and I can be measured. Thus measurement of the Hall voltage provides a measurement of the flux density that surrounds the probe.
   Note also that since

$$V_H = \frac{BI}{nte}$$

the smaller the number of free charged particles there are per unit volume, the bigger the Hall voltage. An insulator would be capable of developing a much larger Hall voltage than a metal can – but, of course, an insulator cannot carry a current. Semiconductors, however, for which the number of charge carriers per unit volume is about 10^6 times smaller than for a metal, *can* develop and maintain relatively large Hall voltages – which is why Hall probes are made from semiconductor material.

The Hall effect and positive charge carriers

Figure 5.24
The polarity of a Hall voltage depends on whether the probe is made of n-type or p-type semiconductor material.

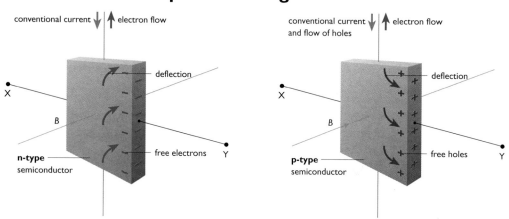

Natural semiconductors have free electrons and free vacancies, or holes, in their energy levels. When an electric field is applied, the electrons move in one direction and the holes move in the other. In an n-type semiconductor, additional free electrons have been added to the crystal, and the dominant means of flow is by electrons. In a p-type semiconductor, additional free vacancies or holes exist, and the material behaves exactly as if it contained positive free charge carriers. (See *Introduction to Advanced Physics*, pages 179–81.)
   Positive charge carriers travel in the opposite direction to electrons. They also, of course, have opposite sign. The result of these two opposites is that in the same magnetic field, electrons and holes experience magnetic deflection in the *same* direction within the Hall probe. If the semiconductor is n-type and carries current due to motion of electrons, then the polarity of the Hall voltage is determined by accumulation of electrons on one face of the Hall probe (Figure 5.24). In a p-type conductor, the holes can be thought of as accumulating on the same face, creating a charge distribution and potential difference with opposite polarity. So the observed polarity of a Hall voltage indicates whether the material is n-type or p-type.

20 Explain why, for measurement of flux density, a Hall probe should be as thin as possible in one of its dimensions. Draw a sketch to show this dimension relative to flux density and current.
21 An external potential difference must be applied to the Hall probe to ensure that there is a current in it. What would you expect to happen to the Hall voltage as this applied potential difference increases?
22 Show that the equation

$$V_H = \frac{BI}{nte}$$

is dimensionally coherent.
23 A Hall probe is used with a voltmeter. Explain how, given a Hall probe, suitable power supply and voltmeter, and a solenoid producing a known flux density at its axis, you would 'calibrate' the probe system in order to measure a range of unknown flux densities.

Flux

Michael Faraday developed ways of representing magnetic fields – using field lines – that were later also applied to gravitational and electric fields. Field lines are a very productive way of visualising phenomena that are quite abstract. One of Faraday's concepts was that of *flux*.

Field lines relate to the directions in which forces act. (In the case of magnetic fields they show the direction of force acting on a small monopole, and *not* the direction of the force acting on a moving charge.) We can also draw field lines close together to represent a stronger field, and further apart for a weaker field. In fact we do this when we draw the field pattern around a magnet – where the field lines are closest together, close to the magnetic poles, the field is strongest. Using this idea, Faraday saw that it was useful to visualise the magnetic field lines passing normally (at right-angles) through different areas. The same number of field lines indicated the same flux, ϕ, whatever the size of the area (Figure 5.25).

Figure 5.25
Two different areas can have the same flux.

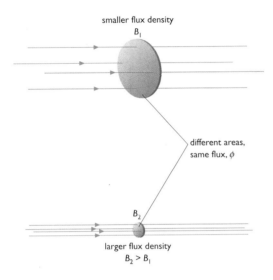

smaller flux density
B_1

different areas, same flux, ϕ

B_2

larger flux density
$B_2 > B_1$

We now define flux from flux density, and say that total flux is the product of the flux density and the normal area:

flux = flux density $\times$ normal area
$$\phi = BA$$

If field lines do not pass normally through an area, we can still work out a value of flux, by considering the component of B that is normal to area A (Figure 5.26). If the angle between the field lines and the area is θ, then the flux is given by

$$\phi = BA \sin \theta$$

The unit of flux is clearly equivalent to $T\,m^2$, which is given its own name in the SI system – the weber, Wb. The concept of flux provides an approach to magnetic field that is extremely useful in understanding magnetic interactions that result in induction of e.m.f. and current, which is the subject of the next chapter.

24 What happens to the flux inside a solenoid when
   a an iron core is added
   b the current is doubled
   c the radius is doubled?
25 What are the differences between flux density and mass density?

Figure 5.26

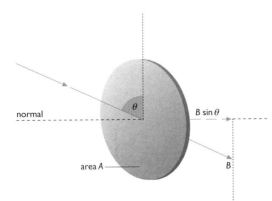

normal

θ

$B \sin \theta$

area A

B

● **Comprehension
and application**

Medical magnetism

Your body is not a magnet. But its hydrogen atoms can be made to align with a strong magnetic field. An MRI machine (Figure 5.27) uses superconducting coils to create magnetic fields that are strong enough to turn the human body into a magnet.

Figure 5.27
The patient in the scanner (background) is subject to a very strong magnetic field. The field becomes non-uniform, so that each point in the patient's body lies in a field of flux density unique to that point. The result is that hydrogen atoms at each point in the body emit radio waves of 'signature' frequencies.

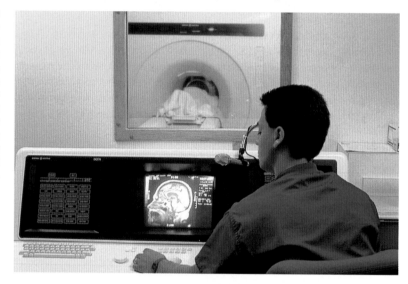

The hydrogen atoms align, but not rigidly with the applied field – their magnetic axes rotate around the field lines, like spinning tops whose axes can rotate around the Earth's gravitational field lines (Figure 5.28). The wobble of the hydrogen atoms is called *precession*. The frequency of precession depends on the applied flux density as well as on the properties of the hydrogen atom.

Figure 5.28
Precession of the magnetic axis of a hydrogen atom compared with precession of a spinning top.

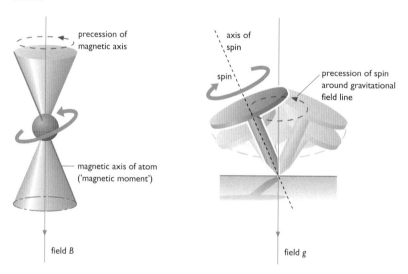

The body is then subjected to a burst of radio waves with a frequency that is equal to the frequency of precession. The energy of the radio waves is taken in by the atoms, causing them to be thrown out of alignment.

The tendency is for the atoms quickly to fall back into alignment. If the arrangement were left unchanged they would lose exactly the same amount of energy as they gained from the radio waves, by emitting waves of the same frequency. However, before this can happen the applied magnetic field is changed, from being uniform to one having gradients of intensity in each of the three dimensions. So the atoms do not fall back to their original frequency of precession, but fall back to a new frequency that matches the new flux density in which they find themselves. The sequence of events is shown in Figure 5.29.

Figure 5.29
Energy loss by an atom as it realigns with the new magnetic field is not the same as that gained from the initial radio waves, and an atom accordingly gives out a frequency that is not the same. Also, because of the variation in flux density from place to place, the frequency of emitted radio waves depends on the atom's location.

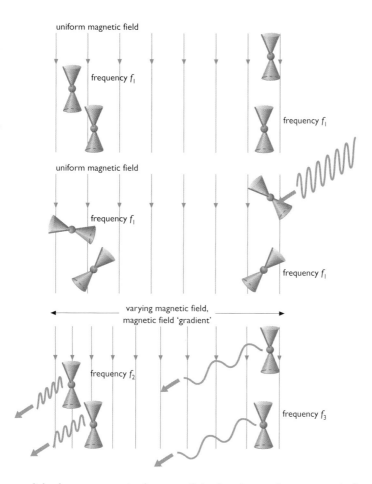

uniform magnetic field

frequency f_1

frequency f_1

uniform magnetic field

frequency f_1

frequency f_1

varying magnetic field, magnetic field 'gradient'

frequency f_2

frequency f_3

Hydrogen atoms precess in a strong magnetic field.

A burst of radio waves provides energy and disturbs the alignment of the hydrogen atoms.

The hydrogen atoms fall back into alignment. To do this they emit radio waves. The frequency of these waves depends on the flux density that *now* surrounds them. Because the magnetic field varies from place to place, hydrogen atoms in different places emit different frequencies. So by detecting the intensities of different frequencies, the densities of hydrogen atoms in different places can be known.

A hydrogen atom in the top of the head, say, does not emit the same frequency of radio wave as a hydrogen atom in the neck. Radio receiving equipment, part of the scanner, can therefore distinguish between waves from the different places. Not only that, but it can examine the intensity of the waves of different frequency, originating in different parts of the body. From these intensities it is possible to compare densities of hydrogen atoms in different places. Since different tissues have different hydrogen densities, the scanner can build up pictures of the tissues – including tumours that would otherwise have to be located by surgical trial and error. In the brain, the technique can distinguish between oxygenated blood and de-oxygenated blood, and so can locate brain activity associated with processes ranging from simple physical actions to deep emotions.

26 Compare magnetisation of the human body during an MRI scan with magnetisation of a piece of iron by an external magnetic field.
27 The field strength required for MRI scanning is in the region of 2 T.
  a Calculate the force on a potassium ion ($q = 1.6 \times 10^{-19}$ C) moving at a speed of $1\,\mathrm{m\,s^{-1}}$.
  b Explain whether you would expect this field to
  i exert forces on thermally moving charged particles in the body
  ii provide energy to such particles, and thereby have a heating effect
  iii exert forces on ions moving during a nerve impulse (see page 89)
  iv magnetise the body permanently.

● **Extra skills task** Information Technology and Communication
  1 Use CD ROMs and the Internet to obtain up-to-date information on brain imaging research, which should include MEG (magnetoencephalography) as well as functional MRI (or fMRI).
  2 Use a DTP package to prepare a brief account of the purposes and methods of the research.
  3 Many people argue that brain imaging technologies provide stronger and stronger evidence that the human brain is an entirely physical organ, while there are many others who believe that the mind and the self are separate from the physical brain. In a group, express your views in turn. Discuss whether the evidence is conclusive for either of these viewpoints. Is it impossible that such conclusive evidence will ever be found?

Examination questions

1 The diagram shows an arrangement in a vacuum to deflect protons into a detector using a magnetic field, which can be assumed to be uniform within the square shown and zero outside it. The motion of the protons is in the plane of the paper.

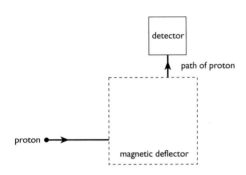

a Sketch the path of a proton through the magnetic deflector. At any point on this path draw an arrow to represent the magnetic force on the proton. Label this arrow F. (2)

b State the direction of the uniform magnetic field causing this motion. (1)

c The speed of a proton as it enters the deflector is $5.0 \times 10^6 \, \text{m s}^{-1}$. If the flux density of the magnetic field is 0.50 T, calculate the magnitude of the magnetic force on the proton. (2)

d If the path were that of an electron with the same velocity, what *two* changes would need to be made to the magnetic field for the electron to enter the detector along the same path? (2)

AQA (NEAB), AS/A level, Module Test PH03, March 1999

2 The diagram shows a slice of p-type semiconducting material that has been connected into a circuit so that a current I flows in the direction shown.

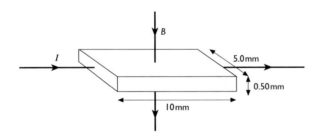

The semiconductor is placed in a magnetic field of flux density 0.050 T at right angles to the surface of the slice. The semiconducting material has a resistivity of $5.0 \times 10^{-2} \, \Omega \, \text{m}$ and a charge carrier density of $4.0 \times 10^{22} \, \text{m}^{-3}$. The potential difference between the ends of the slice is 5.0 V.

a Calculate
  i the current through the slice,
  ii the drift velocity of the charge carriers. (4)

b i Calculate the force exerted on each of the charge carriers due to the presence of the magnetic field.
  ii Indicate the direction of this force on the diagram.
  iii Explain why the presence of the magnetic field will produce across the slice a *steady* potential difference called the Hall voltage.
  iv Indicate on the diagram the two faces across which the Hall voltage is developed.
  v Calculate the magnitude of the Hall voltage. (8)

NEAB, AS/A level, Module Test PH05, March 1998

3 Solar storms consist of protons and helium nuclei emitted from the sun. A particular solar storm hits the Earth's atmosphere at a speed of $1.2 \times 10^6 \, \text{m s}^{-1}$. The particles pass above the northern magnetic pole moving parallel to the Earth's surface. In this region the Earth's magnetic field is directed vertically downwards and has a flux density of $5.8 \times 10^{-5} \, \text{T}$.

a Calculate the radius of the path of a proton from the storm as it passes directly above the Earth's northern magnetic pole. (3)

  mass of a proton $= 1.7 \times 10^{-27} \, \text{kg}$
  charge on a proton $= 1.6 \times 10^{-19} \, \text{C}$

b The diagram shows the initial velocity of a proton and the direction of the Earth's magnetic field. Draw and label the path of the proton and that of a helium nucleus which has the same initial velocity as the proton. State on the diagram the radius of the path of each particle.

  mass of a helium nucleus $= 6.8 \times 10^{-27} \, \text{kg}$
  charge on a helium nucleus $= 3.2 \times 10^{-19} \, \text{C}$

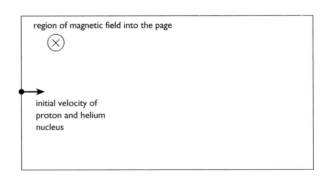

AQA (AEB), A level, Module Paper 8, Summer 1999

4 a The magnitude of the force on a current-carrying conductor in a magnetic field is directly proportional to the magnitude of the current in the conductor. With the aid of a diagram describe how you could demonstrate this in a school laboratory. (4)

b At a certain point on the Earth's surface the horizontal component of the Earth's magnetic field is 1.8×10^{-5} T. A straight piece of conducting wire 2.0 m long, of mass 1.5 g, lies on a horizontal wooden bench in an East–West direction. When a very large current flows momentarily in the wire it is just sufficient to cause the wire to lift up off the surface of the bench.

State the direction of the current in the wire.

Calculate the current.

What other noticeable effect will this current produce? (4)

London, A level, Module Test PH4, June 1997

5 The diagram shows a current flowing through a piece of semiconductor material. The charge carriers are electrons. A magnetic field acts perpendicular to the current in the semiconductor.

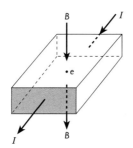

a i State the magnitude of the force experienced by the electron e, due to its movement with velocity v at right angles to a magnetic field of flux density B. (1)
ii Draw an arrow, labelled F, onto the diagram to indicate the direction of the force experienced by the electron. (1)
b Electrons build up on one side of the semiconductor, leaving an equal positive charge on the opposite side. This leads to a potential difference, known as the Hall voltage, V_H, across the piece of semiconductor.
i State the magnitude of the electric field strength set up across the semiconductor, in terms of V_H and the width d of the semiconductor. (1)
ii Draw onto the figure lines of equipotential, labelled P, and electric field lines, labelled E, for the field set up by the Hall voltage.
iii Electrons entering the semiconductor will experience a force due to the electric field. State the magnitude of this force in terms of V_H, d, and the charge e on an electron. (1)
iv The Hall voltage builds up until the forces on the electrons, caused by the magnetic field and by the electric field, are equal and opposite. Show that the Hall voltage is given by:

$$V_H = Bvd$$

where v = the drift velocity of the electrons. (1)
v The semiconductor has a width of 0.83 mm. It is placed in a magnetic field of flux density 1.6 T and the drift velocity of the electrons is 0.048 m s^{-1}. By performing a calculation, suggest a suitable range for a voltmeter to measure the Hall voltage. (2)

AEB, A level, Module Paper 8, January 1998

6 The diagram below shows a beam of electrons which have been accelerated by a potential difference V, travelling in an evacuated tube. A magnetic field acts at right angles to their direction of motion in the shaded region and into the plane of the paper.

a On the diagram draw the path of the electrons in the shaded region. (2)

In the diagram below, a pair of conducting plates, 2.5 cm apart, has been introduced into the shaded region. A potential difference is applied to the plates and is gradually increased until it reaches 400 V when the path of the electrons is a straight line.
b Indicate on the diagram the polarity of the plates. (1)

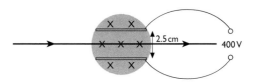

c Calculate
i the electric field strength in the region between the plates,
ii the force on an electron due to this field. (3)
d The magnetic flux density in the shaded region is 1×10^{-3} T. Show that the speed of the electrons must be 1.6×10^7 m s^{-1}. (2)
e Calculate the potential difference V required to accelerate electrons to this speed. (3)

London, A level, Module Test PH4, January 1997

7 a Explain what is meant by a neutral point in a field. (2)

The diagram shows two similar solenoids A and B. Solenoid A has twice the number of turns per metre. Solenoid A carries four times the current as B.

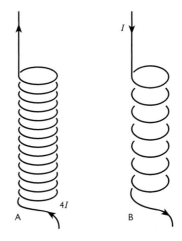

b Draw the magnetic field lines in, around and between the two solenoids. (4)

c If the distance between the centres of A and B is 1 m, estimate the position of the neutral point. Ignore the effect of the Earth's magnetic field. (3)

London, A level, Module Test PH4, January 1998

8 The diagram shows a beam of protons entering a region where a magnetic field acts at right angles to the beam.

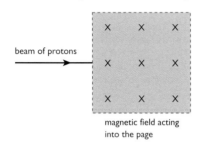

beam of protons

magnetic field acting into the page

a Draw the path of the protons after entering the magnetic field. (1)

b Add to the diagram the path of a beam of protons travelling at a lower speed which enters the magnetic field from the opposite direction. (2)

The fact that a moving charge experiences a force in a magnetic field is used in an electromagnetic pump which moves molten sodium through pipes in a nuclear power station. The pipes are situated in a magnetic field and a direct current is passed through the sodium.

The following diagrams show a pipe of rectangular cross-section containing molten sodium.

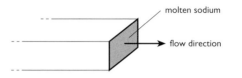

molten sodium

flow direction

cross-section of pipe (molten sodium flows out of page)

c Show on either diagram the direction in which a current would have to flow through the sodium and the orientation of the magnetic field if the molten metal is to flow in the direction shown by the arrow. (3)

Edexcel, A level, Module Test PH4, June 2000

9 A solenoid is formed by winding 250 turns of wire on to a hollow plastic tube of length 0.14 m.

a Show that when a current of 0.80 A flows in the solenoid the magnetic flux density at its centre is 0.0018 T. (2)

The solenoid has a cross-sectional area of $6.0 \times 10^{-3}\,\mathrm{m^2}$. The magnetic flux emerging from one end of the solenoid is $5.4 \times 10^{-6}\,\mathrm{Wb}$ $(\mathrm{T\,m^2})$.

b Calculate the magnetic flux density at the *end* of the solenoid. (2)

c Why is the flux density at the end of the solenoid not equal to the flux density at the centre? (1)

Edexcel, A level, Module Test PH4, June 2000

6 Electromagnetism and alternating current

THE BIG QUESTIONS
- In what ways can electrical changes have magnetic effects, and magnetic changes have electrical effects?
- What useful circuit components rely on such changes?
- What is the relevance of such changes to electricity generation and distribution?
- What is the relevance of such changes to the fundamental nature of electric and magnetic fields, and to the nature of light itself?

KEY VOCABULARY angle of dip   back e.m.f.   capacitive reactance   eddy currents   electromagnetic induction   Faraday's Law   filter circuit   flux linkage   full-wave rectification   half-wave rectification   henry   inductive reactance   inductor   *LCR* circuit   Lenz's Law   mutual inductance   primary coil   resonant circuit   rms values   secondary coil   self-inductance   smoothed d.c.   step-down transformer   step-up transformer   turns ratio

BACKGROUND Physics has always had a dual nature. The study of electromagnetic induction led, on the one hand, to a deeper understanding of the nature of light, and from that to new ways of thinking about space and time. It changed how we think (Figure 6.1). On the other hand, it led to applications of electricity in just about every aspect of the lives we lead, and also to communications technologies based on the electromagnetic spectrum. It changed how we live (Figure 6.2).

Figure 6.1
Physics changes how we think – it provides new models of the world.

Figure 6.2
Physics shapes how we live.

E.m.f. induced by movement

A charged particle moving through a magnetic field experiences a force – a phenomenon covered in the previous chapter. The charged particle might be one of many in a TV electron beam, and as a result of the force the beam is deflected (Figure 6.3a). Or it might be one of many in a current-carrying wire, such that the whole wire experiences force (Figure 6.3b). There is another way in which an electron in a wire can be made to experience force due to a magnetic field; that is, by physical movement of the whole wire (Figure 6.3c).

Figure 6.3
Moving electrons through a magnetic field:
a in a beam,
b as a current in a wire, and
c by physical movement of a wire.

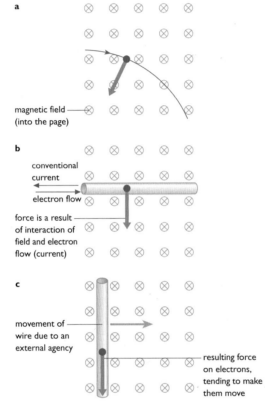

In a beam of electrons, each one behaves independently and moves in a pathway that is an arc of a circle.

In a wire carrying a current, the 'free' electrons are only free to move within the wire. They experience force, resulting in force on the wire: the *motor effect*.

A whole wire, whether or not it is part of a circuit, can be moved through a magnetic field. The 'free' electrons within the wire experience force, resulting in an e.m.f.: this is electromagnetic induction, or the *dynamo effect*.

Figure 6.4 (below)
Electrons moving through a magnetic field experience force. The force can be along the length of the wire, creating an e.m.f. The 'right-hand dynamo rule' shows the direction of the resulting conventional current.

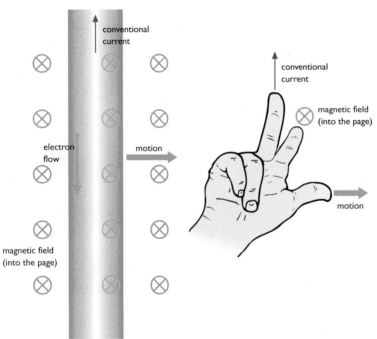

An electron which moves because it is part of a body of metal that moves experiences force just as does, say, an electron in a TV (even though its speed relative to the field is usually much smaller). If the body of metal is a wire, and the length of the wire is perpendicular to its motion and to the field, then the force experienced by the electrons is along the length of the wire (Figure 6.4).

The tendency for electrons to distribute themselves unevenly results in a tendency for electromotive force, e.m.f., to develop in the wire. The creation of an e.m.f. in the moving wire is an example of **electromagnetic induction**. E.m.f. is a voltage. If the wire is part of a circuit, then a current flows as a result of this.

When there is no continuing current in the wire then the induced e.m.f. is identical to a potential difference between the ends of the wire. However, when there is a complete circuit that can carry a current then the potential difference across the wire is less than the e.m.f., because of energy dissipation by heating of the wire. (See *Introduction to Advanced Physics*, Chapter 18.)

Induced e.m.f. and induced current may be in either direction, depending on the relative motion involved. A sensitive ammeter capable of showing positive and negative needle deflections can be used to see this. Such an ammeter, sometimes uncalibrated, with its zero-current position in the centre of its scale, is called a centre-zero galvanometer. Induced current due to motion of a single wire or a single loop can be quite small. The effect is enhanced using loops of conductor, and further enhanced if such coils have many turns. Often, if we simply wave a magnet in the vicinity of a coil then there is enough effect for a meter needle deflection to be seen. Or we can hold the magnet still and move the coil. What we need is relative movement (Figure 6.5).

Figure 6.5
Relative movement of a coil of wire and a magnetic field induces e.m.f., and detectable current in a circuit, which lasts for as long as the movement lasts in the field.

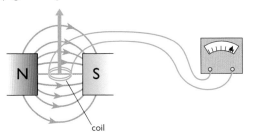

Suppose that the wire is isolated, and an e.m.f. is induced but no current results. In this case e.m.f. and potential difference are the same, and electrons are simply redistributed within the wire. The uneven distribution of charge results in electric force on the electrons. The e.m.f. that is induced grows only until the electric force acting on an electron within the wire is equal to the magnetic force, that is, until

$$\text{electric force} + \text{magnetic force} = 0$$

Where the wire, its motion and the field are mutually perpendicular (Figure 6.6a), we obtain

$$Ee + Bev = 0$$

Note that v is the velocity of an electron through the field. Since its motion is due to the motion of the wire, it is also the velocity of the wire. Electric field strength is related to potential difference across the wire and its length by $E = V/l$, so that

$$\frac{Ve}{l} = -Bev$$

$$V = -Blv$$

and so

$$\text{e.m.f., } \mathcal{E} = -Blv$$

Note that it is essential to use separate symbols, normally E and $\mathcal{E}$, to distinguish between electric field strength and induced e.m.f.

Figure 6.6
Where the velocity of the wire and the flux density are not perpendicular, we calculate the size of the induced e.m.f. by considering the perpendicular components. Induction only takes place where the wire cuts through field lines.

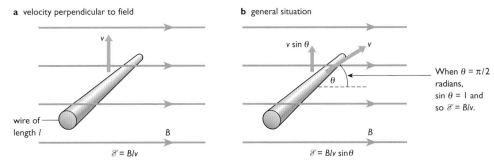

a velocity perpendicular to field

b general situation

$v \sin \theta$ v

When $\theta = \pi/2$ radians, $\sin \theta = 1$ and so $\mathcal{E} = Blv$.

wire of length l B B

$\mathcal{E} = Blv$ $\mathcal{E} = Blv \sin\theta$

The velocity may not always be perpendicular to the field (Figure 6.6b). Where there is an angle θ between the field lines and the velocity,

$$\mathcal{E} = -Blv \sin \theta$$

(In practical use of this formula, the minus sign is often omitted.)

An example of induction due to movement in the Earth's magnetic field

An aeroplane wing is made of metal, and it moves through the Earth's magnetic field. An e.m.f. is induced between the wing tips (Figure 6.7). No current flows, because the wing is not part of a complete circuit. This is an example of a situation in which field and direction of motion of the conductor are not perpendicular. We must use perpendicular components.

Figure 6.7
The velocity of an aircraft flying horizontally is perpendicular to the vertical component of flux density. The aircraft cuts through these vertical field lines. It does not cut through horizontal field lines, and no induction takes place as a result of the horizontal component of flux density.

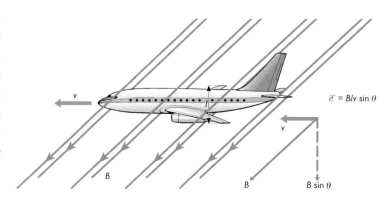

If the flux density of the Earth's field is B, then the vertical component of this is $B\sin\theta$, where θ is the angle between the horizontal and the field lines; θ is sometimes called the **angle of dip** of the field. Angle of dip varies considerably around the world, to a maximum of $\pi/2$ radians at the magnetic poles. If the aircraft's motion is horizontal then it is perpendicular to $B\sin\theta$. Then for an airliner of $30\,\text{m}$ wingspan travelling at $200\,\text{m s}^{-1}$, in a region in which the flux density of the Earth's field is $50\,\mu\text{T}$ and angle of dip is $\pi/6$ radians, the e.m.f. that is induced between the wing tips is given by

$$\mathscr{E} = Blv\sin\theta$$

$$= 50 \times 10^{-6} \times 30 \times 200 \times \sin\frac{\pi}{6}$$

$$= 0.15\,\text{V}$$

Electromagnetic induction due to movement, compared with the motor effect

The electric motor works as a result of interaction of a magnetic field with a current that is driven by an applied potential difference. This could be summarised as:

The motor effect

$$\text{magnetic field} \quad \text{and} \quad \text{current} \quad \rightarrow \quad \text{force (and motion)}$$

Electromagnetic induction requires a magnetic field, but also requires motion that is driven by some other agency (as in the motion of the aircraft, or the motion of a wire that is pulled through the field). The consequence this time is induction of e.m.f., and current if the moving conductor is part of a circuit:

Electromagnetic induction, or the dynamo effect

$$\text{magnetic field} \quad \text{and} \quad \text{relative motion} \quad \rightarrow \quad \text{e.m.f. (and current)}$$

It makes no difference whether the wire (or other conductor) moves through a static field, or whether the wire is still and it is the field that moves. The relative motion must involve the wire cutting through field lines. This is usefully understood in terms of flux, ϕ, where $\phi = BA \sin \theta$ (see Chapter 5). The relative motion results in a wire sweeping out an area while cutting through field lines. The wire cuts flux (Figure 6.8).

Figure 6.8
Flux depends on the intensity of field lines – more strictly called the flux density, B – and on the area through which the wire sweeps. It also depends on the angle between the flux density and the area.

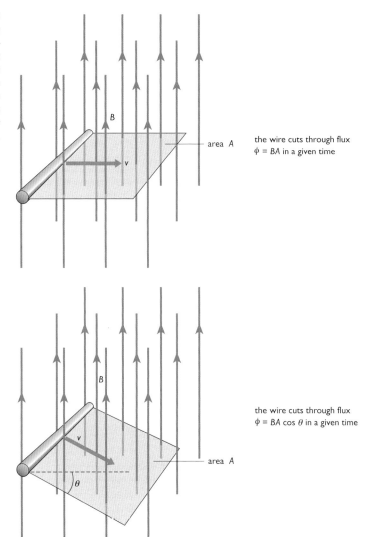

area A

the wire cuts through flux
$\phi = BA$ in a given time

the wire cuts through flux
$\phi = BA \cos \theta$ in a given time

area A

---

1 Make lists of similarities and differences between the motor effect and electromagnetic induction in a moving wire (the dynamo effect).

2 A wire, connected to a galvanometer, is moved vertically in a direction perpendicular to its length at a particular place, and a deflection of the galvanometer needle is seen. No deflection is seen when the wire moves horizontally in the same way.
   a Sketch the wire moving vertically, and the field lines.
   b Sketch the wire moving horizontally, and the field lines.
   c Do your sketches show the only possible arrangements that produce the observed effects?

3 A horizontal wire, connected to a galvanometer, is moved vertically in a direction perpendicular to its length and to the magnetic flux density at a particular place. A deflection of the galvanometer needle is seen.

What will be observed if the wire is turned through $\pi/2$ radians in a horizontal plane, and then moved, again in a direction perpendicular to its own length,
   a horizontally
   b vertically?

4 **a** A 4 m metal pole is held horizontally by a tightrope walker walking towards magnetic 'north' of the Earth at a speed of $0.5 \, \text{m s}^{-1}$, at a location where the total flux density of the Earth's field is $40 \, \mu\text{T}$ and the angle of dip is 0.4π radians. Calculate the e.m.f. induced in the metal pole. Draw a sketch to show its polarity.
   b The tightrope walker falls from high above the ground and reaches terminal velocity of $25 \, \text{m s}^{-1}$, but the pole remains horizontal and perpendicular to the horizontal component of the Earth's flux density. What is the e.m.f. now? Draw a sketch to show its polarity.

Lenz's Law

When a wire moves through a magnetic field then electromagnetic induction can result in current in a circuit, and in the wire itself (Figure 6.9a). So now we have a current-carrying wire in a magnetic field. As a result of the current, the wire experiences a force. The direction of the force opposes the motion (Figure 6.9b).

The direction of the induced current is always such as to result in force that opposes the motion responsible for the induction. This is a statement of what is called **Lenz's Law**, applied to the particular case of induction due to motion. In general, Lenz's Law can be written as:

The direction of an induced current opposes the change that causes it.

Figure 6.9
The current that is induced in a moving wire results in it experiencing magnetic force which opposes its motion.

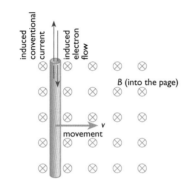

a

B (into the page)

Movement of a conductor relative to a field, cutting flux, induces e.m.f. and current.

b

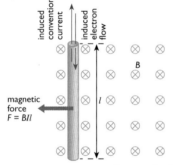

magnetic force
$F = BIl$

The induced current interacts with the field, resulting in a force, given by $F = BIl$. The direction of the force opposes the movement.

a force equal and opposite to $F = BIl$ must be exerted in order to overcome it and sustain the motion of the wire

Figure 6.10 (below)
Work must be done to overcome the magnetic force that a moving wire experiences. Induced current can then result in energy transfer from a circuit by heating and/or working.

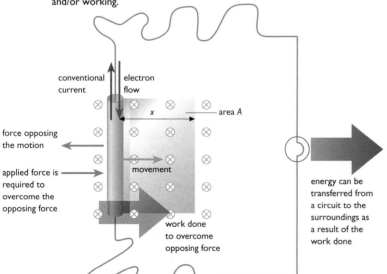

conventional current | electron flow

x — area A

force opposing the motion

applied force is required to overcome the opposing force

movement

work done to overcome opposing force

energy can be transferred from a circuit to the surroundings as a result of the work done

Work must be done by an external agency to overcome this opposing force (Figure 6.10). There must be a source of energy for this – an energy input. The induced current is capable of transferring energy in a circuit (either by heating or by working or both) to the circuit's energy output. Total energy input and output during any period of time are the same. Energy is conserved.

The work that must be done to overcome the opposing force is given by

$$\begin{aligned} \text{work done} &= Fx \\ &= BIlx \\ &= BIA \\ &= \phi I \end{aligned}$$

Lenz's Law applied to induction in coils due to motion

A magnet can be moved towards one end of a coil that is part of a circuit (Figure 6.11). A current is then induced in the coil. As a result of the current, the coil develops a magnetic field. The direction of the induced current is always such as to oppose the motion of the magnet. The coil repels an approaching magnet, but attracts it when it is pulled away.

Figure 6.11
Lenz's Law at work – the polarity of the magnetic field of a coil that results from induced current is always such as to oppose relative motion with the external magnetic field.

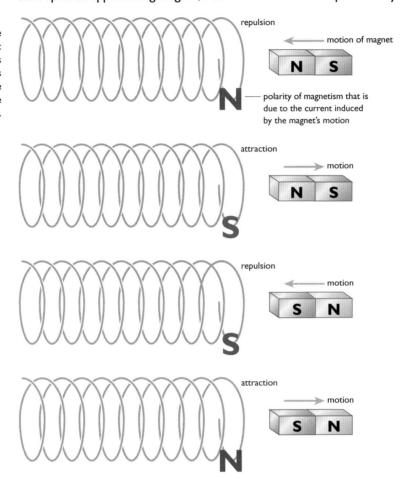

Repulsion or attraction takes place between the end of the coil and the moving magnet, and always opposes the motion of the magnet.

Flux cutting and Faraday's Law

Consider again a wire that moves perpendicularly through a uniform magnetic field (Figure 6.12). It was Michael Faraday (Figure 6.13) who first pictured the wire as cutting through the lines, and used this idea as the basis for investigation of the variables that affect the size of the induced e.m.f. His experimental observations can be summarised as: the size of the induction effect depends on the rate at which flux is cut.

Figure 6.12
Faraday created the concept of flux and discovered that the size of the induced e.m.f. was determined by the rate at which flux was cut.

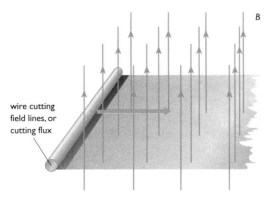

wire cutting field lines, or cutting flux

Faraday's finding was that the more field lines were cut per second, the bigger was the induced e.m.f. For a single wire moving perpendicularly to the field, more field lines (or flux) can be cut in a given time by
- increasing the flux density, B, or
- increasing the area, A, swept by the wire in a given time, either by
  a moving the wire faster, or
  b using a longer wire (or both).

Figure 6.13
Michael Faraday liked to
share his enthusiasm, and
gave Christmas Lectures
for young people at
London's Royal Institution.
The main feature was live
demonstration of newly
explored physical and
chemical phenomena.

Figure 6.13
Michael Faraday liked to
share his enthusiasm, and
gave Christmas Lectures
for young people at
London's Royal Institution.
The main feature was live
demonstration of newly
explored physical and
chemical phenomena.

Faraday discovered that for
a single straight wire the
induced e.m.f. depends on:
• the flux density, B
• the length of the wire
  moving through the
  field, l
• the velocity of relative
  movement, v.

For a single straight wire moving in a uniform field, Faraday's experimental findings can be expressed as:

induced e.m.f., $\mathscr{E} = -$rate of change of flux

$$= -\frac{d\phi}{dt}$$

For a coil of N turns,

$$\mathscr{E} = -N\frac{d\phi}{dt}$$

N is not itself time dependent, and so the quantities $N\,d\phi/dt$ and $d(N\phi)/dt$ are the same. We can therefore write

$$\mathscr{E} = -\frac{d(N\phi)}{dt}$$

The quantity $N\phi$ is sometimes called **flux linkage**. This formula, stating that induced e.m.f. is the same size as the rate of change of flux linkage, applies to all coils. It also applies whether the induction is due to relative physical motion of coil and field, as we have been considering so far, or due to a field that originates from a stationary source but is changing (as in self-induction and mutual induction – see pages 130–134).

The equations

$$\mathscr{E} = -\frac{d\phi}{dt} \qquad \text{and} \qquad \mathscr{E} = -\frac{d(N\phi)}{dt}$$

are both manifestations of **Faraday's Law**, which can be written in words as follows.

• For a single wire or loop:

  Induced e.m.f. is proportional to rate of change of flux.

• For a coil of N turns:

  Induced e.m.f. is proportional to rate of change of flux linkage.

We can show that these findings are consistent with what we already know. We have seen that, for a single wire of length l moving at velocity v, or dx/dt, normally to a field of flux density B,

$$\mathcal{E} = -Blv$$

so that

$$\mathcal{E} = -Bl\frac{dx}{dt}$$

and for our single straight wire, area is swept out at a rate of dA/dt, where

$$\frac{dA}{dt} = \frac{d(lx)}{dt} = l\frac{dx}{dt}$$

So,

$$\mathcal{E} = -B\frac{dA}{dt} = -\frac{d(BA)}{dt}$$

$$\mathcal{E} = -\frac{d\phi}{dt}$$

The minus sign is a manifestation of Lenz's Law. The polarity of the induced e.m.f. is such as to oppose the change of flux.

5 Comment on the statement that 'Lenz's Law is the principle of conservation of energy applied to electromagnetic induction'.

6 For each of the situations shown in Figure 6.14, use Lenz's Law to work out the polarity of the induced e.m.f.

7 A single-turn coil is made in the form of a 'noose' that can be pulled into circular loops of varying area. It is connected to a galvanometer and placed in a field as shown in Figure 6.15.
  a What will happen as the coil area is increased and decreased?
  b What is the effect of changing the speed at which these changes are made?

8 Consider a uniform horizontal field of flux density 8×10^{-2} T.
  a What is the flux in an area, within a vertical plane, described by a square of side 0.05 m?
  b What is the flux linkage between the field and a 50-turn square coil of side 0.05 m placed in the same position?
  c What happens to this flux linkage as the coil turns about a vertical axis that passes centrally through it?
  d Give a formula for the flux linkage in terms of angular orientation.

9 Show that the relationship

$$\mathcal{E} = -\frac{d\phi}{dt}$$

is dimensionally correct.

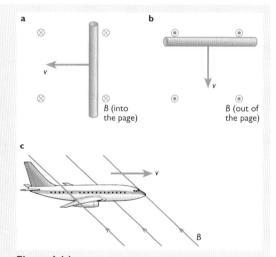

Figure 6.14

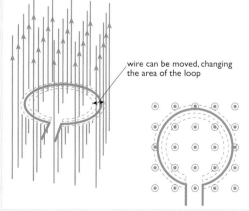

Figure 6.15

Induced eddy currents

We have seen that when a wire moves in a magnetic field, then, providing that it is cutting flux, an e.m.f. is induced. The same process takes place within a sheet of conductor. The induction again obeys Lenz's Law, and the sheet develops its own magnetic field, which interacts with the external field in such a way as to oppose the motion of the sheet. It is possible to feel this by pulling a sheet of copper or aluminium between the poles of a permanent magnet.

The currents induced are circulatory, in the plane of the sheet, such as to produce polarity of the resulting magnetic field which tends to cause repulsion as the sheet (or part of it) enters the field, and attraction as the sheet (or part of it) leaves the field (Figure 6.16). The currents are called **eddy currents**.

Figure 6.16
Lenz's Law provides a reminder of the polarity of induced magnetism in a sheet, and hence of the direction of the induced eddy currents.

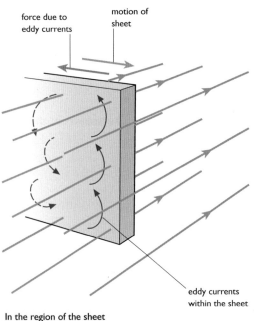

force due to eddy currents

motion of sheet

eddy currents within the sheet

In the region of the sheet that is cutting flux most rapidly, the direction of the eddy currents is such as to produce force that opposes the motion of the sheet.

If the conductor is part of a pendulum system, then the magnetic force resulting from eddy currents opposes the motion of the pendulum, and the swing loses amplitude. The oscillation is damped. In terms of energy, eddy currents produce heating effects within the metal, dissipating energy from the swing and so damping its motion. The presence of slots cut into the metal greatly reduces the effect, by making it impossible for eddy currents to circulate through large areas of the conductor.

The a.c. generator or dynamo

Figure 6.17
E.m.fs induced in the two sides of a rectangular coil add together to provide the total e.m.f.

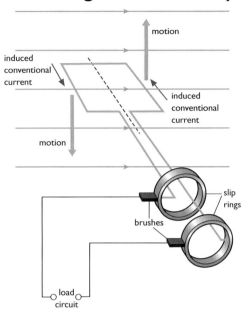

A coil may rotate in a field, so that each side cuts flux. The rotation could be, say, around a horizontal axis that is perpendicular to a horizontal field (Figure 6.17). Then as one side moves up, the other moves down. This means that the two sides of the wire each develop an e.m.f., and these act together to drive current around a circuit.

To prevent twisting of wires as the coil rotates, it must be connected to an external circuit by way of slip rings and 'brushes', which in working dynamos are made of carbon. Carbon brushes are relatively soft so that as they rub against the slip rings they wear slightly. This has the advantage that their shape fits well against the metal to create a good continuous electrical contact. They may be held by springs to ensure that they remain in contact with the rings. Eventually they may wear away and must be replaced.

In a uniform field the sides of the coil are cutting through field lines, i.e. cutting flux, at a high rate when the coil is in a plane parallel to the field lines (Figure 6.18a). But for the instant at which the coil is in a plane that is perpendicular to the field lines, then its sides are not cutting flux at all (Figure 6.18b). So e.m.f. is induced strongly when the coil is in the parallel plane, and not at all when the coil is in the perpendicular plane.

Figure 6.18
Rate of flux cutting depends on relative orientation of coil and field.

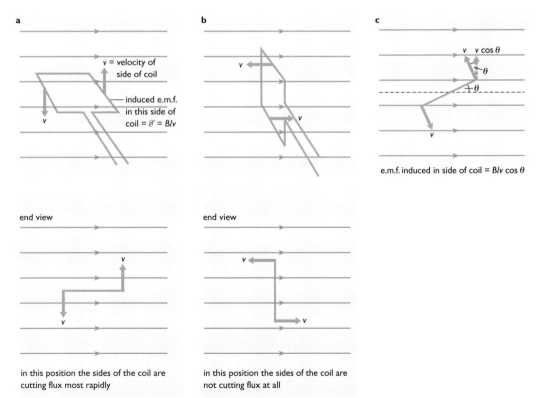

For each turn of each side of a rectangular coil, where field B and velocity v are mutually perpendicular, we have from earlier in this chapter that

$$\mathscr{E} = -\frac{d\phi}{dt} = -Blv$$

But field and velocity are not perpendicular for most of the time. If θ is the angle between the field lines and the plane of the coil (Figure 6.18c), then the component of v that is perpendicular to the field is $v\cos\theta$. Angle θ is time-dependent, and related to the angular velocity, ω, of the coil by $\omega = \theta/t$, or $\theta = \omega t$. In turn, ω is related to the frequency, f, of rotation of the coil by $\omega = 2\pi f$. So, in general,

e.m.f. induced in one side of the coil $= Blv\cos\omega t$ (ignoring the minus sign)

Chapter 1 tells us that $v = r\omega$, where r is the radius of rotation, so

e.m.f. induced in one side of the coil $= Blr\omega\cos\omega t$

A single-turn coil has a wire on each side, and an e.m.f. is induced in each of them, these e.m.fs then adding together. So total e.m.f. for such a coil is given by

$$\mathscr{E} = 2Blr\omega\cos\omega t$$

But $l \times 2r$ is equal to the area of the coil, A, and if the coil has N turns rather than just one, then

$$\mathscr{E} = BAN\omega\cos\omega t$$

Figure 6.19
Graphs of e.m.f. against time and current against time for a resistive circuit are both sinusoidal.

Note that the maximum and minimum values of the e.m.f. are $+BAN\omega$ and $-BAN\omega$. We can write the maximum, or peak, value of $\mathscr{E}$ as $\mathscr{E}_0$. Then, as shown in Figure 6.19,

$$\mathscr{E} = \mathscr{E}_0\cos\omega t$$

Current in an external resistive circuit to which the a.c. generator is connected is proportional to e.m.f., so, as also shown in Figure 6.19,

$$I = I_0\cos\omega t$$

where I is the instantaneous current, the value of the current at any chosen time, t, and I_0 is the peak current.

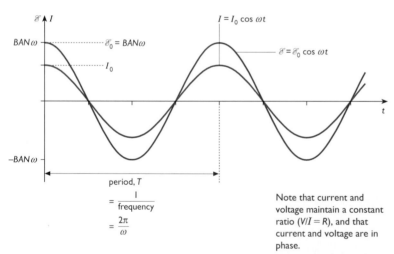

$\mathscr{E}_0 = BAN\omega$

$I = I_0\cos\omega t$

$\mathscr{E} = \mathscr{E}_0\cos\omega t$

period, T
$$= \frac{1}{\text{frequency}}$$
$$= \frac{2\pi}{\omega}$$

Note that current and voltage maintain a constant ratio ($V/I = R$), and that current and voltage are in phase.

10 **a** Give an equation for peak current in terms of flux density, area and number of turns of coil, angular velocity of the coil, and total circuit resistance.
b Sketch a graph of peak current against peak e.m.f. and comment on the significance of the gradient.
11 Why is no net e.m.f. induced in a coil when it sweeps through a uniform magnetic field as shown in Figure 6.20?

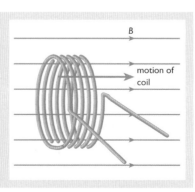

Figure 6.20

Peak and rms values provided by an a.c. generator

In the formula

$$I = I_0 \cos \omega t$$

I is the instantaneous current and I_0 is the peak current. Neither of these tell us much about energy transfer in circuits over a period of time. We know that the power relating to such energy transfer is given by $P = IV$, and since $I = I_0 \cos \omega t$ and $V = V_0 \cos \omega t$, the power output of an a.c. generator can be written as

$$P = I_0 V_0 \cos^2 \omega t$$

This allows us to represent the power instant by instant on a graph of power against time (Figure 6.21a). Note also that power is maximum, and has value P_0, when $\cos^2 \omega t = 1$, that is,

$$P_0 = I_0 V_0$$

Figure 6.21
Power–time graphs for an a.c. generator.

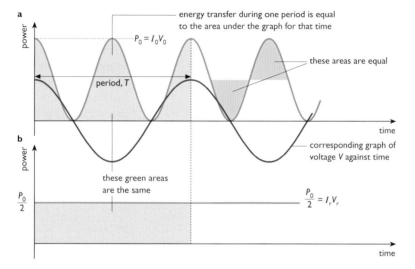

The maximum power output of an a.c. generator is P_0, where

$$P_0 = I_0 V_0$$

This is equivalent to a constant (d.c.) power output of $P_0/2$. The corresponding values of d.c. current and voltage can be called I_r and V_r, and are such that

$$\frac{P_0}{2} = I_r V_r$$

The area under any graph of power against time is equal to energy transferred in the chosen length of time. Note that the graph in Figure 6.21a is symmetrical about its own mid-value. The areas described by the 'top halves' exactly fit into the 'bottom halves' as shown. This means that the area under the power–time graph is the same as the area under a power–time graph for a constant power of $P_0/2$ (Figure 6.21b).

We can define two quantities, I_r and V_r, called **rms values** (rms standing for root-mean-square), which are the constant d.c. current and voltage that would transfer energy at the same rate as our circuit that carries sinusoidal a.c. This 'same rate' is $P_0/2$. So,

$$\frac{P_0}{2} = I_r V_r$$

The energy may be transferred to the circuit's surroundings by resistive heating, and in terms of the circuit resistance, R, we can write

$$\frac{P_0}{2} = I_r^2 R$$

and since $P_0 = I_0^2 R$, we get

$$\frac{P_0/2}{P_0} = \frac{I_r^2 R}{I_0^2 R}$$

which simplifies to

$$I_r = \frac{I_0}{\sqrt{2}} \quad \text{and, similarly,} \quad V_r = \frac{V_0}{\sqrt{2}}$$

12 **a** Current and voltage provided by an a.c. generator are sometimes positive and sometimes negative. Explain why for an a.c. generator connected to a resistor, power can never be negative.
b Explain, using sketch graphs, why the frequency of variation of power in an a.c. generator circuit is twice that of the current and voltage.

13 A circular coil of 20 turns and radius 0.05 m is made to rotate at 20 revolutions per second about an axis that is perpendicular to a field of flux density 10 mT. What is the peak e.m.f. induced?

14 **a** Sketch a graph of output voltage against time for a simple a.c. dynamo.
b On the same axes, sketch graphs to show the output if
i the number of turns, N, of the coil is doubled
ii the number of turns, N, of the coil is halved
iii the angular velocity of rotation, ω, is doubled
iv the angular velocity of rotation, ω, is halved.

Rectification of a.c.

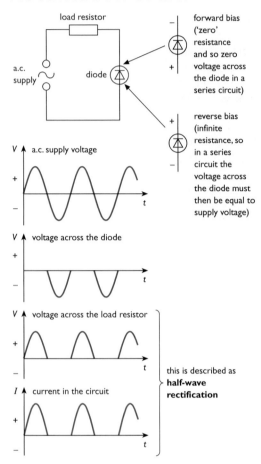

Figure 6.22
A single diode rectifier results in **half-wave rectification**.

For many applications, constant polarity of voltage and constant current direction are required. Such direct current (d.c.) can be created from a.c. by the process of rectification, using one or more diodes. An ideal diode has zero resistance when voltage is applied in one direction (forward bias) and infinite resistance when voltage is applied in the other direction (reverse bias); see *Introduction to Advanced Physics*, page 177. Figures 6.22 and 6.23 show two types of rectification circuit.

A capacitor can be charged when the voltage applied to it is high, and becomes discharged through a resistor when the voltage falls. This is used to provide a voltage that does not vary as strongly as simply rectified a.c. The voltage is described as **smoothed** (Figure 6.24).

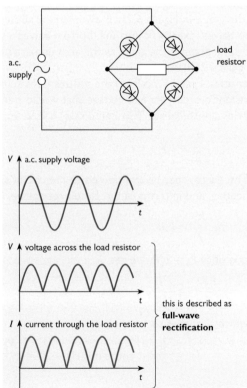

Figure 6.23 (left)
A four-diode circuit, or bridge rectifier, results in **full-wave rectification**.

Figure 6.24 (right)
A capacitor across the load results in **smoothed d.c.**

15 Show that Figure 6.23 is consistent with Kirchhoff's laws.

Induced e.m.f. in a motor

A simple motor consists of a coil rotating in a uniform field. The motion is driven by the interaction of the magnetic field due to the current in the coil with the external magnetic field. But there is more interaction than that. The coil is a conductor moving in a field, and as a result an e.m.f. is induced in it. This e.m.f. opposes the potential difference that is applied to the motor coil, and is called **back e.m.f.** The direction of this e.m.f. is also such as to oppose the change producing it – as predicted by Lenz's Law. The back e.m.f. opposes the rotation of the coil (Figure 6.25a).

Figure 6.25
a For a motor coil rotating freely at a constant speed, the motor turning effect is balanced by the turning effect that is due to the back e.m.f. plus any frictional or viscous resistive force.
b For a motor which has a load – that is, one that is doing external mechanical work – there is increased opposition to motion.

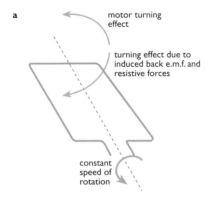

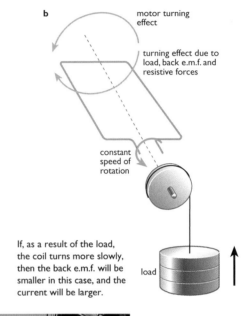

Figure 6.26
Electric motors can lift loads. The effect of the load is to oppose rotation and reduce back e.m.f.

16 a For a simple motor, why must the back e.m.f. always be smaller than the applied potential difference?
b What factors limit the size of the back e.m.f.?
c A body will experience rotational acceleration if subject to a net torque (that is, unbalanced turning effect or couple). When connected to a power supply, a motor quickly accelerates to an equilibrium rotational speed. What factors determine the size of this speed?
d Why does back e.m.f. tend to decrease as the rate of doing work increases?

The size of net current in the coil is limited by the action of the two opposing voltages – the applied potential difference and the back e.m.f. A motor exerts forces on the world around it – that is, a motor does mechanical work. It has a 'load' (Figure 6.26). The load opposes the rotation (Figure 6.25b), and may slow it down, thus reducing back e.m.f. If the coil is prevented from turning by its load, then there is no back e.m.f. The applied potential difference can then drive a current that is large enough to overheat the wires and destroy the motor.

Self-inductance of a coil

A coil lies in its own magnetic field. When the current in the coil is steady, then the magnetic field is unchanging. But if the current changes then the field changes – a magnetic field grows or shrinks in the space that holds the coil. Relative movement of field lines and conducting wire takes place. The wire can be thought of as cutting the flux. The result is electromagnetic induction of an e.m.f. A coil in which such self-induction takes place is called an **inductor**.

The direction of the induced e.m.f. is such as to oppose the change that produces it. That is, the e.m.f. opposes the change in current. For this reason current grows and declines in a coil more slowly than would be the case in a 'non-inductive' circuit with resistance only (Figure 6.27).

Figure 6.27
In a circuit with a coil acting as an inductor, the induced e.m.f. is in the opposite direction to the applied potential difference, and thus opposes the current. The opposition is strongest when the current is changing fastest.

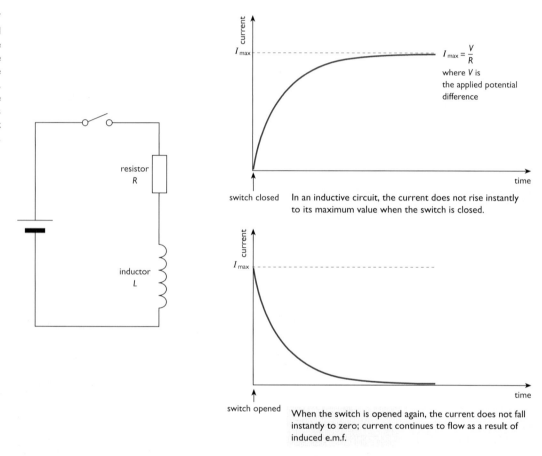

where V is the applied potential difference

In an inductive circuit, the current does not rise instantly to its maximum value when the switch is closed.

When the switch is opened again, the current does not fall instantly to zero; current continues to flow as a result of induced e.m.f.

The size of the induced e.m.f. is proportional to the rate of change of flux, as predicted by Faraday's Law. The flux due to the coil is proportional to the current it carries. So we can say

$$\mathscr{E} \propto -\frac{\mathrm{d}I}{\mathrm{d}t}$$

The constant of proportionality in this relationship depends on complex factors – such as the dimensions of the coil, including length, diameter and number of turns. The minus sign is a consequence of Lenz's Law and the relative directions of the change in current and the induced e.m.f. We can indirectly measure $\mathrm{d}I/\mathrm{d}t$ and investigate its relationship with $\mathscr{E}$ for any particular coil, and so determine the value of the constant of proportionality. This constant is called the **self-inductance** of the coil, for which the symbol L is used:

$$\mathscr{E} = -L\frac{\mathrm{d}I}{\mathrm{d}t}$$

From this equation it seems that a unit $V A^{-1} s$ is suitable for self-inductance, and this is given its own name, the **henry** (H).

Energy storage due to self-induction

We have seen that an effect of self-induction is to reduce rate of increase of current when a circuit is switched on, and to allow current to flow after the inductive circuit is switched off. For a circuit that has resistance as well as inductance, energy may be transferred by the resistor(s) to the surroundings of the circuit after it is isolated from its energy source. The energy can be thought of as having been stored by inductive components of the circuit (Figure 6.28).

Figure 6.28
Energy transfer to and from an inductor.

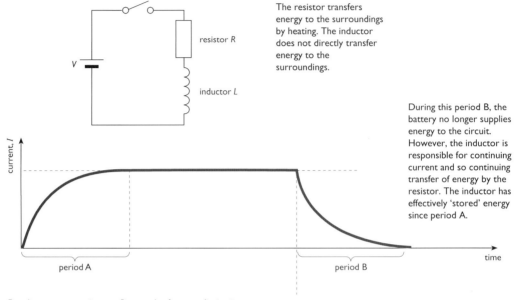

The resistor transfers energy to the surroundings by heating. The inductor does not directly transfer energy to the surroundings.

During this period B, the battery no longer supplies energy to the circuit. However, the inductor is responsible for continuing current and so continuing transfer of energy by the resistor. The inductor has effectively 'stored' energy since period A.

During the initial period A, the presence of the inductor reduces the current. As a result, it reduces the transfer of energy to the surroundings by resistive heating.

For the constant resistance R, a graph of power dissipation against time is the same shape as the current–time graph.

the energy that is not transferred by heating during period A, but would have been transferred by a non-inductive circuit...

... is transferred by heating during period B

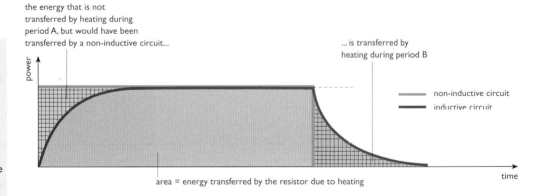

non-inductive circuit

inductive circuit

area = energy transferred by the resistor due to heating

17 **a** A coil of self-inductance 0.5 mH carries a current that is increasing at a rate of 50 mA s⁻¹. What e.m.f. is induced across it?
b What is the polarity of the induced e.m.f. relative to that of the current?
c What tends to happen to such an induced e.m.f. as the current approaches its maximum value?

18 In a circuit with both resistance and inductive reactance, explain what determines the size of the maximum current reached after the circuit is switched on.

We can express this energy storage as:

energy stored by inductor = energy not available for resistive heating during period A

= applied potential difference
× shaded area described by current–time graph during period A

= $\frac{1}{2}LI^2$

= energy available after circuit is broken (period B)

(A derivation of the expression $\frac{1}{2}LI^2$ requires the use of integration.)

Self-induction in a.c. circuits

The relationship between induced e.m.f. and the current in an inductor is given in the form of a differential equation [one that involves a differentiation and relates to rate of change of one variable (in this case current) with another (time)]:

$$\mathscr{E} = -L\frac{\mathrm{d}I}{\mathrm{d}t}$$

We know that, for an a.c. current, instantaneous values are related to the constant peak value by

$$I = I_0 \cos \omega t$$

From this we can plot a graph of current against time (Figure 6.29). The gradient of the graph also varies with time, and we can find a formula for this by differentiation. This gives

$$\frac{\mathrm{d}I}{\mathrm{d}t} = -I_0\omega \sin \omega t$$

and so

$$\mathscr{E} = LI_0\omega \sin \omega t = \mathscr{E}_0 \sin \omega t$$

Figure 6.29
Graphs show that current and voltage are not in phase for an inductor. The induced e.m.f. is largest when current is changing fastest.

We can plot graphs of current against time and induced e.m.f. against time on the same axes (Figure 6.29), and this highlights the phase difference between them. One point that these graphs help to illustrate is that

$$LI_0\omega = \mathscr{E}_0$$

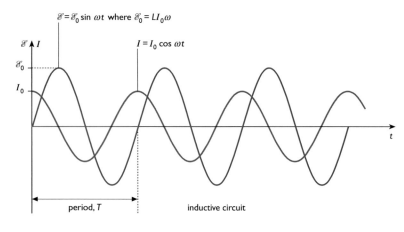

$\mathscr{E} = \mathscr{E}_0 \sin \omega t$ where $\mathscr{E}_0 = LI_0\omega$

$I = I_0 \cos \omega t$

period, T

inductive circuit

Current and e.m.f. are 'out of phase'. The time between the peaks of the two graphs is $T/4$. (The phase difference is a quarter of a revolution, or $\pi/2$ radians – see Chapter 2.)

Very little induction takes place in a simple straight wire resistor, and it can be described as 'non-inductive'. Graphs of potential difference and current against time for such a resistor are relatively simple. The voltage across a resistor – the applied potential difference – drives the current. Current is proportional to it, provided that resistance is constant. Graphs of current against time and voltage against time show a constant voltage/current ratio: the ratio V/I is the circuit resistance in this case.

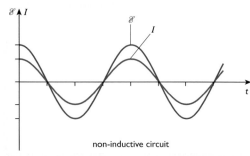

non-inductive circuit

A pure inductor is an idealised coil with no resistance at all. Then there is no dissipation of energy by heating, and no net potential difference across the coil. The applied potential difference is in perfect balance with the induced e.m.f. The relationship between current and voltage is not straightforward because of the phase shift as shown in Figure 6.29. The coil may not have resistance, but we can devise an analogous quantity – a ratio of peak voltage to peak current. This ratio is called **inductive reactance**, X_L:

$$X_L = \frac{\mathscr{E}_0}{I_0}$$

So we can see that

$$X_L = L\omega$$

The unit of this quantity is the VA^{-1}, which we know as the ohm, Ω.

19 Explain why from $\mathscr{E} = -\mathrm{d}\phi/\mathrm{d}t$ it is possible to say that $\mathscr{E} \propto -\mathrm{d}I/\mathrm{d}t$.

20 **a** Sketch a graph of e.m.f. induced in an inductive coil against rate of change of current. What is the significance of the gradient?
b Explain why rate of change of current is difficult to measure.
c How do graphs of e.m.f. against time and current against time make it possible to measure self-inductance?
d How can the data for such graphs be collected?

Mutual inductance of a pair of coils

If induction effects take place within a single coil, then not surprisingly they also take place between one coil and another. An e.m.f. is induced in one coil whenever it lies in the changing magnetic field of another. A coil in which the initial, or primary, changing current results in a changing magnetic field is called a **primary coil**. A coil in which e.m.f. is then induced is called a **secondary coil**. The induced e.m.f. in the secondary coil is proportional to the rate of change of current in the primary coil:

$$\mathscr{E}_s \propto -\frac{dI_p}{dt}$$

In this case the constant of proportionality depends on even more factors than is the case for the single coil. It is influenced not only by the dimensions of each coil, and whether or not they are both wound around a single core, but also by their relative positions. Each pair of coils has a given **mutual inductance**, which must be found by measurement of rate of change of current and of the induced e.m.f.:

$$\mathscr{E}_s = -M\frac{dI_p}{dt}$$

Here M is the mutual inductance and is dimensionally the same as self-inductance. It therefore has the same unit, the henry, H.

Note that when rate of change of current with time, dI/dt, is zero, then e.m.f. is also zero. Current must be changing for e.m.f. to be induced. If the current is alternating, then it is continuously changing, and we can write

$$I_p = I_{p0}\cos \omega t$$

which means that

$$\frac{dI_p}{dt} = -I_{p0}\omega \sin \omega t$$

so that

$$\mathscr{E}_s = MI_{p0}\omega \sin \omega t = \mathscr{E}_{s0}\sin \omega t$$

Figure 6.30
There is a phase difference between primary current and secondary e.m.f. The e.m.f. is largest when the primary current is changing fastest, as predicted by the relationship

$$\mathscr{E}_s = -M\frac{dI_p}{dt}$$

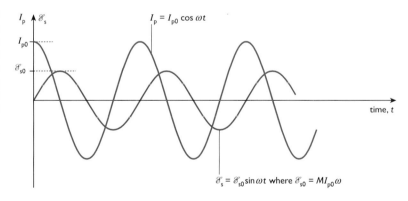

Figure 6.31
A transformer inside a mains adapter.

A pair of coils wound around a single iron core is an effective way of transferring energy from one circuit to another, without any direct electrical connection. It only works on a continuous basis with current that is continuously changing, as a.c. is. Alternating current in the primary coil induces an alternating e.m.f. in the secondary, with a phase difference as shown in Figure 6.30. Such a device is a transformer (Figure 6.31).

The great usefulness of transformers comes from the fact, in the ideal case, that the ratio of voltage across the primary to voltage across the secondary is equal to the ratio of the number of turns in the two coils. That is, for an ideal transformer in which there are no energy losses,

$$\frac{V_p}{V_s} = \frac{N_p}{N_s}$$

This is called the **turns ratio**.

Measured and quoted values of V_p and V_s are usually the rms values.

A transformer can provide a larger output voltage than input voltage, just by having a secondary coil with more turns. This is a **step-up transformer**.

However, the average output power cannot be bigger than the average input power. In fact, it is always less, because energy transfer processes take place within the transformer itself – mostly by heating the core due to eddy currents (Figure 6.32). To reduce eddy currents, transformer cores are usually laminated – made of layers of iron separated by even thinner layers of insulator.

We can, however, imagine a transformer with 100% efficiency. Then,

$$\text{power output} = \text{power input}$$
$$V_s I_s = V_p I_p$$

where I_s and I_p are the secondary circuit and primary circuit rms currents. This tells us that

$$\frac{I_s}{I_p} = \frac{V_p}{V_s}$$

which means that for a step-up transformer, for which $N_s > N_p$ and $V_s > V_p$, the secondary current must be less than the primary current, $I_p > I_s$ (Figure 6.33). That is, if the transformer increases voltage, it must decrease current.

For a **step-down transformer**, the turns ratio formula still applies, but now $N_p > N_s$, so that $V_p > V_s$ and, provided that energy losses are not too large, $I_s > I_p$.

Figure 6.32
Energy transfer by a transformer.

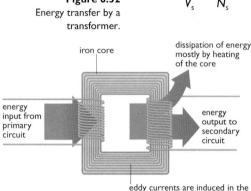

iron core

dissipation of energy mostly by heating of the core

energy input from primary circuit

energy output to secondary circuit

eddy currents are induced in the core, causing some heating

Figure 6.33

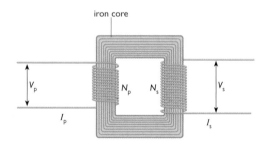

iron core

In the transformer shown

$$N_s > N_p \quad \text{and} \quad V_s > V_p$$

It is a *step-up* transformer. Average power output cannot be greater than average power input, that is

$$V_s I_s \leqslant V_p I_p$$

(They are equal only for a transformer with 100% efficiency). So

$$I_s < I_p$$

(Note that the coils have little resistance and the size of the current in each of them does not depend on the e.m.f. across them in the simple way that it would for resistors.)

21 a Sketch graphs of induced e.m.f. against rate of change of current for a single coil and for a pair of coils.
   b In each case, explain what happens to the induced e.m.f. as the frequency of the a.c. increases.
22 a For a step-up transformer with a turns ratio of $N_p/N_s = 0.02$, what is the minimum ratio between the primary current and the secondary current?
   b Explain why this is a minimum ratio.
23 A transformer for stepping domestic mains voltage of 230 V rms down to 6 V rms has 575 turns on its primary coil.
   a How many turns does it have on its secondary coil?
   b If the maximum current in the secondary circuit is 4 A rms, what is the maximum
   i peak power output
   ii average power output?
   c What is the minimum rms current in the primary circuit?
   d Why will the primary current be bigger than this?
24 A step-down transformer has input and output voltages of 230 V and 12 V. It provides power for twenty 6 W Christmas tree lights, which are connected in parallel, and the primary current is then 0.9 A. What is the transformer efficiency?

Capacitors in a.c. circuits

We know from the definition of capacitance that

$$Q = CV$$

and so

$$\frac{dQ}{dt} = \frac{d(CV)}{dt}$$

Since capacitance of a particular capacitor is not normally time-dependent, this becomes

$$\frac{dQ}{dt} = C\frac{dV}{dt}$$

or

$$I = C\frac{dV}{dt}$$

This gives us a relationship between current and potential difference. It is a differential equation, and not a straightforward relationship. However, it does allow us to compare how the potential difference and the current vary with time in an a.c. circuit with capacitance (Figure 6.34).

If

$$V = V_0 \sin \omega t$$

then

$$\frac{dV}{dt} = V_0 \omega \cos \omega t$$

(You do not need to be able to perform such differentiations, but the process is shown here in order to reveal how the current–time relationship follows from the voltage–time relationship.) Substituting for dV/dt, we get

$$I = CV_0 \omega \cos \omega t = I_0 \cos \omega t$$

Note that

$$I_0 = CV_0 \omega$$

and so

$$\frac{V_0}{I_0} = \frac{1}{C\omega}$$

This ratio of peak voltage to peak current is measured in ohms, Ω. It is called the **capacitive reactance** of the capacitor, X_C.

$$X_C = \frac{1}{C\omega}$$

Figure 6.34
As with an inductive circuit (page 132), there is a phase difference between current and potential difference, but it is not the same. Now the current is largest when the potential difference is changing fastest, as predicted by

$$I = C\frac{dV}{dt}$$

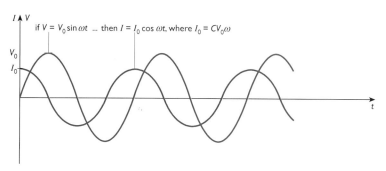

if $V = V_0 \sin \omega t$ ... then $I = I_0 \cos \omega t$, where $I_0 = CV_0 \omega$

25 Sketch graphs of
  a inductive reactance against frequency
  b capacitive reactance against frequency.
  Explain the different types of behaviour.
26 Why are current and voltage always in phase for a purely resistive circuit but not for capacitive or inductive circuits?
27 A capacitor and a pure (zero-resistance) inductor are connected in parallel to a sinusoidal alternating current supply, so that the potential difference across them is always the same. Sketch a graph of this potential difference against time, and on the same axes sketch graphs of the current against time for each of the components.

Phasor diagrams for resistors, inductors and capacitors

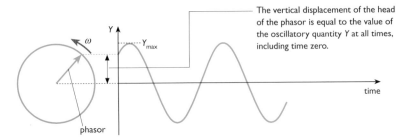

The vertical displacement of the head of the phasor is equal to the value of the oscillatory quantity Y at all times, including time zero.

Figure 6.35
Phasor representation of a quantity Y.

As we saw in Chapter 2 (page 24), a phasor is a useful representation of any oscillating quantity, Y, including current and voltage in an a.c. circuit. The phasor rotates at angular velocity ω (Figure 6.35) and has length Y_{max}. The phasor diagrams for the current and voltage in a.c. circuits containing resistors, inductors and capacitors are described below, and are shown in Figure 6.36.

- *Resistor:* For a resistor, current and voltage are related by

$$I = \frac{V}{R}$$

Their relative values are not time-dependent, and they are always in phase. Their phasors rotate *together*, as shown in Figure 6.36a.

- *Inductor:* For an inductor, current and voltage are related by

$$V = L\frac{dI}{dt}$$

Current I and voltage V are $\pi/2$ radians out of phase, as shown by the rotating phasors in Figure 6.36b. The voltage phasor *leads* the current phasor.

- *Capacitor:* For a capacitor, current and voltage are related by

$$I = C\frac{dV}{dt}$$

Current I and voltage V are again $\pi/2$ radians out of phase, as shown in Figure 6.36c. But this time the voltage phasor *lags* behind the current phasor.

Figure 6.36
Current I and voltage V phasors for components in an a.c. circuit: **a** resistor, **b** inductor, **c** capacitor.

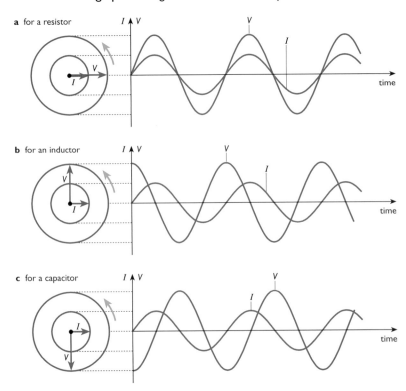

a for a resistor

b for an inductor

c for a capacitor

● **Comprehension and application**

Filter circuits

For a circuit with an a.c. power supply and a *pure resistor*, the resistance is independent of frequency. Current is related to potential difference in quite a simple way, and is independent of frequency (Figure 6.37):

$$I = \frac{V}{R}$$

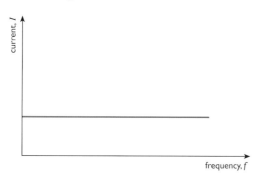

Figure 6.37
In a purely resistive circuit, current is independent of frequency.

For a circuit with an a.c. power supply and a *pure inductor*, reactance is given by

$$X_L = L\omega = \frac{V_0}{I_0}$$

Not only are current and potential difference across an inductor not in phase, but the value of the peak current is dependent on frequency (Figure 6.38):

$$I_0 = \frac{V_0}{L\omega} = \frac{V_0}{L(2\pi f)}$$

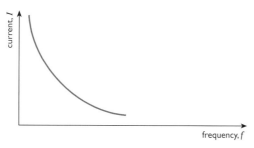

Figure 6.38
In a purely inductive circuit, peak current decreases as frequency increases, for constant voltage.

For a circuit with an a.c. power supply and a *pure capacitor*, reactance is given by

$$X_C = \frac{1}{C\omega} = \frac{V_0}{I_0}$$

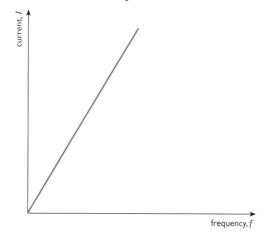

Figure 6.39
In a purely capacitive circuit, peak current increases as frequency increases, at constant voltage.

Again, peak current is dependent on frequency, but the dependence takes a different form (Figure 6.39):

$$I_0 = V_0 C\omega = V_0 C(2\pi f)$$

A circuit that has resistance, inductance and capacitance – known as an **LCR circuit** – is somewhat more complex. For one thing, a resistor, an inductor and a capacitor can be connected in different parallel or series arrangements. The combined effects of the three types of component can be measured not in terms of resistance or reactance but in terms of a new quantity, impedance, that depends on these.

A circuit with all three types of behaviour shows a dependence of peak current on frequency, with a peak of current at a frequency that depends on the detail of the arrangement. The graph takes on the appearance of a resonance curve (see Chapter 2), and the frequency corresponding to the maximum peak current is the resonant frequency (Figure 6.40). The circuit is a **resonant circuit**. By changing variables of the circuit, such as by changing the value of the capacitance, the value of the resonant frequency can be changed.

Figure 6.40
An *LCR* circuit shows resonance, and can be used to maximise response to a chosen frequency.

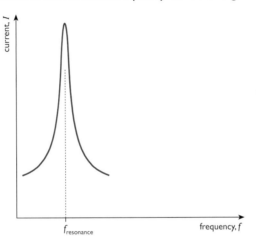

Thus relatively large current can exist in such circuits only at certain frequencies. If voltages at a wide range of frequencies are applied to them simultaneously, they experience the biggest current in response to the voltage closest to the resonant frequency. If, say, a further device – such as a loudspeaker (with amplifier) – is present in the circuit, it will respond most strongly to the circuit's resonant frequency.

This is analogous to a colour filter, on to which a range of frequencies of light falls but only one frequency is transmitted at high intensity. The *LCR* circuit, or resonant circuit, therefore has another name – a **filter circuit**.

Electrons in a TV or radio aerial oscillate at very many superimposed frequencies, because they are absorbing energy from incident waves of very many frequencies. To 'tune in' to a particular station, one frequency range must be chosen. A filter circuit allows this choice.

---

28 a At a frequency of 50 Hz, what is the reactance of
  i a 10 mH inductor
  ii a 100 nF capacitor?
  b At what frequency will they have the same reactance?
29 a What determines the gradient of a graph of inductive reactance against frequency?
  b A filter circuit is effective if it has a sharply peaked resonance curve. Explain how this affects the choice of inductor.

Electric and magnetic interactions and the nature of light

There is an electric field around any charged particle. And we can consider that if it moves there is also a magnetic field. Suppose that it moves in a vacuum, which is the normal environment in most of the Universe. Its fields at a distance from it do not travel rigidly with it. It takes time for the fields to change in response to the movement. The further away from the particle, the longer it takes. The fields travel at a speed that is related to two fundamental constants – the permittivity and the permeability of a vacuum, ϵ_0 and μ_0. James Clerk Maxwell, in a somewhat speculative piece of mathematics, made a prediction about the speed at which the fields spread. This turned out to be stunningly reliable in what it said about the world, and we can write it quite simply:

$$c = \frac{1}{\sqrt{\epsilon_0 \mu_0}}$$

According to Maxwell's calculations, c is equal to $3 \times 10^8 \, \text{m s}^{-1}$. This should be familiar – Maxwell's equations predicted the value of the speed of light, and measurements ever since have agreed with this value.

When the fields of a charged particle interact with the fields of another particle, the possibility exists for energy to pass from one to the other. This happens in a radio transmitter and radio receiving aerial – electrons in one pass energy to electrons in the other. Likewise oscillation of charged particles in a light bulb filament can result in transfer of energy to your retina, where charged particles oscillate to create chemical change and nerve impulses. In a simple two-particle system, one particle loses energy and the other gains it.

There are different ways to think about this, of which the following is one. Suppose that a charged particle oscillates – then the magnetic field oscillates, and the changes spread into space at a speed of $3 \times 10^8 \, \text{m s}^{-1}$. There is a changing magnetic field, and the particle itself lies in this changing magnetic field. There is also an oscillating electric field, which also travels into space at a speed of $3 \times 10^8 \, \text{m s}^{-1}$. We say that the oscillating charged particle is sending electromagnetic radiation into space (Figure 6.41), the frequency of the radiation matching the frequency of oscillation.

The oscillating and rapidly spreading magnetic and electric fields – that is, the electromagnetic radiation – are capable of doing work on charged particles that they come across. Such a charged particle takes energy from the travelling oscillating fields (from the radiation) and itself oscillates with the same frequency. Loss of energy by the fields is described as absorption of radiation.

Since they are capable of doing work, we have to suppose that the oscillating fields, or radiation, carry energy with them from their point of origin. This energy must be supplied to them by their source. Energy is required to emit these travelling oscillating fields, or electromagnetic radiation.

30 Particles in thermal motion in a body of material are accelerating and decelerating. The particles pass energy from one to another, and they also transfer energy out of the body. The body radiates.
a Explain why this causes cooling of the body unless the energy is replaced.
b How might the energy be replaced in
i a heated lamp filament
ii a piece of metal that is in thermal equilibrium with its surroundings?

31 Show that the equation $c = 1/\sqrt{\epsilon_0 \mu_0}$ is dimensionally coherent (having the same dimensions on both sides).

32 Why is it impossible to have
a electrical oscillation without magnetic oscillation
b magnetic oscillation without electric oscillation?

33 **DISCUSS**
Energy is required to generate electromagnetic radiation, and electromagnetic radiation supplies energy when it is absorbed. How do we know that the energy exists during the period between emission and absorption?

Figure 6.41
Electromagnetic interaction spreads out from candle flames and radio transmitters, due to accelerations and decelerations of charged particles.

● **Comprehension and application**

The national grid and mains supply

Figure 6.42 (above)
Thomas Edison poses with a product of his research. He was perhaps the first industrialist to realise the importance of systematic research and development for creation of new consumer products.

Thomas Alva Edison (Figure 6.42) was a central character in the booming economy of the United States in the late 19th century. He was a believer in industry and a believer in innovation, but he was not a scientist. The new science of his day included Maxwell's equations and the discovery of a wide range of electromagnetic radiations, debate about the nature of matter, the existence of atoms, the identity of electrons, and so on. Edison knew little of this. He was a person who knew that new technologies provided a way to make money, and he created the first industrial laboratory. The developments in which he at least played a part included the light bulb and sound recording. One of his great schemes was the provision of city-wide electrical supplies. The Edison Electric Light Company used d.c. generators, acting very much like batteries in circuits. But in order to deliver enough power to light very many streets and houses, with very many energy transfer appliances connected in parallel, a large current was necessary. Large currents in wires produce strong heating effects, and rapid energy dissipation. One attempted solution was to use high voltage so that current could be as small as possible for a given power. This made the supply more dangerous. Even so, Edison's d.c. system needed to have generators close to the consumers, so that the wires could be as short as possible.

Others, notably George Westinghouse and Nikola Tesla, saw the problem, and they saw a solution. Use of a.c. rather than d.c. allowed transformers to be used. These could step voltage up and current down, close to the generator, and power could then be transmitted over longer distances at the low current, greatly reducing heating losses. Near to the location of the consumer, the high voltage could be stepped down, and the total current could be stepped up to provide power to very many parallel circuits.

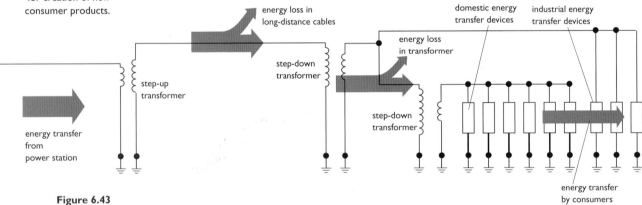

Figure 6.43
Schematic representation of energy transfer by an a.c. distribution system.

Such a distribution system is shown in Figure 6.43. A power station might provide power at a rate of a few hundred megawatts. The output from the generators may have an rms voltage of 20 kV. This is then stepped up to 400 kV for long-distance 'distribution'. A transformer can step this down to 275 kV for district distribution, and another transformer can take this to 33 kV for local use. Industrial use, including supply for electric trains, might require this voltage, but 33 kV is dangerously high for use where direct contact with people is possible, so for domestic use a further transformer steps the voltage down to 230 V (in Europe).

34 a If rms value of domestic mains voltage is 230 V, what is the peak voltage?
b If its frequency is 50 Hz, write down an equation for voltage against time, in which these are the only variables with unspecified value.
c Sketch a graph of the voltage against time, with numbered axes.
35 A commercial consumer requires power at 10 MW, which is supplied through 1 km cable of resistance 0.01 Ω m⁻¹. The voltage as measured by the consumer is 33 kV.
a What is the current in the consumer's circuit when at full power?
b What is the voltage drop along the length of the cable?
c What is the rate at which the cable dissipates energy?
d What are the equivalent values of current, voltage drop and power loss if the consumer potential difference is 230 V?

● **Extra skills task**   Application of Number
Voltages used in a commercial electricity distribution system are as shown in Table 6.1.
Transformers are used between each stage. Assume that each one has an efficiency of 80%.

Table 6.1

Stage	Voltage/kV	Typical cable length/km	Typical cable resistance/$\Omega\,m^{-1}$
generator output	20	0.01	1.0×10^{-3}
long-distance distribution	400	50	5.0×10^{-3}
urban area distribution	275	10	5.0×10^{-3}
local distribution	33	5	1.0×10^{-2}
domestic use	0.23	0.5	1.0

Make a poster-size schematic diagram of a simple system with a power station feeding a long-distance distribution cable at a power of 100 kW, one urban area distribution cable, and several local distribution systems in parallel, each one serving many domestic systems. Use the diagram to show a comparison of the voltage, current and thermal dissipation of energy for each stage of transmission from power station to consumer. Use bar charts with clear scales, and include all of your calculations on the poster. Show the percentage of the input power of 100 kW that is delivered to consumers.

● ## Examination questions

1 The first graph shows how the reactance X_C of a capacitor C varies with the frequency f of an applied alternating voltage. The diagram shows the circuit which was used to obtain data for the graph.

The second graph shows the trace on the oscilloscope when the measurements for point P on the first graph were being made. The Y-amplification setting on the oscilloscope was $5\,V\,cm^{-1}$.

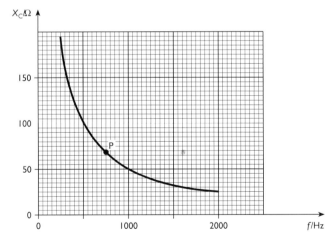

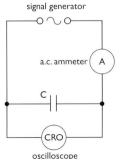

signal generator

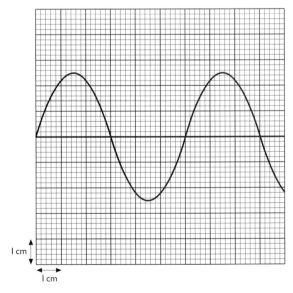

a i Determine the root mean square (r.m.s.) voltage across C when the measurements for point P were being obtained. (3)
ii What was the reading on the a.c. ammeter when the measurements for point P were being made? (3)

b Draw, on the axes of the second graph, a graph to show the corresponding variation of current with time in the circuit. It is not necessary to give a value for the maximum current. (2)

c Use information from the first graph to calculate the capacitance of the capacitor C in the circuit diagram. (3)

AEB, A level, Paper 2, Summer 1998

2 a The switch in the circuit below is closed at time $t = 0$. The graph shows how the current I in the circuit subsequently varies with time t.

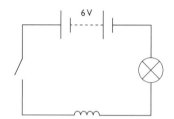

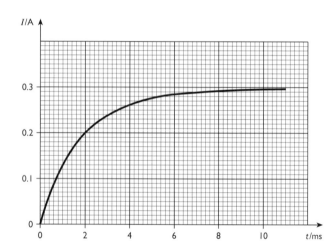

Explain, in terms of Faraday's and Lenz's laws, why the current does not rise immediately to its maximum value. (4)

b Use information from the circuit diagram and the graph to determine the self-inductance of the coil. (3)

c Assuming that the battery and the coil have negligible resistance, determine the final resistance of the lamp. (2)

d The air-cored *plane circular coil* used as the inductor in the circuit has 600 turns and radius 0.050 m. The permeability of free space (air) is 1.26×10^{-6} H m^{-1}. Calculate the final magnetic flux density through the coil. (2)

AEB, A level, Paper 2, Summer 1998

3 A transformer is required to produce an r.m.s. output of 2.0×10^{3} V when it is connected to the 230 V r.m.s. mains supply. The primary coil has 800 turns.

a Calculate the number of turns required on the secondary coil, assuming the transformer is ideal. (2)

b The output from the transformer is to be rectified and smoothed.

i State the names of the components needed for these processes.

1 Rectification

2 Smoothing (1)

ii Calculate the maximum possible smoothed output voltage when the transformer is used in this way. (1)

c The transformer suffers from *eddy current* losses.

i Explain how *eddy currents* arise. (4)

ii State the feature of transformers designed to minimise eddy currents. (1)

AQA (AEB), A level, Paper 2, Summer 2000

4 The a.c. mains supply of root-mean-square (r.m.s.) voltage 230 V is to be converted into a much lower steady d.c. voltage for use with a high resistance load.

a Draw a diagram of a suitable circuit. Your circuit should include a transformer, a full-wave rectifier using diodes, and a capacitor. Add arrows to your diagram to show the route of the current through the rectifier during one half-cycle. (6)

b If the output of the transformer in **a** is 12 V r.m.s. explain why the steady d.c. output of the circuit you have drawn in **a** is not 12 V. The diodes in the circuit can be considered ideal. (2)

OCR (Oxford), A level, 6843, June 1999

5 a In the context of the operation of an electric motor, explain what is meant by the term *back e.m.f.* (2)

b The coil wound on the armature of a small d.c. electric motor consists of 100 turns of wire with a total resistance of 2.0 Ω. The area enclosed by the coil is 2.5×10^{-3} m^2. The coil is connected to a 10 V battery of negligible internal resistance. The armature rotates in a radial magnetic field of flux density 0.50 T produced by an electromagnet.

i Explain why it is an advantage for the armature to rotate in a radial magnetic field rather than a uniform one. (2)

ii The armature rotates at 75 rad s^{-1}. Calculate the current [in A] in the coil. (3)

c With the rotation of the armature opposed by frictional forces only, the strength of the field of the electromagnet is slightly reduced. It is observed that the speed of rotation of the armature increases. Suggest, using electromagnetic principles, why this increase occurs. (2)

OCR (Oxford), A level, 6843, June 1999

6 The diagram shows a 0.050 H inductor, of negligible resistance, connected to an a.c. source. The instantaneous current, I (in mA), in the circuit at time t is given by the expression

$$I = 2.0 \sin(1000 \pi t)$$

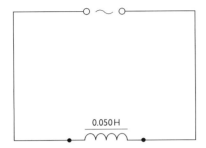

a Show that the frequency of the a.c. source is 500 Hz. (1)

b Calculate the peak voltage [in V] of the a.c. source. (3)

OCR (Oxford), A level, 6843, June 1999

7 The diagram shows the principle of one type of residual current device (R.C.D.) that utilizes mutual induction. The function of the device is to switch off the mains supply when, owing to a fault, current flows from the live wire to earth.

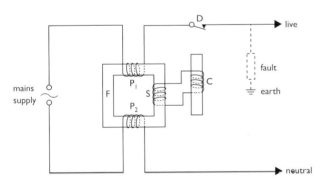

Three coils, P_1, P_2 and S, are wound on a laminated iron former, F. Coils P_1 and P_2 are connected into the live and neutral wires respectively of the supply. Coil S is in series with a relay coil C.

The currents in P_1 and P_2 flow in the same sense when viewed from the left of the diagram. Relay switch D is opened if C carries a current.

a What is meant by *mutual induction*? (1)

b Explain:

i why, in normal use, D experiences no electromagnetic force; (2)

ii why D opens when a fault occurs. (3)

c State and explain what would occur if a solid iron former were used instead of a laminated iron former. (2)

OCR (Oxford), A level, 6843, June 1999

8 The relationship between E, the instantaneous voltage output of an a.c. source, and time, t, is given by the expression:

$$E = E_0 \sin(\omega t)$$

a Using the axes below, sketch a graph of E against t for one cycle of the output of the source. Add suitable labels to show the significance of E_0 and ω. (3)

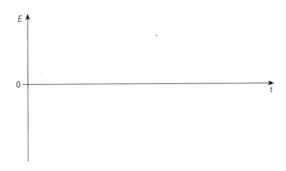

b The root-mean-square (r.m.s.) voltage of the source is 6.0 V. The source is connected to a 0.50 H inductor in a circuit of negligible resistance. The r.m.s. current in the circuit is 3.8 mA. Calculate the frequency [in Hz] of the source. (3)

OCR (Oxford), A level, 6843, March 1999

9 a The primary coil, P, of an ideal transformer has N_p turns, and is connected to an alternating voltage supply. The secondary coil, S, which has N_s turns, is connected to a resistor. The r.m.s. currents in P and S are I_p and I_s respectively.

Starting with appropriate voltage and power relations, derive an expression for the ratio I_p/I_s in terms of N_p and N_s. (2)

b The primary coil of a transformer is connected to the 230 V mains supply. The transformer has two secondary coils. One secondary coil is to provide 350 V and deliver a current of 50 mA. The other secondary coil is to provide 6.0 V and deliver a current of 2.0 A.

i The transformer is fitted with a single fuse connected in the primary circuit. Show that a suitable rating for this fuse is 150 mA. (2)

ii Under unusual circumstances, only the 6.0 V secondary circuit is used. It is found that the current in this circuit can rise to 5.0 A, which is sufficient to cause the 6.0 V coil to start to melt. Explain why the single 150 mA fuse in the primary circuit will now prove inadequate. (1)

iii Suggest how the fitting of an additional fuse will protect the transformer from damage should the event outlined in **ii** occur. (1)

UODLE, A level, Paper 4, March 1998

10 The armature of a small d.c. electric motor consists of 75 turns of fine wire each enclosing an area of $2.5 \times 10^{-4} \, \text{m}^2$. The resistance of the armature is $8.0 \, \Omega$. The armature rotates in a radial magnetic field of flux density $0.40 \, \text{T}$.

  a The armature is connected to a $12 \, \text{V}$ d.c. supply. Initially the armature is prevented from rotating. For this condition, calculate the torque [in N m] acting on the armature. (3)

  b The armature is now released and allowed to rotate freely. Explain, in terms of electromagnetic principles:

    i why the current in the armature progressively decreases as the angular velocity of the armature increases; (3)

    ii why a maximum angular velocity is eventually reached. (2)

    UODLE, A level, Paper 4, March 1998

11 **a** Describe the motion of the conduction electrons in a mains cable in which there is an r.m.s. current of $5.0 \, \text{A}$. (2)

  b The cable has a cross-sectional area of $0.20 \, \text{mm}^2$ and the number of conduction electrons per unit volume is $1.0 \times 10^{29} \, \text{m}^{-3}$. Calculate the maximum speed v_{max} of the electrons. State any assumption you make.

    Explain how you could use a knowledge of this maximum speed to calculate the amplitude of oscillation of the electrons in the cable. (6)

    London, A level, Synoptic Paper PH6, June 1998 (part)

12 Magnetic flux density B varies with distance beyond one end of a large bar magnet as shown on the graph below.

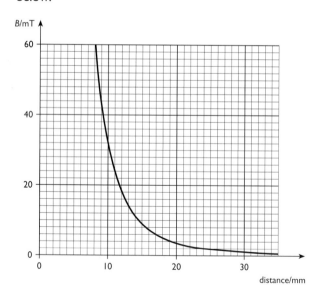

A circular loop of wire of cross-sectional area $16 \, \text{cm}^2$ is placed a few centimetres beyond the end of the bar magnet. The axis of the loop is aligned with the axis of the magnet.

  a **i** Calculate the *total* magnetic flux through the loop when it is $30 \, \text{mm}$ from the end of the magnet.

    ii Calculate the total magnetic flux through the loop when it is $10 \, \text{mm}$ from the end of the magnet. (3)

  b The loop of wire is moved towards the magnet from the $30 \, \text{mm}$ position to the $10 \, \text{mm}$ position so that a steady e.m.f. of $15 \, \mu\text{V}$ is induced in it. Calculate the average speed of movement of the loop. (3)

  c In what way would the speed of the loop have to be changed while moving towards the magnet between these two positions in order to maintain a steady e.m.f.? (1)

    London, A level, Module Test PH4, June 1999

13 **a** A large solenoid is $45 \, \text{cm}$ long and has $72 \, \text{turns}$. Calculate the magnetic flux density inside the solenoid when a current of $2.5 \, \text{A}$ flows in it. (2)

A small solenoid is placed at the centre of the large solenoid as shown. The small solenoid is connected to a digital voltmeter.

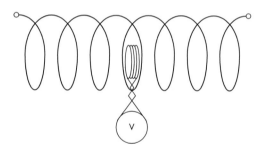

  b State what would be observed on the *voltmeter* when each of the following operations is carried out consecutively.

    i A battery is connected across the large solenoid.

    ii The battery is disconnected.

    iii A very low frequency alternating supply is connected across the large solenoid. (5)

    London, A level, Module Test PH4, January 2000

7 Nuclear energy

THE BIG QUESTIONS

- What are the differences between a nuclear bomb and a nuclear reactor?
- What are the past, present and future of nuclear technologies?
- Where does the energy come from?

KEY VOCABULARY

barn   binding energy per nucleon   chain reaction   control rods   coolant   criticality
critical mass   delayed neutrons   depleted (fuel rods)   enriched (uranium)
fast neutrons   fast reactor   fuel rods   mass defect   moderator   pressure vessel
pressurised water reactor   shielding   subcriticality   supercriticality
thermal neutrons   thermal neutron absorption cross-section   thermal reactor
total binding energy   tunnelling

BACKGROUND

There is no way of generating the electricity we want without having an environmental impact. When nuclear fission was first used to heat steam to turn power station turbines in the 1950s (Figure 7.1), this was seen as a huge step forward – the 'modern way' to cheap smoke-free electricity. Einstein's name could be used to show just how modern this energy resource was – since the energy is made available by loss of the total mass of the particles involved in fission, exactly according to the equation that had become so famous, $E = mc^2$. Every nation in the world that could afford to build nuclear power stations did so.

Fission itself produces no smoke or carbon dioxide to pollute the atmosphere. Within the rods of fuel in the power stations, uranium nuclei split into smaller nuclei. The quantities are very small, but these smaller nuclei are highly radioactive, and the 'spent' fuel rods contain this high-level radioactive waste (Figure 7.2). To this day, nobody has solved the problem of what to do with this waste, and it remains in 'temporary' storage. Further radioactive waste is produced by the flow of neutrons that emerge from nuclear reactors. These neutrons can enter nuclei in surrounding materials, changing their structure and making them unstable.

Nuclear power stations continue to provide us with electricity, but their waste provides us with a headache that won't go away.

Figure 7.1 (left) In the 1950s nuclear power promised a bright new future, with cheap and clean energy for all.

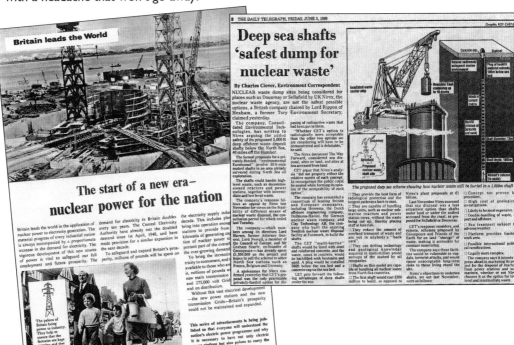

Figure 7.2 (right) Nuclear power stations have created a new kind of waste. Fission products are the chief components of high-level nuclear waste, and problems of long-term storage have not been solved.

Nuclear size and density

This chapter considers the behaviour of assemblies of nucleons (Figure 7.3). A fundamental question we should ask before going too far is: 'Do nucleons merge together or just stick together?'

Figure 7.3
Which is the truer picture?
Only in the second of these
two possibilities does

volume of nucleus =
number of nucleons
× volume of each nucleon

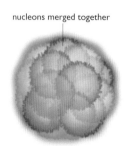

nucleons merged together

The density of the nucleus is greater than the density of an individual nucleon.

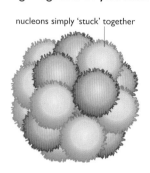

nucleons simply 'stuck' together

The density of the nucleus is the same as the density of each individual nucleon.

If the nucleons are merged together, even partly, when in a nucleus, then they occupy a volume that is less than their combined individual volumes. If they do not merge but stick together, rather like a cluster of Velcro-covered balls, then the volume of the nucleus is the same as the total combined volumes of the nucleons. In the latter case:

volume of nucleus = number of nucleons, A × volume of each nucleon, v

That is, for the whole nucleus, of radius r, we can write

$$\tfrac{4}{3}\pi r^3 = Av$$

For a single nucleon, of radius r_0, we can simply say that

$$\tfrac{4}{3}\pi r_0^3 = v$$

which would mean that

$$\tfrac{4}{3}\pi r^3 = A\tfrac{4}{3}\pi r_0^3$$
$$r^3 = Ar_0^3$$

and so

$$r = r_0 A^{1/3}$$

Figure 7.4 (below)
Bombardment of nuclei by
alpha particles provides
information about
maximum nuclear size.

By testing the truth of this equation we can find out the answer to our original question. This requires that we have a way of comparing nuclear radii.

One way to predict nuclear size is to fire alpha particles at the nucleus, as in the original experiment of Ernest Rutherford and his colleagues (see *Introduction to Advanced Physics*, page 60). If a radioactive isotope of known decay energy is used as the source of the alpha particles and they then travel through a vacuum, their initial kinetic energy is known.

If they are to approach a nucleus then alpha particles must 'climb the potential (or potential energy) hill' associated with it (Figure 7.4). They may do this thanks to their initial kinetic energy. We know that an alpha particle can approach a nucleus directly towards its centre, and rebound without being absorbed by it. This tells us that nuclear radius must be less than the distance of closest approach.

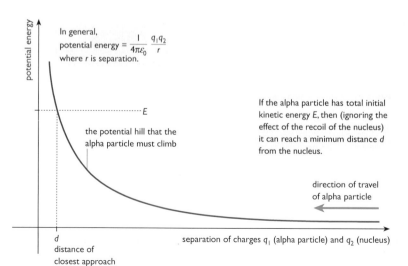

potential energy

In general,
potential energy $= \dfrac{1}{4\pi\varepsilon_0}\dfrac{q_1 q_2}{r}$
where r is separation.

E

the potential hill that the alpha particle must climb

If the alpha particle has total initial kinetic energy E, then (ignoring the effect of the recoil of the nucleus) it can reach a minimum distance d from the nucleus.

direction of travel of alpha particle

d
distance of closest approach

separation of charges q_1 (alpha particle) and q_2 (nucleus)

If we ignore the recoil of the nucleus, which has a minor effect provided that the nucleus is very much larger than the alpha particle, then at the shortest distance of approach to a nucleus, all of the initial kinetic energy of the alpha particle has become potential energy. We can write

$$\text{KE lost, } E = \text{PE gained}$$

$$= \frac{1}{4\pi\epsilon_0}\frac{q_1 q_2}{d}$$

(The expression on the right-hand side is given in Chapter 3.) Here d is the distance of closest approach, q_1 is the alpha particle charge (equal to twice the proton charge), q_2 is the nucleus charge (equal to the proton charge e multiplied by the number of protons), and ϵ_0 is the permittivity of a vacuum. So substituting for q_1 and q_2

$$\text{KE lost, } E = \frac{1}{4\pi\epsilon_0}\frac{2eZe}{d} = \frac{1}{4\pi\epsilon_0}\frac{2Ze^2}{d}$$

where e is the proton charge and Z is the number of protons in the nucleus. If we assume that the alpha particle keeps its energy and does not pass any to the nucleus (though there will be some recoil of the nucleus so this is an approximation), then the predicted distance of closest approach is given by

$$d = \frac{1}{4\pi\epsilon_0}\frac{2Ze^2}{E}$$

where E is initial alpha particle kinetic energy.

An energetic alpha particle might have an initial energy of 8.0 MeV. For a gold nucleus, as an example, $Z = 79$. The value of $1/(4\pi\epsilon_0)$ is $9 \times 10^9\,\text{N}\,\text{m}^2\,\text{C}^{-2}$ (from Chapter 3). So, for a gold nucleus, converting MeV into joules ($1\,\text{eV} = 1.6 \times 10^{-19}\,\text{J}$) so that we have a consistent SI equation, we would expect

$$d = 9 \times 10^9 \times \frac{2 \times 79 \times (1.6 \times 10^{-19})^2}{8 \times 10^6 \times 1.6 \times 10^{-19}}\,\text{m}$$

$$= 2.8 \times 10^{-14}\,\text{m}$$

This tells us that the gold nucleus must have a radius less than this distance.

Electron diffraction by nuclei provides a method of making measured comparisons of nuclear radii. A diffraction pattern produced by a beam of electrons passing amongst nuclei shows a central maximum, a first minimum, and so on (Figure 7.5). The angle at which the first minimum,

Figure 7.5
Diffraction patterns for 420 MeV electrons diffracted by carbon and oxygen nuclei.

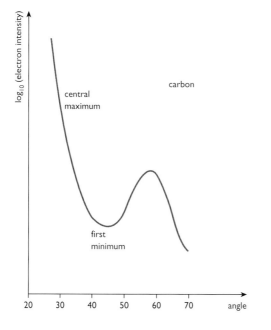

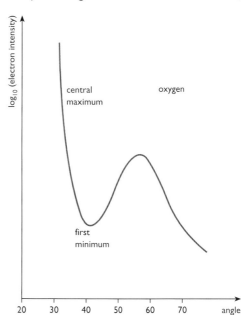

for example, occurs depends on nuclear radius, and thus provides a means of measurement of the radius. Data obtained from such electron diffraction, also called electron scattering, are given in Table 7.1. These data are used to generate a graph of r against $A^{1/3}$, as in Figure 7.6. The straight line shows that the relationship $r \propto A^{1/3}$ proposed above is reliable, and therefore confirms that when nucleons cluster together their total volume (and therefore their density) experiences little change. It thus confirms that nuclei all have similar density, and that nucleons 'stick' together rather than 'merge' together in nuclei (see Figure 7.3).

Table 7.1
Electron diffraction data.

Nuclide	Radius, r/fm	Nucleon number, A	$A^{1/3}$
$^{1}_{1}$H (proton)	$1.20 = r_0$	1	1.00
$^{4}_{2}$He	1.92	4	1.32
$^{12}_{6}$C	2.81	12	2.29
$^{16}_{8}$O	3.15	16	2.52
$^{28}_{14}$Si	3.62	28	3.03
$^{40}_{20}$Ca	4.19	40	3.42
$^{51}_{23}$V	4.27	51	3.71
$^{88}_{38}$Sr	4.93	88	4.45
$^{115}_{49}$In	5.35	115	4.86

$1 \, \text{fm} = 10^{-15} \, \text{m}$

Figure 7.6
r against $A^{1/3}$ using electron diffraction data.

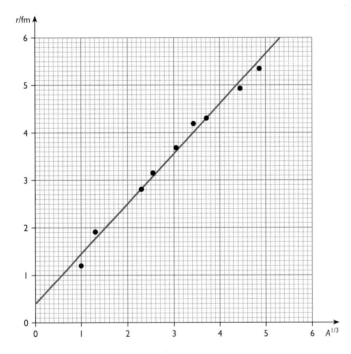

1 Explain why alpha particle distance of closest approach only provides an approximate and maximum value for the radius of a nucleus.

2 a When an alpha particle travels along a line that runs through the centre of a nucleus and hits it, then, if the nucleus is much larger than the alpha particle, the alpha particle returns along the same pathway with its speed almost unchanged. Explain this in terms of kinetic energy and momentum conservation. (The collision is elastic.)
 b Explain what happens when the alpha particle hits a helium nucleus in the same way.

3 Suppose that analysis of the electron diffraction data had shown that $r \propto A$. What conclusion would we have had to make about nucleons and nuclei?

Mass and energy

Einstein had already shown through his theory of special relativity that the energy possessed by a simple and independent body (one not experiencing interactions, either internally or with other external bodies) has two components – energy due to its motion (kinetic energy) and energy due to its mass (mass-energy). Energy and mass are not entirely separate quantities but are inter-convertible (see Figure 28.8 in *Introduction to Advanced Physics*, and Figure 7.7 below). Energy, such as kinetic energy or photon energy, can become mass; and mass can become energy – in the form of kinetic energy or energy enabling the creation of photons of electromagnetic radiation. To convert mass in kilograms into energy in joules we use the formula $E = mc^2$, where c is the speed of light, so c^2 is $9 \times 10^{16} \, \text{m}^2\text{s}^{-2}$. So we can say:

$$\text{energy in joules} = \text{mass in kilograms} \times 9 \times 10^{16}$$

Figure 7.7
A PET scan – based on positron emission tomography. A beta$^+$-emitting radioactive substance is put into the brain (through a drink). The positrons emerge into the surrounding material, meet electrons, and both electrons and positrons cease to exist. This is mutual annihilation, and involves conversion of the masses of the particles into energy that is carried away by gamma ray photons, which can be detected from outside the brain to produce pictures like this one.

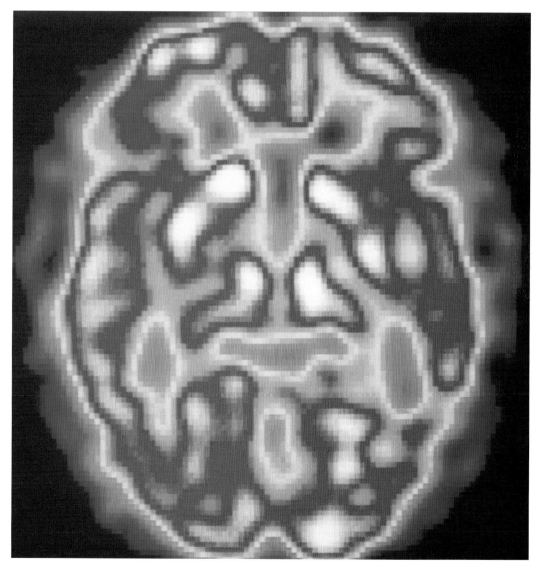

Units of mass and energy

Atoms are small and the kilogram is not a convenient unit for measuring their masses, so instead we can use the atomic mass unit, u. One atomic mass unit (1 u) is equal to one-twelfth of the mass of a neutral atom of carbon-12. An atom of carbon-12 contains 12 nucleons (six protons and six neutrons all of similar mass) and the masses of the electrons are very small in comparison. One atomic mass unit is *approximately* equal to the mass of a nucleon.

Particle physicists deal with conversions between mass and energy as a matter of routine, for example in the analysis of particle tracks in detectors (Figure 7.8). So they use the same units for both mass and energy. The unit that is most convenient for their work on mass-energy is not the atomic mass unit but the MeV – millions of electronvolts (see Table 7.2).

Just as we can say that

$$\text{energy in joules} = \text{mass in kilograms} \times 9 \times 10^{16}$$

so we can also use a conversion factor to convert mass in atomic mass units into mass-energy in MeV:

$$\text{energy in MeV} = \text{mass in u} \times 931.3$$

Figure 7.8
Output from the ALEPH detector at CERN, showing tracks of a 'jet' of particles. Many particles can be created because of the high energy of the colliding electron and positron.

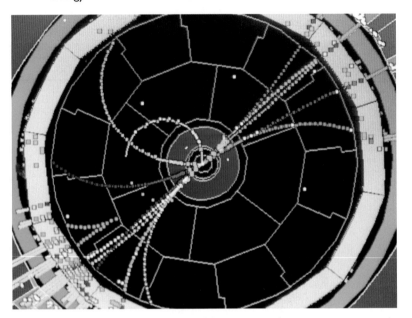

Table 7.2
Some masses in different units.

Particle	Mass/kg	Mass/u (relative atomic mass)	Mass-energy/MeV
neutral atom of carbon-12	1.993×10^{-26}	12.0000	11 200
neutral atom of hydrogen-1	$1.673\,560 \times 10^{-27}$	1.0078	938.6
proton	$1.672\,649 \times 10^{-27}$	1.0073	938.1
neutron	$1.674\,954 \times 10^{-27}$	1.0087	939.6
electron	$9.110\,953 \times 10^{-31}$	0.0005	0.511

4 Estimate the energy, in joules, available from your own body mass.

5 Use data from Table 7.2 to calculate the conversion factor for changing atomic mass units to kilograms.

6 a Estimate your body mass in atomic mass units.
  b Estimate how many nucleons you are made of.

7 A proton in a particle accelerator can gain as much as 10^{12} eV of energy.
  a What is the equivalent quantity in atomic mass units?
  b By what factor is it bigger than the mass of a stationary proton measured in MeV?

8 a Is the difference between the mass of a neutron and the mass of a proton bigger or smaller than the mass of an electron?
  b How does your answer to **a** explain that a β^- particle emitted from a nucleus can be very fast?

Nuclear mass defect and binding energy

The relative atomic mass of carbon-12 is 12.000000 u. Our knowledge of atomic structure tells us that the atom is made up of six protons, six neutrons and six electrons. But the sum of the separate masses of these constituents is significantly more than 12.000000 u. The mass of the separate constituents of the atom is more than the mass of the atom itself. In coming together, they must have lost mass. This does not appear to happen when mixing flour, eggs and milk, when mass is effectively conserved. Behaviour at the nuclear level is rather different.

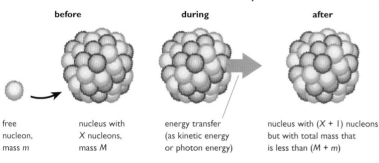

before — free nucleon, mass m

nucleus with X nucleons, mass M

during — energy transfer (as kinetic energy or photon energy)

after — nucleus with $(X + 1)$ nucleons but with total mass that is less than $(M + m)$

Adding an extra nucleon to a nucleus does not, in general, increase total mass by the mass of the nucleon. Instead, the mass of the new nucleus is less than the mass of the nucleon and the original nucleus (Figure 7.9). In the process of combining nucleon and nucleus, total mass decreases and energy is made available. This is usually in the form of either kinetic energy of the particles (or particle) involved or photon energy.

Figure 7.9 If twelve nucleons could be brought together to create a carbon nucleus, then we would find not only that total mass would decrease significantly, but also that energy would be 'released'. This energy might take the form of kinetic energy of the new nucleus, or photons might be created and would carry energy away, or both. The 'disappeared' mass is called the **mass defect** of the nucleus. It is equivalent to the energy released as predicted by the formula $E = mc^2$.

The **total binding energy** of the nucleus is the change in energy experienced by the system of nucleons when they assemble as a nucleus. Since energy is transferred outside the system, the system loses energy. Total binding energy of nucleons is a negative quantity of potential energy, given by

$$\text{total binding energy} = \text{mass defect of nucleus} \times c^2$$

If the nucleus were to break apart again, into twelve nucleons, it would have to gain an amount of energy equal in size to its binding energy. Until the nucleus receives that energy, its nucleons remain bound together, hence the name 'binding energy'.

Figure 7.10
An analogy.

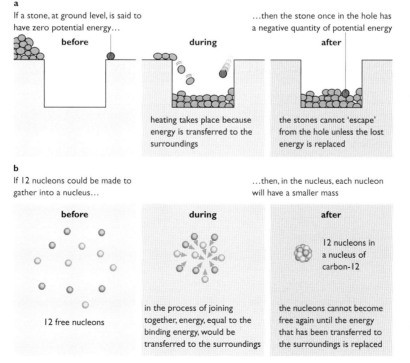

a

If a stone, at ground level, is said to have zero potential energy...

...then the stone once in the hole has a negative quantity of potential energy

before

during — heating takes place because energy is transferred to the surroundings

after — the stones cannot 'escape' from the hole unless the lost energy is replaced

Some stones lie on a building site. A builder comes along and shovels them into the foundations of a new house. The stones have less potential energy in the hole than they do when at ground level. They will need to gain energy to escape from the hole, and until they do gain energy they are trapped. Their energy in the hole is negative relative to their energy when at ground level.

b

If 12 nucleons could be made to gather into a nucleus...

...then, in the nucleus, each nucleon will have a smaller mass

before — 12 free nucleons

during — in the process of joining together, energy, equal to the binding energy, would be transferred to the surroundings

after — 12 nucleons in a nucleus of carbon-12. the nucleons cannot become free again until the energy that has been transferred to the surroundings is replaced

Likewise with nucleons – when they come together into a nucleus they lose energy. Their energy is negative relative to their energy when they are free. Until they regain this energy, they are trapped. The energy they all need to escape from each other is equal to the total binding energy of the nucleus.

Figure 7.11
Graph of energy versus distance from the centre of a nucleus, for a proton (red line) and a neutron (green line).

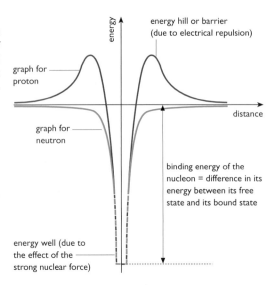

energy

energy hill or barrier
(due to electrical repulsion)

graph for proton

distance

graph for neutron

binding energy of the
nucleon = difference in its
energy between its free
state and its bound state

energy well (due to
the effect of the
strong nuclear force)

A proton approaching the nucleus must 'climb over' (or 'tunnel through') the energy barrier before it can fall into the nucleus. Neutrons do not experience electrical repulsion and so for them there is no hill to climb. So in general nuclei can absorb neutrons more easily.

The stones and the hole in the ground (Figure 7.10) provide an analogy for nucleons and binding energy. However, there is an important difference between nucleons and stones. Protons have to be forced close together. Work must be done to overcome the electrical repulsion between them, before the strong nuclear force takes over and pulls them together – into the 'hole'. The protons must first climb over or **tunnel** through the 'hill' that surrounds the 'hole' (Figure 7.11). Shapes in ground levels – holes and hills – are useful analogies which describe the potential energies that nucleons possess at different places.

9 People sometimes talk about 'pure energy' in relation to the formula $E = mc^2$.
  a When mass is converted to energy, how is the energy carried away?
  b Is there such a thing as pure energy?
10 Is the mass of a nucleus of oxygen-16, $^{16}_{8}O$, more than or less than the mass of eight free protons and eight free neutrons?
11 Use the following data in this question:

  proton mass = 1.007 277 u
  neutron mass = 1.008 665 u
  mass of helium nucleus = 4.002 784 u

  a What is the relationship between the energy released when four nucleons come together to form a helium nucleus and the energy that must be supplied to separate them again?
  b What is the mass defect of the helium nucleus in atomic mass units?
  c What is its total binding energy in MeV?
12 Comment on the statement that: 'Total binding energy and mass defect are both negative quantities.'
13 Investigate the relevance of the term 'tunnelling', and explain its use, regarding
  a alpha emission
  b STM (scanning tunnelling microscope) imaging.

Binding energy per nucleon

A very large nucleus has more binding energy than a very small one simply because it has been created by the assembly of more nucleons. Comparison of total binding energies of different nuclei shows this trend, but a more useful comparison is made in terms of **binding energy per nucleon** – the total binding energy divided by the number of nucleons. For example, the binding energy per nucleon of a carbon-12 nucleus is its total binding energy divided by twelve.

When we plot nuclei on to a chart of binding energy per nucleon against size of nucleus we reveal a very clear pattern (Figure 7.12). Iron nuclei, and others of similar 'medium' size, have the most binding energy per nucleon. Nucleons in mid-sized nuclei are particularly well bound together.

Figure 7.12
Mid-sized nuclei are more tightly bound together than either small ones or large ones.

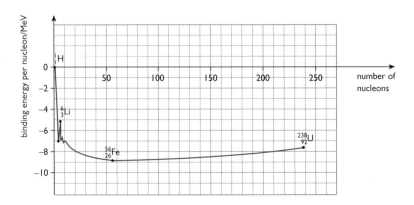

Fission, fusion and changes in binding energy per nucleon

The size of binding energy per nucleon increases when a large nucleus breaks into two mid-sized nuclei (Figure 7.13a). This is nuclear *fission* (see *Introduction to Advanced Physics*, Chapter 28). The increase in binding energy makes the nucleons more tightly bound in the two daughter nuclei and involves release of energy. For example, in the fission of a nucleus of uranium-236 into iodine-137 and rubidium-97, the nucleons are more tightly bound in the daughter nuclei than in the parent nucleus. The binding energy per nucleon has increased in size.

In lithium-6, binding energy per nucleon is particularly small. This means that breaking a lithium-6 nucleus apart is much easier, nucleon for nucleon, than breaking an iron nucleus apart. It also means that if two lithium-6 nuclei were to join or fuse together then their binding energy per nucleon would greatly increase in size (Figure 7.13b). A relatively large amount of energy would be released. A nucleus of carbon-12 would be created. This would be an example of *fusion* (see *Introduction to Advanced Physics*, Chapter 28) of small nuclei into a larger one:

$$^6_3Li + ^6_3Li \rightarrow ^{12}_6C$$

The nucleons are more tightly bound in the carbon nucleus than they are in the lithium nuclei. Fusion of the lithium into carbon would release energy:

binding energy per nucleon of a nucleus of lithium-6 $= -5.076\,\text{MeV}$
binding energy per nucleon of a nucleus of carbon-12 $= -7.424\,\text{MeV}$
change in binding energy per nucleon $= (-7.424) - (-5.076)$
change in total binding energy $= -2.348 \times 12$
$= -28.176\,\text{MeV}$

Figure 7.13
Any process that increases the size of binding energy per nucleon transfers energy outwards.

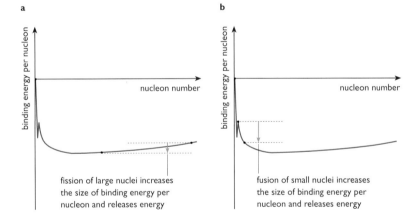

a

b

fission of large nuclei increases the size of binding energy per nucleon and releases energy

fusion of small nuclei increases the size of binding energy per nucleon and releases energy

Since all of the energies involved have negative values, the signs are sometimes omitted. While binding energies are always negative quantities of potential energy, some writers, including some writers of examination questions, omit the negative signs.

Fission of large nuclei, like fusion of small ones, makes energy available. In both cases the energy is available as kinetic energy, carried by the particles themselves, and as energy that is carried away by photons. These are the energies that constitute available 'nuclear energy'.

---

14 Which would you expect to be easier (requiring less energy), the removal of a nucleon from an iron atom or from a lithium atom?

15 Deuterium is an isotope of hydrogen. A deuterium nucleus has one proton and one neutron.
a Explain why deuterium is chemically indistinguishable from hydrogen-1.
b The following are the masses of a proton, a neutron and a deuterium nucleus, in atomic mass units:

proton   1.007 277 u
neutron   1.008 665 u
deuterium nucleus   2.013 553 u

i Calculate the mass defect of the deuterium nucleus, in atomic mass units.
ii Calculate the total binding energy and binding energy per nucleon for the deuterium nucleus, in MeV.

16 The process of fusion to create a deuterium nucleus, as in question 15, can be written as:

$$^1_1H + ^1_0n \rightarrow ^2_1H + \text{photon}$$

a Assuming that the original particles are stationary, the deuterium nucleus and the photon travel apart in opposite directions with equal and opposite momenta. Draw a sketch of this.
b Using the de Broglie relationship (see *Introduction to Advanced Physics*, page 269), write down an expression for the photon momentum.
c Using the simple product of nuclear mass and velocity for the momentum of the nucleus, write down an expression for the momentum of the deuterium nucleus.
d Write down an algebraic equation to show the momentum equality (ignoring relativistic effects).

Fission as an energy resource

Fission is a quite rare event in nature. However, large nuclei can be induced to fission by absorption of neutrons. This is the basis of all working nuclear power stations. It is, unfortunately, a messy business (Figure 7.14). The product nuclei of the fission, though they have more binding energy per nucleon than the parent nuclei, are neutron-rich relative to stable nuclides of similar size. They are therefore almost always beta⁻ emitters (see Figure 7.15 and *Introduction to Advanced Physics*, Chapter 28). These radioactive products of fission create enormous disposal problems, and in catastrophic accidents they can be released into the environment. Also, fission of a large nucleus usually releases a few neutrons from it, along with the creation of the daughter nuclei. These neutrons may be absorbed by nuclei in surrounding material, making those nuclei unstable and the material radioactive. The physical structure of a nuclear reactor becomes radioactive, and when its working life is finished expensive decommissioning is necessary to protect the environment.

Figure 7.14
In a mine in New Mexico, a trial project is being run to test ways of storing high-level nuclear waste which contains material with very long half-lives.

Fission products increase in quantity in the fuel of a reactor. Many have short half-lives, and energy from their decay contributes to the energy that is needed to generate electricity. But many have long half-lives, and they contribute to the radioactivity of spent nuclear fuel. This radioactivity makes the spent fuel very hazardous. Although nuclear power stations produce very low volumes of such waste, disposing of it where it cannot affect the living environment is a big problem that has still not been solved.

Figure 7.15
Daughter nuclei of fission processes are formed very quickly from the parent nucleus, with nucleon structures that are very variable. They are almost always neutron-rich and so undergo decay such that neutron number decreases. This means that they usually decay by beta⁻ emission. (For detailed charts of nuclides, see *Introduction to Advanced Physics*, pages 295 and 296.)

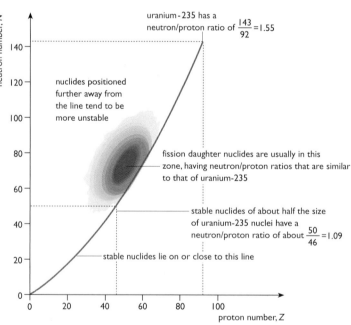

uranium-235 has a neutron/proton ratio of $\frac{143}{92} = 1.55$

nuclides positioned further away from the line tend to be more unstable

fission daughter nuclides are usually in this zone, having neutron/proton ratios that are similar to that of uranium-235

stable nuclides of about half the size of uranium-235 nuclei have a neutron/proton ratio of about $\frac{50}{46} = 1.09$

stable nuclides lie on or close to this line

17 Use a graph of binding energy per nucleon against nucleon number to show why fusion of iron cannot release energy from the nuclei into their environments.

18 Why would you expect a nucleus with more binding energy per nucleon to be more stable?

19 If free nucleons are, somehow, assembled to create a lithium nucleus and an iron nucleus, in which case does each nucleon lose most mass-energy?

Fission and neutrons

Fission involves much more drastic nuclear change than radioactive emission does. Following radioactive emission the daughter nucleus, even in the case of alpha emission, is only a little smaller than the parent nucleus, whereas fission of a large nucleus creates two mid-sized nuclei both of which must be called daughter nuclei. A single nucleus becomes two.

Fission was not discovered until more than 40 years after radioactivity, and six years after the discovery of neutrons. There are no materials in the environment that fission without first experiencing some other significant change to trigger the process. This is because nuclei that fission are very unstable – if fission is going to happen to a nucleus it will usually happen soon after the nucleus forms.

In 1938 it was found that, when a sample of uranium was bombarded by a stream of neutrons, the sample then contained elements with much smaller nucleon number. The neutrons were inducing uranium nuclei to split, or to fission.

The liquid drop model of the nucleus is useful here (Figure 7.16). The nucleus that is induced to fission by absorbing a neutron is in an excited state – with a large excess of energy compared with a ground-state nucleus. A large drop of water on a very hot surface is also in an excited state – and is likely to vibrate or wobble violently and soon break up into smaller droplets. In both the liquid drop and the nucleus, the instability is considerable and break-up happens quickly.

Figure 7.16
The liquid drop model of fission of an excited nucleus. Fission events take place very quickly. The sizes of the daughter nuclei and the number of neutrons produced varies from one event to another.

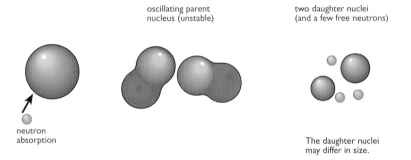

oscillating parent
nucleus (unstable)

two daughter nuclei
(and a few free neutrons)

neutron
absorption

The daughter nuclei
may differ in size.

Uranium-235 and plutonium-239 are nuclides that can be bombarded by neutrons in order to induce fission. These are the fuels for power stations or for bombs. In both cases, absorption of a neutron produces a slightly larger nuclide – uranium-236 and plutonium-240 respectively. It is these larger nuclides that are excited and unstable and which fission. The daughter nuclides produced by fission of the same parent nuclide can vary in size and structure – the high instability makes fission a very rapid process, and there is no time for nucleons to gather into particular patterns. As well as the daughter nuclei, fission produces free neutrons; again the number is not fixed, but is usually between one and four. These are some examples:

$$^{235}_{92}\text{U} + ^{1}_{0}\text{n} \rightarrow ^{236}_{92}\text{U*} \rightarrow ^{86}_{36}\text{Kr} + ^{147}_{56}\text{Ba} + 3^{1}_{0}\text{n}$$

$$^{235}_{92}\text{U} + ^{1}_{0}\text{n} \rightarrow ^{236}_{92}\text{U*} \rightarrow ^{104}_{42}\text{Mo} + ^{130}_{50}\text{Sn} + 2^{1}_{0}\text{n}$$

$$^{239}_{94}\text{Pu} + ^{1}_{0}\text{n} \rightarrow ^{240}_{94}\text{Pu*} \rightarrow ^{100}_{40}\text{Zr} + ^{137}_{54}\text{Xe} + 3^{1}_{0}\text{n}$$

$$^{239}_{94}\text{Pu} + ^{1}_{0}\text{n} \rightarrow ^{240}_{94}\text{Pu*} \rightarrow ^{111}_{44}\text{Ru} + ^{125}_{50}\text{Sn} + 4^{1}_{0}\text{n}$$

The asterisks (*) show that the larger nuclides are unstable.

20 DISCUSS
The liquid drop model and the 'shovel of stones' model (or potential well model) are very different ways of thinking about nuclei. Are they incompatible? Explain.

21 DISCUSS
Should alpha emission be considered to be just a particular example of fission?

22 What is the net change in total number of neutrons in each of the fission reactions shown above?

Chain reaction

Fission can be triggered by bombarding suitable material – a material such as uranium which contains nuclei that can undergo fission – with a stream of neutrons. Some of the neutrons are absorbed by nuclei of the material, in some cases making slightly larger nuclei that are unstable and then undergo fission.

We can see from the chart of nuclides (page 154) that the fissioning nuclei have a larger neutron:proton ratio than do stable nuclei of about half the size. So fission tends to produce daughter nuclei that are very neutron-rich compared with stable nuclei of similar size. These initial daughter nuclei are so neutron-rich that they eject neutrons immediately, only occasionally with a significant delay. So the fission process is induced by neutrons, but produces more neutrons than it takes in.

The newly freed nucleons are then available to induce more fission. In a sample of uranium or plutonium, the possibility then exists of a **chain reaction** – just one initial fissioning nucleus providing the neutrons that are needed to induce more fissions, which then provide further neutrons, and so on (Figure 7.17).

Figure 7.17
The principle of the chain reaction. Neutrons that are set free by the fission process are then absorbed by other nuclei, causing them to fission.

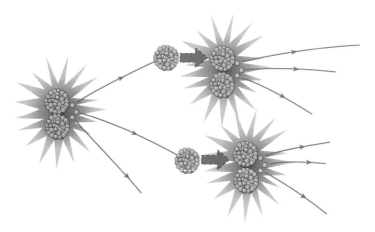

Figure 7.18
Different neutrons produced by fission events (at the green spots) in a sample of uranium meet different fates. It is the proportion that induce further fission that matters to the maintenance of a chain reaction.

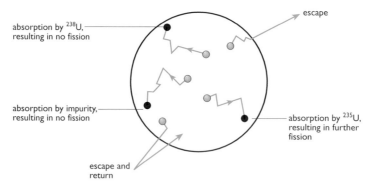

absorption by ^{238}U, resulting in no fission

escape

absorption by impurity, resulting in no fission

absorption by ^{235}U, resulting in further fission

escape and return

Most natural uranium consists of two isotopes, uranium-238 (99.3%) and uranium-235 (0.7%).

Since each fission using uranium-235 as fuel creates an average of about 2.5 free neutrons, the possibility exists for the chain reaction to grow. As the rate of fission grows, the rate of release of energy grows. An uncontrolled chain reaction has the potential for huge and rapid energy release, as in a bomb. The fate of the ejected neutrons (Figure 7.18) determines how the reaction proceeds:

• If exactly one neutron from each fission event goes on to be absorbed by a nucleus and induce fission, then fission in the material will continue at a constant rate – such a situation is called **criticality** (Figure 7.19).
• If more than one neutron from each fission event induces a further fission event, then the chain reaction grows – this is **supercriticality**.
• If less than one neutron induces further fission, then the chain reaction slows down and dies – this is **subcriticality**.

Figure 7.19
Critical and supercritical
chain reactions.

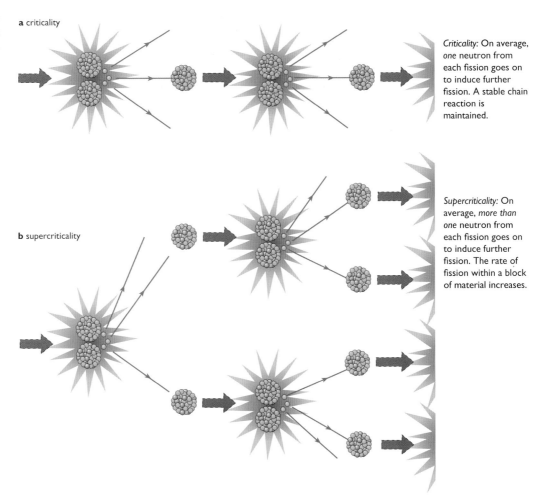

a criticality

Criticality: On average, *one* neutron from each fission goes on to induce further fission. A stable chain reaction is maintained.

b supercriticality

Supercriticality: On average, *more than one* neutron from each fission goes on to induce further fission. The rate of fission within a block of material increases.

The bomb and critical size

A chain reaction can become subcritical if too many neutrons escape through the surface of the sample before being absorbed by nuclei and inducing fission. A smaller sphere has a larger ratio of surface area to volume than a larger sphere has. The rate of escape relative to the number of neutrons is bigger in a small sphere than in a large one. The mass of a sphere of uranium-235 or plutonium-239 required to maintain a critical chain reaction is called the **critical mass** (Figure 7.20).

Figure 7.20
Critical mass is a matter of shape as well as size.

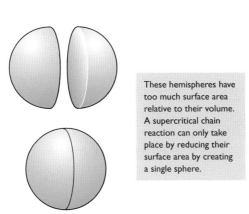

These hemispheres have too much surface area relative to their volume. A supercritical chain reaction can only take place by reducing their surface area by creating a single sphere.

The same mass of a material in a different shape, such as a long cylinder, has a larger surface area relative to volume than the sphere, and so, on its own, is unable to sustain a supercritical chain reaction. It can, however, be made to sustain a critical chain reaction when positioned close to other such cylinders – as in a nuclear reactor.

Figure 7.21
The first fission bomb test, in the desert of south-western USA, 0.025 s after detonation. A half-sphere of white-hot air surrounds the explosion, while the mechanical and thermal blasts begin to whip dust from the ground.

To explode a fission bomb (Figure 7.21), two (or more) part-spheres of material smaller than the critical mass can be thrown together, and the presence of a few neutrons is enough to begin the supercritical chain reaction. The energy release is then so rapid that the material vaporises. The separate part-spheres must be brought together very quickly indeed – using chemical explosives – or the material will disperse and the chain reaction will die down. In the bomb dropped on Hiroshima in 1945 at the end of World War II, only 1.3% of the uranium-235 fissioned before the bomb vaporised itself – releasing energy equivalent to 15 000 tonnes of TNT. The remaining uranium and the daughter nuclides, most of them highly unstable, were dispersed into the environment, so that the effects of ionising radiation killed thousands more people in the days and the years that followed (Figure 7.22).

Figure 7.22
Sadako's Monument, Hiroshima, Japan.

Sadako Sasaki was two years old when the atom bomb was dropped on Hiroshima. She was indoors at the time, and though her house was destroyed, it protected her from both the mechanical blast and the thermal blast that burnt thousands of others. But ten years later, she died of leukaemia – cancer of the blood cells. Many other children in Hiroshima and Nagasaki died in the same way. During her illness, Sadako made hundreds of paper birds. In 1958 a monument was built in Hiroshima's Peace Park, 'To the children of the atomic bomb', featuring a statue of Sadako. Now millions of paper birds from all over the world are placed around the monument each year.

23 Use the formulae for a sphere, i.e.

volume $= \frac{4}{3}\pi r^3$

area $= 4\pi r^2$

to show that the ratio of surface area to volume is greater for a small sphere than for a large one.

24 a Sketch a graph of number of free neutrons against time for a material in which a stable critical chain reaction is taking place.
  b How would the graph be different
  i for subcriticality
  ii for supercriticality?
  c What happens to the liberated neutrons that do not trigger further fission?

The thermal fission reactor

Fuel and neutron energies

A bomb can be made with plutonium or uranium. For uranium, the important fissioning isotope is uranium-235. But only 0.7% of natural uranium is uranium-235. The rest is uranium-238, which behaves differently. It *can* absorb neutrons, but when it does it only rarely undergoes fission. However, uranium-238 nuclei have a much greater tendency to absorb **fast neutrons** than slower ones. Uranium-235 nuclei, on the other hand, are more likely to absorb neutrons and undergo fission when the neutrons are slower.

Neutrons are emitted, during and following fission events, with high energies (up to a few MeV) – and these are fast neutrons. A natural mixture of 99.3% uranium-238 and 0.7% uranium-235 is useless as a bomb. Absorption of fast neutrons by uranium-238 makes a chain reaction (which must involve uranium-235) impossible.

To make a chain reaction it is necessary to take natural uranium and greatly increase the proportion of uranium-235. Material for which this has been done is called **enriched** uranium. The enrichment process involves separation of the two chemically identical isotopes. It can be done by centrifuge – spinning material around so that the heavier isotopes tend to accumulate at the larger radii of the revolutions. It is difficult and time-consuming, and therefore expensive, to achieve a significant level of separation.

If the neutrons are slowed down, then far more are absorbed by the uranium-235. With less energetic neutrons, a chain reaction is possible in natural uranium. Such neutrons are called **thermal neutrons**, because their speed depends on the temperature of the surrounding material. That is, they have energies that are comparable with those of the surrounding atoms. The energies of thermal neutrons are in the region of 0.01 to 0.10 eV.

Plutonium can sustain a chain reaction with fast neutrons. Reactors that use plutonium and/or heavily enriched fuel, with fast neutrons, are called **fast reactors**; while those that use natural uranium or fuel that is only a little enriched, with thermal neutrons, are called **thermal reactors**. Working civil (non-military) reactors are all thermal reactors, of which the most common type is the **pressurised water reactor**.

Maintaining the critical chain reaction

The thermal fission reactor consists of fuel, moderator, control rods and a cooling system. The reactor is designed to support a stable critical chain reaction. The moderator, control rods, their physical arrangement with the fuel rods and with each other, and delayed neutrons combine to achieve this. In order to act as an energy source, the cooling system transfers energy from the reactor to a boiler where water is turned to hot steam, which can drive turbines, just as in a conventional power station.

The fissile material, or nuclear fuel, in tubes called **fuel rods**, is surrounded by the **moderator**. This is a material in which fast neutrons produced by fission collide with nuclei but are not often absorbed by them. The collisions transfer energy from the neutrons to the moderator nuclei. The neutrons return into the fuel with lower energies – as thermal neutrons. **Control rods** are rods of material such as boron or cadmium that are particularly good at absorbing neutrons. Rods can be inserted by variable amounts into the fuel–moderator core, where their effect is to mop up neutrons to prevent them returning to the fuel. So control rods can slow down the chain reaction. By dropping rods into the reactor it can be shut down very quickly – which is important for reactor safety.

While most neutrons in the reactor are the direct products of fission, some are emitted by daughter products at some time after fission (with effective half-lives up to a few seconds). This delayed production of neutrons limits the rate at which a chain reaction can grow or decline. The chain reaction cannot speed up suddenly because it takes a relatively long time for the **delayed neutrons** to build up. It cannot slow down quickly (unless control rods are inserted) because it takes a long time for delayed neutrons to disappear from the system. Delayed neutrons are very important to chain reaction stability.

Figure 7.23
A feedback loop – in general, negative feedback tends to result in stability of a system.

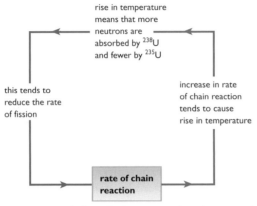

rise in temperature means that more neutrons are absorbed by ^{238}U and fewer by ^{235}U

this tends to reduce the rate of fission

increase in rate of chain reaction tends to cause rise in temperature

rate of chain reaction

The presence of a high proportion of uranium-238 means that, if the temperature of the reactor rises significantly, the rate of absorption of neutrons by uranium-238 increases. This tends to slow down the chain reaction if it increases above the desired rate. That is, it provides some level of negative feedback (Figure 7.23).

A primary **coolant** material flows through pipes between the reactor core, from which energy must be transferred, usually to a secondary cooling system. The primary coolant is subject to a very high exposure to free neutrons, and though the material is chosen to be relatively poor at absorbing neutrons (see next section), some absorption takes place, and the coolant may become radioactive. By using a double system – primary and secondary cooling circuits – the radioactive primary coolant is contained within the shielded reactor building. The primary coolant might be water, carbon dioxide, helium or liquid sodium. The secondary cooling system provides the steam that turns the turbines.

In a pressurised water reactor (Figure 7.24), the primary coolant water must not boil even though its temperature can reach 325 °C, and so pressure has to be about 1500 times greater than atmospheric pressure. The reactor is encased in a steel **pressure vessel**. The reactor and the primary cooling system are contained within a thick concrete building that provides **shielding** – protection of the surroundings from the flow of gamma radiation and neutrons emerging from the reactor.

Figure 7.24
Pressurised water reactor.

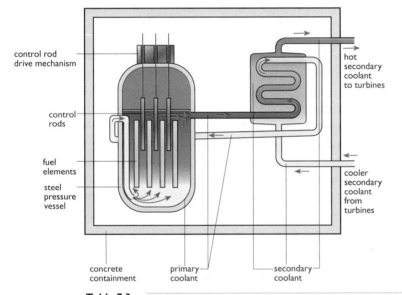

control rod drive mechanism

control rods

fuel elements

steel pressure vessel

concrete containment

primary coolant

secondary coolant

hot secondary coolant to turbines

cooler secondary coolant from turbines

Choice of material for moderator and control rods

The tendency for the nuclei of material to absorb neutrons is very dependent on neutron energy. However, if we specify the energy of the neutrons, for example by dealing only with thermal neutrons, then we can quote a value for each element which indicates its neutron-absorbing behaviour. Such a value is called **thermal neutron absorption cross-section** and is measured in **barns** (as in 'as big as a barn door' – an expression from the American West relating to a barn door as a shooting target). A barn is a measure of the effective absorbing area that a nucleus presents to a neutron, and is equal to 10^{-28} m^2. Some values are given in Table 7.3.

Table 7.3
Some values of thermal neutron absorption cross-section

Nuclide	Thermal neutron absorption cross-section/barn
hydrogen-1	3.32×10^{-2}
hydrogen-2 (deuterium)	5.30×10^{-4}
carbon-12	3.40×10^{-3}
boron-10	759
oxygen-16	2.70×10^{-4}
cadmium-114	2450
uranium-235	694
uranium-238	7.59

A moderator must take energy away from neutrons, slowing them down but absorbing as few of them as possible. In any collision between bodies, one of which can be considered to be moving and the other stationary, momentum and energy conservation mean that maximum energy is transferred from one body to the other if they are the same size. A pea colliding with another pea loses a greater proportion of its initial kinetic energy than a pea colliding with a cannonball. A neutron colliding with a hydrogen nucleus (without fusing with it) loses a greater proportion of its initial kinetic energy than a neutron colliding with a nucleus of uranium-238 (without being absorbed).

So a moderator should have small nuclei, and these should have a low tendency to absorb neutrons, as given by Table 7.3. Water (hydrogen and oxygen nuclei) and carbon are good moderator materials.

Older gas-cooled reactors use graphite as the moderator. Pressurised water reactors use the primary coolant water as the moderator. This has the advantage of allowing reactors to be relatively small.

The chief criterion for the control rods, on the other hand, is that they should absorb neutrons. So they are made of material with high neutron absorption cross-section. Boron-10 and cadmium-114 are suitable materials.

25 a Describe what a moderator does to neutrons produced by fission events.
b Describe the effects of the neutrons on the moderator material.
c Use momentum and energy conservation to show that, in collisions of a moving body of mass m, more of its initial kinetic energy is transferred to a stationary body of mass m than to a stationary body of mass M, where $M \gg m$.
26 Make a list of similarities and differences between a bomb and a reactor.
27 No nuclear reactor has ever become a nuclear bomb. Explain why it doesn't happen.
28 After the Chernobyl disaster (see *Introduction to Advanced Physics*, page 79), boron was amongst the materials dropped from helicopters on to the hot wreckage. Why boron?

Depletion of fuel and nuclear waste

In time, the proportion of uranium-235 in a fuel rod decreases, or becomes **depleted**, and the proportion of daughter products increases. This tends to suppress the chain reaction, and the canisters must be removed for reprocessing (Figure 7.25) – which involves removing the daughter products and enhancing the proportion of uranium-235.

Most daughter nuclides are highly neutron-rich, and highly unstable. That is, they have short half-lives. Their decay almost always happens within the reactor – and the energy released by the decay contributes to the energy that is transferred to the coolant. But the nuclides into which the daughter products decay are usually also radioactive, but not so very unstable. These nuclides usually have longer half-lives, and they are present in large quantities when the fuel is taken for reprocessing. A typical example of a daughter nuclide of fission is iodine-137, and the nuclide then goes through a series of decays. These are shown in brief, with their half-lives:

$$^{137}_{53}\text{I} \xrightarrow[30\,\text{s}]{} {}^{137}_{54}\text{Xe} \xrightarrow[3.4\,\text{min}]{} {}^{137}_{55}\text{Cs} \xrightarrow[33\,\text{years}]{} {}^{137}_{56}\text{Ba}$$

Enough of these materials have long half-lives to mean that this component of spent fuel remains radioactive for very long times. Fortunately the total quantities of material involved are very small; nevertheless the waste must be disposed of where it cannot escape into the environment.

Another component of high-level radioactive waste consists of isotopes of uranium, especially uranium-238, which has taken no part in fission. Both uranium-238 and uranium-235 have extremely long half-lives, and so such waste will remain radioactive for a very long time indeed. The waste also contains other isotopes of large nuclei, created as a result of neutron absorption by uranium and subsequent radioactive decay.

The favoured plan is to embed the material in glass and bury it deep underground – but the high-level waste produced by the world's nuclear power stations is still in temporary storage because no sites can be found for its burial.

Figure 7.25
These are 'ponds' at Sellafield, N.W. England, where spent fuel rods from nuclear power stations are stored until they can be reprocessed. Radioactivity causes heating, so the water is there to keep the fuel cool as well as to absorb radiation.

All of the materials of the reactor, pressure vessel and shielding absorb neutrons, and so changes take place to nuclei within them. These nuclei usually become unstable as a result, and the materials become radioactive. This is a relatively modest level of radioactivity, compared with that of the depleted fuel, but it still presents disposal problems that must be solved. Closing down a nuclear power station at the end of its working life – decommissioning – requires that these materials are kept away from the environment. For example, material that cannot be moved away from the power station site may have to be encased in concrete. Decommissioning of nuclear power stations is therefore expensive and this is one reason why world use of nuclear power is decreasing.

Figure 7.26
Disposal of low-level waste.

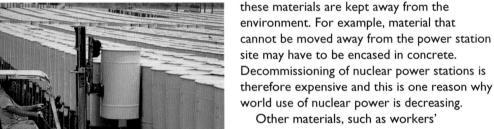

Other materials, such as workers' protective clothing, have very little or no radioactivity above background levels. This is low-level waste, but as a precautionary measure it must still be disposed of with care (Figure 7.26).

Reactors as sources of useful radioisotopes

Material can be inserted within the shielding of a reactor to expose it deliberately to the flow of neutrons. The absorption of neutrons then produces radioactive isotopes, which can be used as tracers in medicine and industry, and to kill cancers. They can also kill bacteria in food, but public concern over use of ionising radiation for this purpose means that irradiation of food does not take place in many countries, including the UK.

Some examples of the production of useful radioisotopes by neutron absorption are as follows:

$$^{31}_{15}P + ^{1}_{0}n \rightarrow ^{32}_{15}P + \gamma$$

$$^{59}_{27}Co + ^{1}_{0}n \rightarrow ^{60}_{27}Co^*$$

$$^{32}_{16}S + ^{1}_{0}n \rightarrow ^{32}_{15}P + ^{1}_{1}p$$

Others are produced less directly, such as by neutron absorption followed by beta⁻ decay:

$$^{130}_{52}Te + ^{1}_{1}n \rightarrow ^{131}_{52}Te + \gamma$$

$$^{131}_{52}Te \rightarrow ^{131}_{53}I + ^{0}_{-1}\beta + \overline{\nu}$$

Phosphorus-32 is a beta⁻ emitter with a half-life of 14.3 days. Cobalt-60 is formed in an excited state (hence the asterisk; see *Introduction to Advanced Physics*, page 294) and falls to its ground state by emission of a gamma ray, with a half-life of 10.7 minutes. Iodine-131 is a beta⁻ and gamma emitter with a half-life of 8.0 days. These half-lives are long enough for material to be transported from the reactor to be used as industrial or medical tracers, but not so long that the activity of a given sample is too low for practical use.

29 a Apart from carrying energy away from the reactor, what is the function of the coolant water in a pressurised water reactor?
 b Why is it pressurised?
30 a How is the flow of neutrons into the environment minimised?
 b What would be the effect of escaped neutrons on material in the environment?
31 Why is reactor decommissioning expensive?
32 Suggest why the pattern of increasing half-lives shown in the example on page 161 for $^{137}_{53}I$ is likely to be a common pattern in nuclides in reactor waste.
33 a Explain why a sample of an isotope with a long half-life has a relatively low activity.
 b Why does this present a problem where an isotope is to be injected as a tracer into a hospital patient?

34 a Use the graph on page 152 to estimate the energy made available per nucleon during fission of a uranium-235 nucleus.
 b What is the corresponding energy made available per nucleus of uranium-235?
 c What is the corresponding energy per kilogram of uranium?
 d How long does it take for a 1 GW (1 GW = 1 gigawatt = 10^9 W) coal-fired power station to supply this much energy?
 e What is the mass of daughter products, approximately, produced from 1 kg of fissioned uranium-235?
 f Estimate the mass of daughter products produced by one year's operation of a 600 MW thermal output nuclear power station.
 g Explain briefly why the electrical output of a power station is less than the thermal output from the reactor.

Fusion as an energy resource

Unfortunately, fusion of nuclei is difficult to arrange, because of the electrical repulsion between the protons. This, as we have seen, creates a potential barrier around nuclei. The energy that a nucleus needs in order to merge with another nucleus is large. Such energies can only be achieved by accelerating nuclei and then colliding them, or by giving them enough thermal energy to overcome the repulsion. The latter requires extremely high temperature, 4.5×10^7 K, at which the material is a plasma. (Its electrons are unable to remain bound to nuclei.) Part of the problem here is the need to contain material that is so hot that it has dramatic effects on any other material it touches. Also, on touching other material, energy is transferred and the plasma tends to be cooled. If this happens only a little then it is possible that continuing fusion will provide energy and maintain the high temperature. But if the plasma is in closer contact with another material, then it becomes impossible to maintain the temperature that is needed for fusion to continue.

The best answer is to use magnetic fields to contain the plasma's charged particles. Unsurprisingly, at such high temperatures holding a plasma in a stable arrangement and then allowing fusion to take place is very difficult indeed. Research and development programmes exist to try to achieve this (Figure 7.27) but such programmes have already occupied the time and efforts of more than one generation of scientists and complete success is still a long way away.

However, fusion has already been applied for military use. The 'hydrogen bomb' is a fusion bomb, in which the very high temperatures are achieved by using a fission explosion as the first source of energy. Hydrogen bombs have existed ever since the furious early days of the 'Cold War' that followed World War II. They are extremely powerful, politically and literally.

Figure 7.27
Torus-shaped experimental research reactors called 'tokamaks', such as the JET prototype reactor, have been used in attempts to contain a plasma and allow fusion to occur.

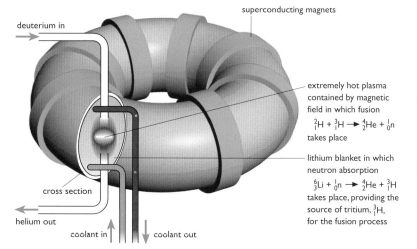

superconducting magnets

deuterium in

extremely hot plasma contained by magnetic field in which fusion

$$^2_1H + ^3_1H \rightarrow ^4_2He + ^1_0n$$

takes place

lithium blanket in which neutron absorption

$$^6_3Li + ^1_0n \rightarrow ^4_2He + ^3_1H$$

takes place, providing the source of tritium, ^3_1H, for the fusion process

cross section

helium out

coolant in   coolant out

Collisions of helium nuclei and neutrons, the products of fusion, with the lithium blanket cause heating. The fission of lithium in the blanket has some cooling effect but not enough to balance the heating. The net heating of the blanket allows energy transfer from it by a coolant, making energy available for generation of electricity.

Nuclear fusion in stars

Nuclear fusion is the source of energy for the Sun, and therefore for our lives. Inside the Sun, nuclei are so close together and so energetic that they can sometimes overcome their electrical repulsion and get close enough together for the strong nuclear force to take over. They fuse together into larger nuclei. The most important fusion processes are the ones involving very small nuclei – mostly of isotopes of hydrogen. These processes increase binding energy per nucleon, and so make energy available. Energy is carried away by photons – providing the light by which we live. A typical fusion reaction is:

$$^3_1H + ^2_1H \rightarrow ^4_2He + ^1_0n + Q$$

Processes of fusion release energy until the created nuclei become as big as those of iron. Then if two iron nuclei, say, fuse together the total binding energy per nucleon decreases. Fusion of pairs of nuclei of iron (and of other nuclei of similar size) does not make energy available but takes energy from its surroundings. Very large stars (see Chapter 8) eventually have enough such mid-sized nuclei for the energy-absorbing behaviour of iron nuclei to be significant. For these stars, fusion of iron is linked to their ultimate cause of death.

35 Why is there more iron in old stars than in young stars?

• **Comprehension and application**

The Manhattan Project

Figure 7.28
Lisa Meitner – a refugee from the Nazis. In the 1930s, a quarter of German physicists were sacked or fled.

The person who played the major part in understanding the production of mid-sized nuclei by the impact of neutrons on uranium was called Lisa Meitner (Figure 7.28), who was in exile in Sweden having had to leave her home country of Germany. This was the time of Nazi power. Through the years of Nazi rule in Germany in the 1930s very many scientists in Germany lost their jobs – in most cases because they were Jewish. Because of this and because of threats to their lives in the developing Holocaust, most of them left Germany and fled to the USA or Great Britain.

Albert Einstein had moved to America. He was a lifelong pacifist – opposing all war. But in 1939 he wrote a famous letter to the President of the USA:

> Sir,
> Some recent work by E. Fermi and L. Szilard, which has been communicated to me in manuscript, leads me to expect that the element uranium may be turned into a new and important source of energy in the immediate future. . . . I believe therefore that it is my duty to bring to your attention the following facts and recommendations.
>
> This new phenomenon would also lead to the construction of bombs, and it is conceivable – though much less certain – that extremely powerful bombs of a new type may thus be constructed. A single bomb of this type, carried by boat and exploded in a port, might very well destroy the whole port together with some of the surrounding territory. However such bombs might very well prove to be too heavy for transportation by air.
>
> I understand that Germany has actually stopped the sale of uranium from Czechoslovakian mines which she has taken over. That she should have taken such early action might perhaps be understood on the ground that the son of the German Under-Secretary of State, von Weizsacker, is attached to the Kaiser-Wilhelm Institute in Berlin where some of the American work on uranium is now being repeated.
> Yours very truly,
> > A. Einstein

In the 1920s science had been carried out without frontiers. Nobody dreamt that there might be any reason why new ideas about atoms and the quantum behaviour of light and matter should be kept secret. But war changed that. By 1940 there was a fear that Hitler's Nazis would develop a nuclear bomb and win World War II. In fact the Nazis did operate a programme to investigate the possibilities of using nuclear processes as the sources of energy for destruction. But it seems to have been a relatively half-hearted project, compared to what happened in the USA. Nazi politicians were happy to invest in technologies such as rockets and jet engines, but had little time for more abstract science.

In the USA the politicians not only learned of the potential of nuclear weapons – that they could be enormously more powerful than chemical explosives – they also learned that to develop such weapons would require a huge programme of research and development. The programme that they set up was called the Manhattan Project, and though work took place at a number of locations, in 1942 a whole new town grew up at Los Alamos in the southwestern USA. Scientists gathered there from all over the USA, from the UK, and from other countries as well. Many scientists who had fled from Germany became involved.

Through the years up to 1945 war continued in Europe, South-East Asia and the Pacific. Very many buildings in central London were destroyed, Coventry could be seen burning for miles around, and in one night in 1944 tens of thousands of people died, mostly burned to death, when British and American bombers attacked the city of Dresden. All of this was done using chemical explosives.

By the summer of 1945 the war in Europe had ended, but war continued in the Pacific. A nuclear bomb was tested and another one was made ready, and without delay this was dropped on the Japanese city of Hiroshima. (It was dropped from an aircraft – Einstein had been wrong about the necessary size of an atomic bomb.) A little later, as soon as a second bomb was ready, it was dropped on Nagasaki. Thousands died – burnt or buried under rubble, and thousands more died later of the effects of ionising radiation. Japan surrendered. The Manhattan Project had ended the war. This is one reason why, since then, politicians have taken science seriously.

36 Make a list of the differences between chemical explosives and nuclear bombs.

37 The Manhattan Project developed the atomic bomb. The 'hydrogen bomb' was developed over the following years. Describe the fundamental difference between the first atomic bombs and later hydrogen bombs.

● **Extra skills task** Communication

Discuss the following. (You might find it useful and interesting to join with students of history and religious studies for this.)

1 Why was work in physics in the 1950s so different from work in physics in the 1920s? Is the way that physics is done now more like it was in the 1920s or 1950s?

2 Enough nuclear bombs to destroy the world still exist. Do they have any influence on your life?

3 Some say that science is about truth, and that it is up to politicians to decide how new understanding is used. Do you agree with that? Who else might be involved?

4 If you had been a physicist in 1942, would *you* have joined the Manhattan Project?

● ## Examination questions

Rest masses in u:

electron	0.000 549
proton	1.007 276
neutron	1.008 665
α-particle	4.001 508
$^{235}_{92}$U nucleus	234.995 1

$1 \, u = 931 \, MeV$

1 One possible fission reaction caused by the absorption of a slow neutron by the nucleus of an atom of uranium-235 is

$$^{239}_{92}U + {}^{1}_{0}n \rightarrow {}^{90}_{36}Kr + {}^{144}_{56}Ba + 2{}^{1}_{0}n + Q$$

where Q represents the energy released by the fission process.

The diagram represents part of a fuel rod suspended in a cylindrical cavity within the moderator in a nuclear reactor. The gap between the fuel rod and the moderator is the passageway for the liquid coolant.

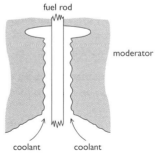

fuel rod

moderator

coolant   coolant

a Explain why the fuel rods heat up. (3)

b Explain how most of the energy released in the fuel rods reaches the coolant. (2)

c Part of the energy released is transferred from the fuel rod directly to the moderator and then back to the coolant. How does this happen? (3)

d Explain, with reference to the example of the fission reaction given above, the essential part played by the moderator in the maintenance of a chain reaction. (3)

London, A level, Module Test PH3, January 2000

2 The diagram shows the essential components of a thermal fission reactor.

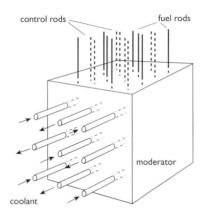

control rods   fuel rods

moderator

coolant

a Explain the function of the moderator. (2)
b Explain the function of the control rods. (2)
c The energy released is transferred to the boiler of a turbine generator. Describe how energy is transferred from the fuel rods to the boiler. (4)

London, A level, Module Test PH3, June 1996

3 This question is about nuclei and nuclear fission.
a Small nuclei, with atomic number less than about 20, generally have roughly equal numbers of neutrons and protons in the nucleus. However, larger nuclei have more neutrons than protons. Explain why this is to be expected, given that the Coulomb force is long range while the strong nuclear force is short range. (3)
b The fission of a uranium-235 nucleus, induced by a neutron, results in two smaller nuclei plus a few neutrons. For example, possible fission products may be lanthanum and bromine nuclei and three neutrons, as follows:

$$^{235}U + n = {}^{146}La + {}^{87}Br + 3n$$

i Explain why neutrons are to be expected as products of fission, in the light of the issues involved in **a** above. (1)
ii The total rest mass of the fission products on the right of the equation is 0.20 u less than the rest mass of uranium plus neutron on the left. Calculate the energy released in the reaction, and state in what form it is released. (2)
iii What is the practical importance of the fact that a fission induced by one neutron may produce more neutrons? Explain briefly. (2)

IB, Standard level 3, Paper 430, November 1998

4 a i What is meant by the term *thermal neutron* in a nuclear reactor?
ii Describe what happens when a thermal neutron strikes a uranium-235 nucleus. (4)

b i A spherical mass of material of radius R undergoes fission. Assume that fission is caused only by neutrons which are absorbed by the material. Show that

$$\frac{\text{rate of loss of neutrons}}{\text{rate of production of neutrons}} \propto \frac{1}{R}$$

ii Use the relationship in part **b i** to explain why there is a minimum mass of fissile material required to establish a self-sustaining chain reaction. (4)
c i State the function of the moderator in a thermal nuclear reactor.
ii Name *one* example of a material suitable for use as a moderator.
iii Explain why fuel rods in a thermal nuclear reactor are replaced before all the uranium-235 has been used. (3)

NEAB, AS/A level, Module Test PH06, March 1998

5 a i Explain what is meant by the *binding energy* of a nucleus.
ii Calculate the binding energy per nucleon of a nucleus of the uranium isotope ${}^{235}_{92}U$. (6)
b Use the liquid drop model of the nucleus to describe how fission occurs and explain qualitatively why energy is released when a ${}^{235}_{92}U$ nucleus undergoes fission as a result of being struck by a slow moving neutron. (4)

NEAB, AS/A level, Module Test PH06, March 1998

6 a i What is meant by the binding energy of a nucleus?
ii Using information from the data at the start of these questions, calculate the binding energy of an α particle in units of MeV.
iii Sketch a graph of binding energy per nucleon [y-axis] against nucleon number [x-axis] for the naturally occurring nuclides, indicating approximate scales on the axes of your graph.

On the same axes mark a point representing the ${}^{4}_{2}He$ nucleus. (8)
b Comment on the relative position of the ${}^{4}_{2}He$ nucleus in the graph referred to in part **a iii** and on its significance in the formation of α particle clusters. (2)

NEAB, AS/A level, Module Test PH06, June 1997 (part)

7 Spent fuel rods from a nuclear reactor contain highly radioactive fission products and plutonium, which is also highly radioactive.
a Explain why most of the fission products in a spent fuel rod emit β^- particles.
b Explain why only a minority of the fission products in a spent fuel rod would emit neutrons.
c Describe how plutonium-239 is produced in a fuel rod in a nuclear reactor. (5)

NEAB, AS/A level, Module Test PH06, June 1997 (part)

8 a Explain why the coulomb repulsion between protons in nuclei does not cause all nuclei to be unstable. (2)

b A $^{239}_{94}Pu$ nucleus may oscillate and/or undergo fission when it absorbs a neutron. The amplitude of the oscillation depends on the kinetic energy of the incident neutron. The sequence of diagrams below shows how a nucleus may oscillate following a collision from a *thermal* neutron.

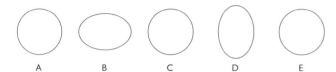

A   B   C   D   E

The binding energy of the $^{239}_{94}Pu$ nucleus changes as it moves from its initial spherical shape in A to its elongated form in B.
State and explain how
i electrostatic potential energy and
ii surface energy
contribute to the change in binding energy as the $^{239}_{94}Pu$ nucleus changes shape. (4)

c Draw a sequence of diagrams, similar to those in part **b**, to show what happens to the $^{239}_{94}Pu$ nucleus when it is hit by a very fast neutron. (2)

AQA (NEAB), AS/A level, Module Test PH06, March 1999

9 Natural uranium consists of 99.3% $^{238}_{92}U$ and 0.7% $^{235}_{92}U$. In many nuclear reactors, the fuel consists of enriched uranium enclosed in sealed metal containers.

a i Explain what is meant by *enriched uranium*.
ii Why is enriched uranium rather than natural uranium used in many nuclear reactors? (2)

b i By considering the neutrons involved in the fission process, explain how the rate of production of heat in a nuclear reactor is controlled.
ii Explain why all the fuel in a nuclear reactor is *not* placed in a single fuel rod. (5)

AQA (NEAB), AS/A level, Module Test PH06, March 1999

10 The nuclear reaction represented by the equation

$$^{235}_{92}U + ^{1}_{0}n \rightarrow ^{94}_{39}Y + ^{139}_{53}I + 3^{1}_{0}n$$

takes place in the core of a nuclear reactor in a power station.

a Explain how this process may lead to a chain reaction. (1)
b Explain why, in practice, a moderator is required in the reactor core. (2)
c Making reference to the nuclear equation, explain why control rods are needed. (3)

d The fission of a nucleus of $^{235}_{92}U$ produces, on average, 200 MeV of energy. The nuclear power station has an overall efficiency of 22% and produces 3000 MW of electrical power.
Calculate
i the total power output [in W] of the reactor core,
ii the total energy output of the reactor core during one year,
iii the number of $^{235}_{92}U$ fissions needed to maintain the reactor's output for one year,
iv the mass [in kg] of Uranium-235 which undergoes fission during this period. (8)

OCR (Cambridge), A level, 4834, November 1999

11 a i State the conditions in the Sun that allow nuclear fusion to take place.
ii Explain why these conditions are necessary. (4)

b The total mass of the protons in a star at its formation was 2.0×10^{30} kg. Energy is produced in the star by the fusion of protons. The net outcome of the fusion process is that 6.4 MeV are produced for each proton in the star. Assume that the star radiates energy at a constant rate of 3.9×10^{26} W during its lifetime.
i Calculate the number of protons in the star at its formation.
ii Estimate the lifetime of the star. (4)

c i State the equation for the reaction which is most likely to create energy from fusion on a practical scale on Earth.
ii Suggest how energy could be extracted from the JET prototype reactor. (4)

OCR (Cambridge), A level, 4834, June 1999

12 A nuclear power station has an electrical power output of 2000 MW. It converts 25% of its fission energy to electrical energy. The fission of a Uranium-235 nucleus by a neutron releases 3.0×10^{-11} J.

a Calculate
i the power [in MW] from fission of nuclei in the reactor,
ii the number of uranium nuclei fissioned in one second,
iii the mass [in kg] of Uranium-235 which is fissioned during one year's continuous operation. (6)

b Describe how, in the reactor core,
i the speed of the neutrons is controlled,
ii the number of neutrons present is controlled. (5)

c State how the fuel rods are disposed of after they have been removed from the reactor. (3)

UCLES, A level, 4834, June 1998

8 Starlight and the expanding Universe

THE BIG QUESTIONS
- How can we study stars with so little light to provide data?
- How can we measure distances in space?
- Why is the Big Bang theory the preferred account of the Universe?

KEY VOCABULARY

Big Bang   Big Crunch   black dwarf   black hole   CCD   Cepheid variable
closed Universe   cosmic background radiation   cosmological principle   critical Universe
geocentric Universe   gravitational collapse   gravitational lensing   heliocentric Universe
Hertzsprung–Russell diagram   Hubble's Law   inflation (of Universe)   light-year
Local Group   Local Supercluster   luminosity   neutron star   Olbers' paradox
open Universe   parsec   quasar   red giant   red shift, z   singularity
steady-state theory   supernova   topology   white dwarf

BACKGROUND

Scientists build concepts of the unfamiliar from concepts that are already familiar. So we visualise the travel of light as something similar to the travel of ripples on a pond, or we picture atoms as tiny balls. With the Universe as a whole we have a real problem – the Universe, in its completeness, is not like anything else that is familiar (Figure 8.1).

Figure 8.1
The Universe is not just 'out there', but is here as well. We are not just observers of it, but part of it.

168

Stars

Detecting starlight

How can we know about stars? They appear to us as tiny specks of light in the night sky. Their light cannot compete, on our retinas, even with the scattered light of the Sun so we don't see them during daylight. In our attempts to understand them, they don't seem to give us much to go on. We have to apply a high level of ingenuity to find out anything about them at all.

All we have from stars is their electromagnetic radiation. However, this can be from all parts of the spectrum. So we can make observations using detectors that are sensitive to different parts of the spectrum.

Radio telescopes can be land-based, because radio waves of many wavelengths can penetrate our atmosphere. This means that it is relatively easy to make large radio telescopes, which also has the advantage of providing images with high resolution (Figure 8.2), as given by the Rayleigh criterion (see *Introduction to Advanced Physics*, page 233). Radio telescopes at two locations a long way apart – detecting radiation from the same image – provide excellent resolution.

Telescopes that provide magnification for visible light can also be based here on the Earth's surface. The best of these telescopes have large apertures, such as the recently installed Gemini mountain-top telescopes – one in Hawaii to watch the northern sky, and one in Chile to watch the southern sky. Both of these collect light using mirrors that are 8.1 m across. If telescopes are fixed then they rotate with the Earth and do not remain focused on the same part of the sky. So most telescopes are able to move to compensate for the Earth's spin, so that a single area of sky can be studied for long periods. When light from sources has low intensity, this is essential to allow the creation of images.

Images are not normally produced by photographic film, but by **CCDs** (charge-coupled devices). These are small sandwiches of silicon dioxide, in rows and columns to create pixels, and from which incident light causes photoelectric emissions (Figure 8.3). The emitted electrons are collected on electrodes and stored there until a whole row of pixels has been illuminated. Pixels can have sides of length 15 µm or smaller, and can be arrayed in lines of many thousands. Each one can store half a million or more electrons. In an array of 2048×2048 pixels there are more than four million of them, each providing 2 bytes worth of information. So a single image from such an array requires 8 Mb of memory capacity.

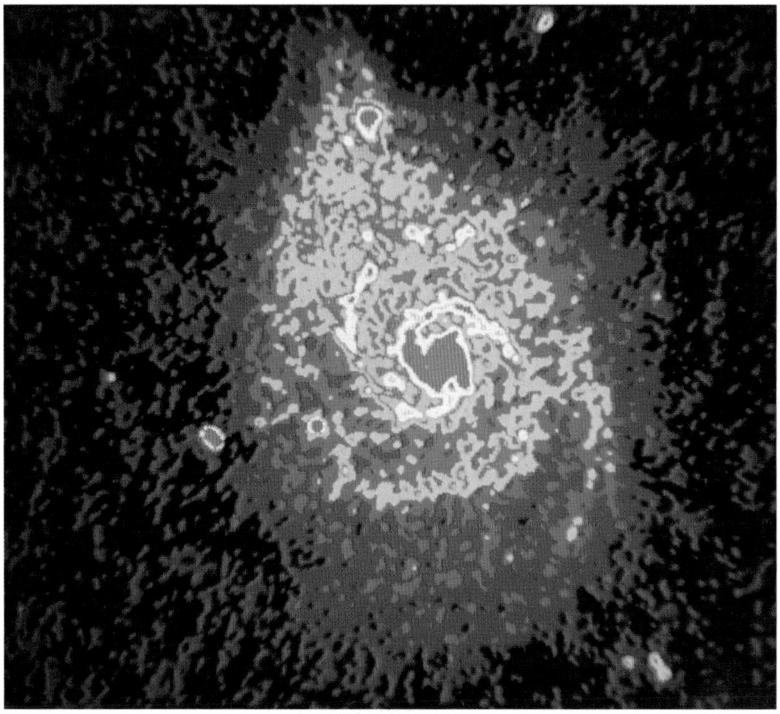

Figure 8.2 (below) Radio image ($\lambda = 0.21$ m) of galaxy M51, or the Whirlpool Galaxy, which is about 20 million light-years from us. The image was made with the Very Large Array of radio telescopes in New Mexico. Use of an array increases resolution.

The colours have been added to map the intensity of radiation (red for high intensity, purple for low).

Figure 8.3 (right) The structure of a CCD.

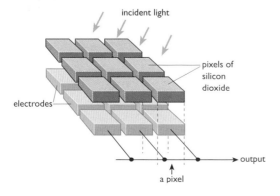

Incident light causes photoelectric emission in the silicon dioxide layer of each pixel. The emitted electrons are collected and stored on positive electrodes. At intervals, electrons are allowed to flow off each electrode in turn. This creates a current–time graph providing a measure of the light intensity that has struck each pixel.

Figure 8.4
A picture made by combining images created by visible light and by X-rays, of a cluster of galaxies with a cloud of gas that is emitting X-rays. The X-ray image was made using detectors carried by a satellite.

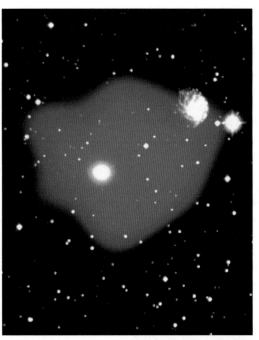

To read an image, the voltages applied to the electrode associated with each pixel are varied, so that the collected electrons are allowed to flow to a terminal. The graph of current against time reproduces the pattern of distribution of electrons that was caused by the original exposure to light.

CCDs are very useful in data collection in astronomy, because they can respond to a wide range of electromagnetic radiation, and their response is in electrical form. They are also very sensitive to low intensities of light.

To examine wavelengths of the spectrum other than radio or visible, it is necessary to launch the detectors into space. X-rays, ultraviolet and infrared radiations from distance sources are much more detectable in space than they are from the bottom of our absorbing atmosphere. An X-ray image produced by a satellite detector is shown in Figure 8.4.

Even for visible wavelengths, there is benefit in placing the telescope above the atmosphere. There the light is not subject to the variable refraction that causes stars to twinkle noticeably. For telescopes that are intended to produce sharp images of distant objects, refraction and scattering present problems. The Hubble Space Telescope (HST), in orbit above the atmosphere, has an aperture of 2.4 m, and does not suffer from the effects of atmospheric refraction and scattering. Figure 8.5 shows a typical HST image.

Figure 8.5
An HST image of a black hole.

This is a Hubble Space Telescope image produced by light from a disc of material falling rapidly into a supermassive black hole in the galaxy M87, 50 million light-years from Earth. The region of space in the image is several light-years across. Measurement of blue shift and red shift of light from the material allows calculation of the speed of its spin, and from this the mass of the invisible black hole at the centre can be estimated – it's about 3 billion times more massive than the Sun, and getting bigger and bigger as it swallows surrounding material.

Spectra, surface temperature and chemical composition

Having obtained the light, the intensity of the radiation can be measured. The next requirement may be to analyse it by spreading it into a spectrum (Figure 8.6). This can be done with a diffraction grating, and the process makes absorption lines directly visible (see *Introduction to Advanced Physics*, Chapter 27). This provides remarkably straightforward information about the elements that exist in the outer layers of a star. The spectrum also gives a reliable indication of the temperature of the source.

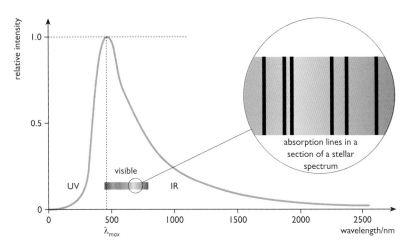

Figure 8.6
The spectrum of a
typical star.

absorption lines in a
section of a stellar
spectrum

We see light that escapes
from the surface of a star.
Its overall spectrum
provides information about
its surface temperature.
The wavelength of the
highest intensity, λ_{max}, is
related to surface
temperature quite simply:

$$\lambda_{max}T = 2.898 \times 10^{-3}\,\text{m K}$$
(note the unit:
metre kelvin)

Absorption of some
wavelengths tells us which
elements are present in the
outer layers of a star, and
in what proportions.

Luminosity and distance

The brightness of the light that we see provides further clues. Of course, a strong influence on the brightness that we detect is the distance of the star from Earth. Although stars are very large, we can, from our position on Earth, reliably consider them to be point sources. The light from all point objects obeys the inverse square law. So to know about the actual **luminosity** of a star, defined as the amount of energy a star radiates every second (its total power output), we need to know how far away it is. We use the equation

$$I_{earth} = \frac{L}{4\pi d^2}$$

where I_{earth} is the observed intensity of radiation (power per square metre) as detected on Earth, L is the luminosity or total power of radiation emitted by the star, and d is the distance of the source (star) from the observer on Earth. I_{earth} can be measured directly.

For stars that are relatively local, to measure d we can use the 'parallax effect' (Figure 8.7) that results from the Earth's shifting position as it orbits the Sun. We can look at the direction to a star at six-monthly intervals. More distant stars do not show such a parallax effect, so we can use them as 'fixed points' to allow measurement of angles.

Figure 8.7 (below)
Measuring distances to
stars.

Using parallax to measure distances to local stars
In radians, $\theta \approx r/d$. But r is known, and θ can be measured
from six-monthly observation of star position relative to the
background of more distant stars. So d can be calculated.

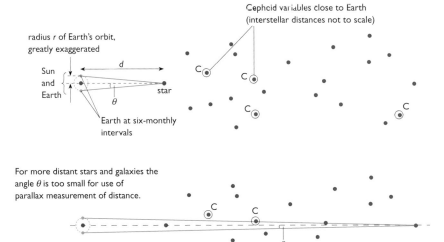

Cepheid variables close to Earth
(interstellar distances not to scale)

radius r of Earth's orbit,
greatly exaggerated

Sun
and
Earth

d

star

θ

Earth at six-monthly
intervals

For more distant stars and galaxies the
angle θ is too small for use of
parallax measurement of distance.

**Using Cepheid variables to measure
distances to stars and galaxies**
Parallax measurement of distance to Cepheid
variables that are relatively close to Earth allows us
to find their mean luminosity from $I_{earth} = L/4\pi d^2$.
From this we can see that Cepheid variables follow
a pattern – the period of variation of their
brightness is related to their luminosity, as
discovered by Henrietta Leavitt in 1908. By
assuming that all Cepheid variables obey this
pattern, we can then observe the periods and so
find the luminosities of more distant ones. From this
we can find their distance, again using $I_{earth} = L/4\pi d^2$.
This was how Edwin Hubble measured the distance
to Andromeda in the 1920s, and showed that
Andromeda is a galaxy separate from our own.

For more distant stars there is another guide to distance. One type of star called a **Cepheid variable** varies periodically in brightness (observed intensity). It turns out that the period of the variation is related to the mean luminosity of such a star. We know this by looking at Cepheid variables that are relatively close to us, for which the independent parallax method can be used to determine distance, and from this we can work out their luminosity. We can then apply knowledge of the luminosity–period relationship to more distant Cepheids to determine *their* mean luminosity. We measure the intensity of the light, I_{earth}, that we detect. The inverse square law can then be used to calculate distance (see Figure 8.7).

The Hertzsprung–Russell diagram

Having found out how to determine both the luminosity of a star and its surface temperature, we can develop a useful way of classifying stars by plotting a graph using these variables. Such a graph is called a **Hertzsprung–Russell diagram** (Figure 8.8). Temperature is plotted on the *x*-axis. The normal Cartesian convention is flouted here, as astronomers habitually plot temperature with value *decreasing* along the axis. Temperature is related to colour – cooler stars are redder, hotter stars are bluer (see Figure 8.9). Relative luminosity is plotted on the *y*-axis. Because of the very wide range of temperatures and stellar luminosities, logarithmic scales are used.

Figure 8.8
The Hertzsprung–Russell diagram.

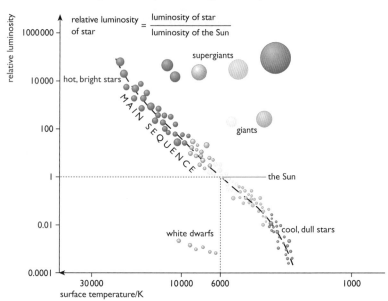

The location of an individual star on such a graph lets us establish a loose system of classification and naming, as shown in Figure 8.8. Stars are not scattered randomly across the chart but most fall into a pattern called the Main Sequence.

Fusion in stars makes energy available to create radiation, consuming mass at an awesome rate. The Sun, for example, loses a mass of 4.5 million tonnes every second. Also, heavier nuclei are created from smaller ones, so that the composition of a star changes (see the next section). As the composition changes, the star's surface temperature and luminosity change, and the star moves off the Main Sequence of the Hertzsprung–Russell diagram. Eventually, in a course of events dependent on the star's size, the star dies and its material is scattered, often violently, back into space.

Figure 8.9
Stars have colour.

It is a limitation of human vision that normally prevents us from seeing the colour of stars. Our wavelength-sensitive cells (cones) only work with relatively bright light. For weak sources of light we rely on rod cells, of which our retinas have only one kind. Our retinas respond to weak sources in the same way whatever the colour. Photography, however, can reveal star colour. This image was taken with a long exposure time, and the camera did not track across the sky to compensate for the rotation of the Earth, so each star appears as a circle of light.

1 a In producing images of stars using ground-based optical telescopes, what are the advantages of using long exposure times?
   b What problem arises due to the spin of the Earth, and how can this problem be solved?
2 List the nature of the evidence that can be collected to determine
   a the chemical composition of a star
   b the surface temperature of a star
   c the luminosity of a star that is not a Cepheid variable
   d the luminosity of a Cepheid variable.
3 a Use the Rayleigh criterion, θ (in radians) $= 1.22\lambda/D$, where θ is the least angular separation of resolvable points, λ is wavelength and D is aperture, to calculate the minimum angle resolvable by the Hubble Space Telescope for green light.
   b If two galaxies in a cluster 1000 million light-years away can just be resolved, what is the minimum distance between them?
   c How can we only find minimum distance and not absolute distance between them? (Draw a sketch to help you to answer this.)
4 a State an appropriate unit for luminosity.
   b Write down an equation relating the intensity of light reaching the Earth from a star (ignoring any effects of the atmosphere) to its luminosity.
   c Explain why knowledge of value of luminosity of a star allows us to calculate its distance from us.
   d Hence explain why Cepheid variables are so important in the development of our ideas of cosmology.

Stellar evolution

A star is formed from a cloud of gas, mostly hydrogen, and dust that is initially spread over a huge volume, but which is pulled together by its own collective gravity. This **gravitational collapse** of the cloud creates a body of large density, and the loss of gravitational potential energy in the process is very large indeed. The result is that the original particles acquire high kinetic energy, so that the collisions between them are very violent. Atoms lose their electrons. Not only that, collisions take place in which electrical repulsion of nuclei is no longer strong enough to keep them apart. They can become close enough together for the strong nuclear force to take effect, so that they merge. Fusion takes place, with hydrogen as the principal raw material. This begins the process of conversion of mass to energy, and much of the released energy takes the form of photons which begin to stream from the new star (Figure 8.10).

Figure 8.10

Starbirth: this is the Orion Nebula, containing clouds of gas (mainly hydrogen). The gas is ionised by radiation from the hot, young stars in the centre of the picture.

Every star then exists in a state of slowly evolving stability. On the one hand there is the tendency for the material to continue to collapse under gravity. On the other hand there is the tendency for the violent thermal activity and the emission of radiation resulting from fusion to blow the material apart. These two opposing tendencies reach equilibrium. The detail of this equilibrium depends on the mass of material involved.

In general, the more massive the star, the greater is the gravitational pressure and so the higher is the rate of energy release by fusion required for equilibrium. So big stars 'use up' their supply of fusing nuclei more quickly than do smaller stars, such that big stars have shorter lives. For example, the Sun has a total expected lifespan of about 10 billion years (it's now about half-way through), a star three times as big has a life expectancy of only 0.5 billion years, while a star with half the mass of the Sun might last for 200 billion years.

Very massive stars experience several stages of fusion in their cores. First hydrogen fuses to helium, then helium to carbon, and so on, creating larger and larger nuclei. Thus such large stars in 'later life' can have shells or layers, like layers of an onion, with heavier nuclei towards their centres. It was in this way that the nuclei of the heavier elements of our world were made.

Our own star, the Sun, is very 'average' in many ways, but is one of a minority of stars that exists alone. Most stars have partners, creating binary systems (see Figure 3.33). The effect of one star on another can affect their development, and especially their ultimate deaths. However, here we consider stars as independent bodies.

It is not only the life expectancy of a star that depends on its mass, but also the way in which it dies. Older stars have outer layers in which hydrogen is the fuel for fusion, while in inner layers helium is the fuel, and for massive stars there may be further layers beneath that as we have mentioned. Most stars, and this will include the Sun, become **red giants** after the end of their equilibrium phase. The process is started by cooling in the inner core, resulting in reduced thermal pressure and 'radiation pressure' and so causing gravitational collapse of the hydrogen shell. But the gravitational collapse provides energy for heating the shell, and so the rate of fusion in the shell increases. This makes the shell expand enormously. The outermost surface of the star becomes cooler, and its light becomes redder, but the larger surface area means that the star's luminosity increases. Meanwhile the gravitational collapse affects the core as well, and ultimately the process of fusion of helium (or larger nuclides) in the core cause the outer shell to expand further and thin, leaving the hot superdense core as a **white dwarf**. Slowly this cools, and becomes a **black dwarf**.

For stars that are several times more massive than the Sun, death may be even more dramatic. A core of carbon is created by fusion of helium, and once this core is sufficiently compressed then fusion of the carbon itself takes place, triggered suddenly throughout the core. The rapid release of energy makes the star briefly as bright as a galaxy – as bright as 10 billion stars. The star explodes – a **supernova** – and its material spreads back into the space around.

In even larger stars, fusion of carbon can continue more steadily, producing still larger nuclides and ultimately creating iron nuclei. The iron nuclei also experience fusion, but these fusions are different – they are energy-consuming. The central core of the star collapses under gravity. This increases temperature but cannot now greatly increase the rate of fusion, so collapse continues. Outer layers also collapse around the core, compressing it further. It becomes denser than an atomic nucleus – protons and electrons join together to create neutrons. Meanwhile, the collapse of the outer layers heats these, increasing the rate of fusion so that, suddenly and dramatically, the star explodes as a supernova event. This spreads the material of these layers into space, leaving a tiny hot body behind – a **neutron star**. And if this supernova remnant is massive enough, its gravity continues to pull the matter towards a single point – a **singularity** – with a surrounding gravitational field that is so strong that nothing, not even light, can escape from it. It becomes a **black hole**.

5 Explain whether you would expect a star, such as the Sun, to remain at the same point on a Hertzsprung–Russell diagram permanently.

6 Sketch intensity–wavelength graphs (spectra) for a red giant and a high-temperature white dwarf, on the same axes.

7 Human lifetimes are less than those of trees and stars. Would you be able to work out details of how trees are formed and die from a year's observation of a forest?

The Milky Way and other galaxies

Before the 1920s, people had seen fuzzy features in the night sky and some had suggested that they were clouds of gas, or nebulae, in between the stars. But it turned out that we are in a vast gathering of stars, a galaxy, and many of these 'clouds' were outside this. They are galaxies separate from our own. Our home galaxy, called the Galaxy (note the capital 'G') or the Milky Way, is one amongst very many.

We in fact seem to live in one of the largest of a cluster of 30 galaxies called the **Local Group**, each galaxy containing many stars. To give an indication of what 'many' means here, there are about 10^{11} stars in our Milky Way. The Sun is one of them.

The Hubble Space Telescope inspects tiny sectors of the sky, and in each one it finds more and more galaxies. There are bigger clusters, often with thousands of galaxies, and superclusters, which are made up of hundreds of clusters. The Local Group is a cluster that is part of the **Local Supercluster**, which also holds the Virgo and Coma clusters, both of these containing huge numbers of galaxies. All these bodies are entangled in each other's gravitational fields.

Measuring space

To arrive at our present view we have had to learn to measure distances in space, as already described earlier in this chapter. The units that people have used make their own interesting story. The kilometre is sadly inadequate as a measure of distances in space, even for relatively local distances such as the radii of orbits of moons. Because astronomy developed so far before the modern form of the SI system was created, the habit of using megametres (metres $\times 10^6$), gigametres (metres $\times 10^9$), terametres (metres $\times 10^{12}$), petametres (metres $\times 10^{15}$) and exametres (metres $\times 10^{18}$) never developed. Astronomers have their own units, including the light-year and the parsec. These are defined as follows:

- The **light-year** is the distance travelled by light in one year. To three significant figures it is equal to 9.46×10^{15} m.
- The **parsec** is the distance from which the Earth and Sun would have an angular separation of one second of arc (Figure 8.11). It is 3.2616 light-years or 3.0857×10^{16} m. Parsec is abbreviated to pc. For consideration of far space, even the parsec is too small, so physicists and astronomers use the kiloparsec (kpc) and the megaparsec (Mpc).

Figure 8.11
The concept of the parsec developed from the use of parallax methods to measure distances to stars.

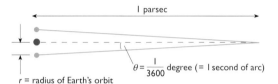

One parsec (or 1 pc) is the distance from the Earth at which the angle subtended by the radius of the Earth's orbit around the Sun is 1/3600 of a degree (also known as one second of arc).

$\theta = \dfrac{1}{3600}$ degree (= 1 second of arc)

r = radius of Earth's orbit

Table 8.1 (below)
Some astronomical distances.

The Milky Way, or the Galaxy of which our Sun is a member, has a diameter of about 30 kpc. The diameter of the Local Supercluster (Figure 8.12), which includes not only the Local Group of galaxies but also other clusters, is 30 Mpc, but it is still described as 'local'. Compared to the whole of the observable Universe, it is quite small. Some typical distances are given in Table 8.1.

radius of Earth's orbit around the Sun	1.5×10^8 km
distance to nearest star from the Solar System	1.33 pc
diameter of the Galaxy or Milky Way	30 kpc
distance to the nearest neighbour galaxy (Andromeda)	770 kpc
distance to the Virgo cluster	19 Mpc
mean diameter of the Local Supercluster	30 Mpc
distance to the edge of the observable Universe	100 Gpc

8 What is the radius of the Earth's orbit in
  a parsecs
  b light-years
  c light-minutes?
9 What is the mean diameter of the Local Supercluster in kilometres?
10 What is the ratio of
  a the diameter of the Solar System to that of the Milky Way
  b the diameter of the Milky Way to that of the Local Supercluster?
11 What is the ratio of the approximate time taken for light to cross the Local Supercluster, 10^8 years, to
  a your lifetime so far
  b the 10 000 years since the last Ice Age receded
  c the million years or so since the first humanoids lived on the Earth?

Figure 8.12
The Local Supercluster is a gathering of galaxies and clusters of galaxies. It includes the Local Group, of which our Galaxy is a part. To see the Local Supercluster from outside, you would have to travel for almost 10^8 years at the speed of light.

Note that the galaxy shown on page 169 is about 2×10^7 light-years, or 6 Mpc, from us and is therefore within the Local Supercluster.

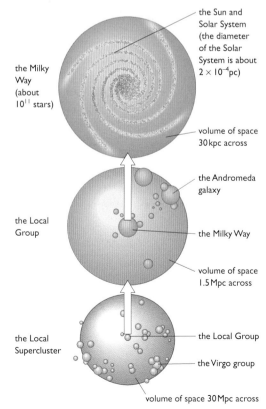

the Sun and Solar System (the diameter of the Solar System is about 2×10^{-4} pc)

the Milky Way (about 10^{11} stars)

volume of space 30 kpc across

the Andromeda galaxy

the Milky Way

volume of space 1.5 Mpc across

the Local Group

the Virgo group

the Local Group

the Local Supercluster

volume of space 30 Mpc across

The Universe – and the revision of ideas

The Universe with no centre

The stars at night are awesome to see. Perhaps it was the night sky that first inspired people to want to know more. They found patterns that stayed the same from one human lifetime to the next, but they also discovered that some of the specks of light are not part of the fixed pattern but wander in their own way through the sky. From this came concepts of orbit, and the wandering specks were called planets. At first the orbits were believed to be centred on the Earth – and this concept is called the **geocentric Universe**. Later, people identified other centres – first the centre of the orbits of the planets – and saw space as being centred on the Sun, the **heliocentric Universe**. Now we can also visualise the centre of the orbit of our Galaxy, which is one amongst billions of observable galaxies, suggesting that the Universe has no centre at all, but is 'acentric'. The idea that the Universe has no single centre, and that therefore one position in the Universe is the same as any other for observing large-scale phenomena, is called the **cosmological principle**.

Olbers' paradox

In our search for different ways to model the Universe of which we are a part, we might picture it as something like the spaces we are used to – like rooms. If we suppose that the walls, floor and ceiling are infinitely far apart, then we have set up a model of the Universe. Rooms are familiar. They are full of air through which we move, and we could picture a room as containing a permanent three-dimensional framework in which events take place. Assumptions that we make when using such a model for the Universe are as follows:

A Events (changes) can take place within the Universe, but the Universe itself is unchanging.
B The Universe has sources of light (galaxies of stars) scattered throughout it and moving within it.
C The Universe is infinitely large.

But if we apply such a model to the Universe we get into trouble. One spot of trouble is called **Olbers' paradox**.

Heinrich Olbers was born about 30 years after Newton's death, but very much at a time when Newton's ideas had a *wow!* factor due to their newness. (In the same way, about 50 years after Einstein's death, relativistic ideas are still regarded as amazing and even sometimes described as 'weird'.) Olbers contemplated a problem that had already been pointed out – that if the Universe is an infinite space scattered with stars, continuing in all directions without end, then it must contain an infinite number of stars. Which means that wherever you look into the sky, you are looking towards a star, and behind that are more stars. Light should reach you from all directions of the sky – and in all directions the night sky should be as bright as a surface of a star (Figure 8.13).

Figure 8.13
The problem with an infinite Universe.

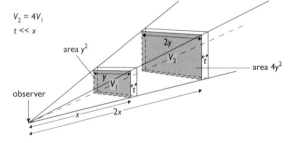

$V_2 = 4V_1$
$t << x$

Light from an individual star or galaxy obeys the inverse square law, so more distant light sources are fainter. To the observer, a star in V_1 is four times as bright as an identical star in V_2. However, if light sources are in general quite evenly scattered through space, a more distant layer or shell of space contains more light sources than a nearer one. If the density of stars is the same in volumes V_1 and V_2, then V_2 contains four times as many stars as V_1. This exactly compensates for the inverse square law effect, so the total brightness of the two volumes of stars is the same. This means that, on the basis of assumptions A, B and C listed above, more distant regions of space should be just as bright as near ones. If space is an infinite and unchanging framework in which the stars are situated, there would be an infinite number of such layers or volumes, each one as bright as any other. The sky would be flooded with light, night and day. It isn't. This is Olbers' paradox.

A mismatch between an argument that sounds sensible, and what we actually see, is always interesting. Olbers' suggestion was that the light was absorbed by clouds of gas in space. But the gas would be heated by this absorption, and would eventually reach the same temperature as the stars, and glow brightly itself. The solution doesn't work.

The problem must lie in one or more of our assumptions:

12 a What does the inverse square law predict about light from a distant star?
b Why is this not enough to explain why the sky is not continuously very bright?

A Changes can take place within the Universe, but the Universe itself is unchanging. – The Big Bang theory of the Universe (see later in this chapter) abandons this assumption, and in doing so solves Olbers' paradox.
B The Universe has sources of light scattered throughout it. – Our observations support this, so the problem and solution do not seem to lie here.
C The Universe is infinitely large. – We cannot know this. There are limits to how far we can see. The Big Bang theory claims that the Universe had a beginning, in which case there has not been time for light to travel from an infinite distance away.

Red shift and Hubble's Law
Red shift

Figure 8.14
Red shift.

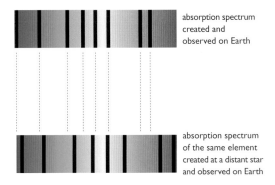

absorption spectrum created and observed on Earth

absorption spectrum of the same element created at a distant star and observed on Earth

Before seeing how the Big Bang theory solves Olbers' paradox we need to take another look at some observations. If you look at a single star it might be tempting to suppose that its light doesn't tell us very much. But we have seen that we can learn about chemical composition from absorption spectra, absolute luminosity and distance from Earth, and surface temperature. From absorption spectra we can also detect and measure **red shift**.

Red shift results in an increase in observed wavelength, towards the red end of the visible spectrum, that occurs when the source and the observer are moving apart (Figure 8.14). Blue shift is seen when the source moves towards the observer (or the observer moves towards the source). Both red shift and blue shift are examples of the Doppler effect (Figure 8.15). You can read about the principles of Doppler shift in *Introduction to Advanced Physics*, pages 283 and 453.

Figure 8.15
Summary of the Doppler effect for a moving source.

Wave A is emitted when source is at position A.
Likewise, wave B is emitted at position B, and so on.

wave A
wave B
wave C
wave D
motion of source
v
observer 1
$\lambda_{obs} = \dfrac{c + v}{f}$
A B C D
λ_{obs}
observer 2
$\lambda_{obs} = \dfrac{c - v}{f}$

- For observer 1, the detected wavelength is *greater than* that emitted (red shift).
- For observer 2, the detected wavelength is *less than* that emitted (blue shift).
- For observer 3, the detected wavelength is *the same as* that emitted.

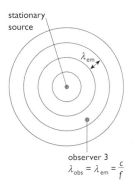

stationary source
λ_{em}
observer 3
$\lambda_{obs} = \lambda_{em} = \dfrac{c}{f}$

The ratio of relative speed of source and observer, v, to the speed of light, c, provides a simple way of thinking about movement on a cosmological scale. This ratio is simply called the red shift, z, of the source:

$$z = \frac{v}{c}$$

v is sometimes called 'speed of recession'. Red shift is a ratio of speeds, and is therefore dimensionless. We can show how z is related to the increase in observed wavelength, as follows.

We know that, for the Doppler effect in general, for a source that is moving away from the observer,

$$\text{observed wavelength, } \lambda_{obs} = \frac{c + v}{f}$$

where c is the speed of light and v is the speed of movement of the source. We then assume that if we make laboratory measurements on the same light – which might, for example, be the light corresponding to a line of the absorption spectrum of an element such as helium – then we will obtain an answer that is the same as the wavelength emitted by the source, and so

$$\text{emitted wavelength, } \lambda_{em} = \frac{c}{f}$$

The difference between these wavelengths is the wavelength shift, $\Delta\lambda$:

$$\Delta\lambda = \lambda_{obs} - \lambda_{em}$$

The ratio of this to the emitted wavelength is

$$\frac{\Delta\lambda}{\lambda_{em}} = \frac{\lambda_{obs} - \lambda_{em}}{\lambda_{em}}$$
$$= \frac{(c + v)/f - c/f}{c/f}$$
$$= \frac{c + v - c}{c}$$
$$= \frac{v}{c}$$

Thus

$$\text{red shift, } z = \frac{v}{c} = \frac{\Delta\lambda}{\lambda_{em}}$$

Note that red shift is constant for a given source, depending on its speed of recession. This means that the wavelength change $\Delta\lambda$ is proportional to the emitted wavelength λ_{em}, so that the change is larger for red light than for blue.

Hubble's Law

Immanuel Kant lived in the 18th century, and is still thought of as one of the world's great philosophers. He was also an astronomer, and he looked at a faint object in the sky that we now call M31, or Andromeda. He suggested that this was not a star but a vast collection of stars, which he called an 'island universe', but this idea did not become accepted science until the 1920s.

People looked at other faint and fuzzy objects in the sky, and called them *nebulae*, from the Latin word for clouds, and debated whether they were clouds of gas or more 'island universes'. Edwin Hubble began to provide the answer in 1924, when he studied Cepheid variable stars in Andromeda.

Figure 8.16
The Virgo cluster contains several thousand galaxies. Together with the Local Group in which we live, and other objects, it forms part of the Local Supercluster. At 19 Mpc away from Earth, it is a relatively close neighbour. Cepheid variable sources within Virgo have been studied and hence its distance is quite reliably known. For more distant objects Cepheid variables are too faint to provide reliable information, and so other methods must be used.

Figure 8.17
At present, if the Hubble Space Telescope cannot detect it, then it is not practically observable at visible wavelengths. Something in the region of 5×10^{10} galaxies are detectable to the Hubble Space Telescope. Not that anybody has counted them all – the number is an estimate based on the density of galaxies in each sector of the sky.

It turned out that Andromeda was a long way away – about two million light-years, or 700 kpc. Andromeda could not be a cloud of gas in relatively local interstellar space. To be seen at all at such a distance it had to be a very bright object indeed – a collection of vast numbers of stars – a galaxy like our own, rather than some part of the Milky Way. There are more of these galaxies, billions of them, as can be seen in images produced by large ground-based telescopes (Figure 8.16) and by the orbiting space telescope named after Hubble (Figure 8.17).

From study of red shifts we can measure the speeds at which galaxies and clusters are moving away, or receding, from us. What we find seems (though not with complete certainty) to be a proportionality relating speed of recession to distance, known as **Hubble's Law**:

$$v = Hd$$

H is the Hubble constant. Since distance, d, to other galaxies is measured in megaparsecs, Mpc, and speed of recession in km s^{-1}, the value of the Hubble constant is usually given in km s^{-1} Mpc^{-1}. Estimates of its value vary from 40 to 100 km s^{-1} Mpc^{-1}, the large uncertainty being due to the difficulty in measuring distances to light sources that are much outside the Local Group (Figure 8.18).

Figure 8.18 (below)
The gradient of the graph is the Hubble constant.

13 **a** For a source of light moving towards the observer,

$$\lambda_{obs} = \frac{c - v}{f}$$

Use this to relate blue shift, z, to v and c for such a source.
b Blue-shifted galaxies do exist. Explain why these are much more likely to be observed at relatively short distances from us.

14 **a** Sketch a graph of red shift, z, against distance of galaxies from us. What is the estimated value, with units, of the gradient of the graph?
b If one galaxy is in orbit around another at high speed, both at considerable distance from us, how could red shift measurement detect this?
c Comment on the positions of these two galaxies on the graph of part **a**.

15 Does the Universe observable to the Hubble Space Telescope contain a mole of stars? (A mole contains 6.0×10^{23} 'particles'.)

16 Assuming that $H = 55$ km s^{-1} Mpc^{-1}, give its value in SI units.

17 How was it possible to conclude that Andromeda is outside our Galaxy?

18 If, for a distant cluster of galaxies, $z = 0.89$, what is the distance to the cluster?

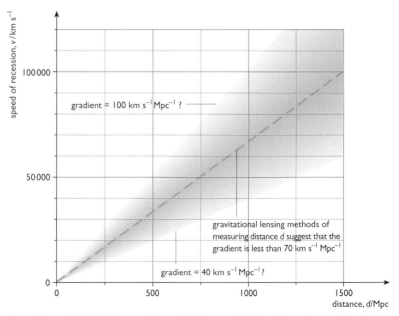

According to Hubble's Law, speed of recession is proportional to distance from us. But there is considerable uncertainty in the value of the constant of proportionality, the Hubble constant, H. More measurements of distances d greater than a few Mpc are needed in order to obtain a more reliable value of H.

The Big Bang

The red shift phenomenon tells us that the Universe itself is not a passive background to events. It is itself changing. More precisely, it is expanding. The **Big Bang** theory is a coherent set of ideas based on this expansion, and on the idea that the Universe developed from 'nothing' (or very nearly nothing).

The Big Bang is accepted as the most likely account of the history of the Universe, but it has always been controversial. An alternative theory – called the **steady-state theory** – suggested that expansion of the Universe does not necessarily mean that it is growing from an initial vanishing point, but that it is possible that new space, and matter, is continuously created throughout the Universe and that the Universe had no beginning. Observations of distant objects seem to falsify this. When we look at distant objects we are looking back in time. We can look back a long way. We find that objects a long way away (which are also objects a long time ago) are fundamentally different to nearer objects. This evidence points away from the steady-state theory and towards the Big Bang theory. The Big Bang has other supporting evidence in the form of the microwave background radiation.

Models of expansion

The expansion of the Universe is not easy to understand. There are not many models to help us. Our familiar world provides models, but we have to be particularly careful not to put too much faith in these. We are used to expansions that have centres – as in the case of growth of bacterial colonies, or chemical explosions, say. Cosmologists have looked around for suitable models for describing expansion without a centre, or expansion for which every point has an equal claim to being the centre, and have discovered the humble loaf of bread and the party balloon (Figures 8.19 and 8.20).

Figure 8.19
As the yeast cells grow and multiply they produce the carbon dioxide gas that makes the bubbles in the bread, and as the bubbles form the bread 'rises'. If the bread has no edges (and at this point it becomes a more unfamiliar sort of bread) then any one yeast cell has an equal right with all others to claim to be at the centre of the process of expansion.

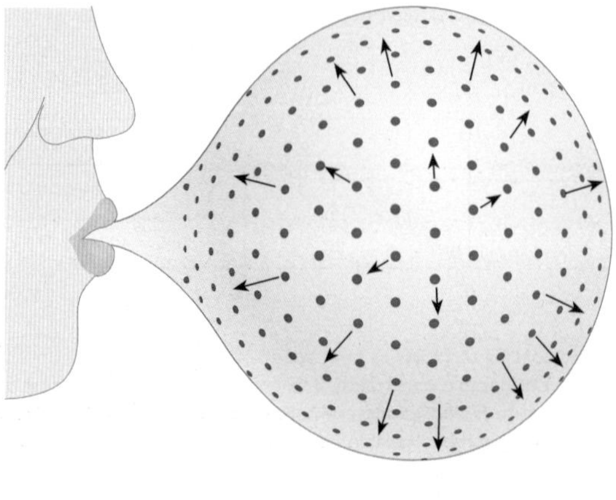

A yeast cell observing other yeast cells in rising dough lives in an expanding universe.

The further away another cell is, the faster it recedes.

Any cell will have the same impression of the expansion.

Figure 8.20
If a balloon is covered in a spotted pattern, then as the balloon expands all the dots become further apart. No dot is at the centre of the expansion, and the expanding 'space' has no edges. The mathematics of shapes is called topology (see page 185), and provides a useful source of models to help people to understand a very complex Universe.

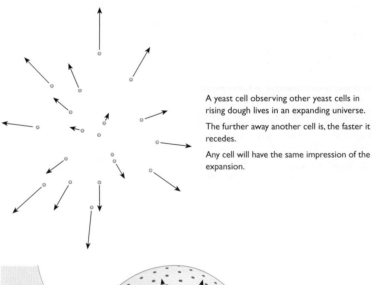

Spots on an expanding balloon all move further away from each other.

All spots have the same status as viewing points. There is no one centre to the expansion.

Microwave background radiation

Figure 8.21
Robert Wilson and Arno Penzias in 1978 when they received the Nobel Prize for Physics.

We have learned to use different parts of the electromagnetic spectrum in many different ways, and prominent amongst these is for communication. In 1965, two American physicists Arno Penzias and Robert Wilson, working for a telecommunications company, were developing a system for receiving radiation with wavelength of a few centimetres. The wavelengths are much longer than those of, say, visible light, but they are short relative to the radio range, and so communications specialists called them microwave wavelengths. The high frequency of microwaves allows them to carry information at high rate, and so they are much used for telecommunications. (In microwave ovens, radiation with a very specific wavelength is particularly absorbed by water molecules, causing them to vibrate with increasing amplitude.)

The two physicists (Figure 8.21) were experiencing difficulty. Their microwave receiver provided, as its output, a continuous background noise, whatever signals they transmitted to it. They considered all possibilities, even that it had something to do with the droppings of the pigeons which were roosting in the large shed with the microwave equipment. The pigeons were shot. The noise continued.

Eventually, others suggested that the noise was nothing to do with the specifics of the equipment, nor with local conditions (such as the presence of pigeons, or of the Earth's surface or anything on or above it). Instead they said it was the effect of radiation that was ever-present in the Universe, arriving at the detectors from all directions in the sky. They suggested that the radiation provided evidence in support of the Big Bang theory – that it could be explained as radiation emitted by the very young Universe that has existed throughout the Universe ever since. Now it is called the **cosmic background radiation**.

Figure 8.22 (below)
The front page of *The Independent* on 24 April 1992 proclaimed the evidence for the Big Bang. Never before had a piece of abstract science reached the front pages.

How the universe began

The cosmic background radiation, however, presented one problem for the Big Bang theory. It seemed to be incredibly uniform. Its intensity never varied, from time to time or from place to place. Yet the Universe that we now live in is full of non-uniformity in the density of matter – from the level of particles to the level of superclusters of galaxies. How could a perfectly smooth early Universe become lumpy? There was no explanation for this, but satellite observations in 1992 led to great excitement (Figure 8.22). These found very tiny but very significant variations in the intensity of the cosmic background radiation. We are lumps in a lumpy Universe because lumpiness seems to have existed at an early stage (see Figure 8.25, page 184).

The Big Bang theory claims that the radiation has been travelling since the Universe was young, and as the Universe has expanded a lot, so have the waves of radiation. They were initially much shorter wavelength waves emitted by *very* energetic material. In time, wavelength has increased, as if the source has cooled, so that it now appears to be radiation that would be emitted by material with a temperature of about 3 K.

19 Why may there be no difference between radiation from a very old hot source and radiation from a present-day cold source?

The Big Bang and Olbers' paradox

20 Describe briefly the evidence to support the Big Bang theory, in terms of:
a Olbers' paradox
b microwave background radiation
c red shifts.

The sky is not intensely bright day or night. This is a very simple observation that is at odds with old assumptions about the Universe. This mismatch between observation and assumption is Olbers' paradox. We solve the paradox if we accept that the Universe is not just a passive or unchanging framework in which events happen, but that it had a beginning and is still changing. The Big Bang theory supposes that the Universe is in a state of continuing change. And in a Universe that is expanding from a beginning, we can only see light from sources that are close enough to us for the light to have reached us in the time available since the beginning.

The age of the Universe and the Hubble constant

By making a crude assumption that the rate of expansion has always been the same, we can use Hubble's Law to estimate the age of the Universe, T:

$$v = \frac{d}{T}$$

where v is the average speed with which two bodies in the Universe have moved apart throughout time T, and d is the present-day separation. We also know from Hubble's Law that

$$v = Hd$$

So,

$$T = \frac{1}{H}$$

Taking

$$H = 55\,\text{km}\,\text{s}^{-1}\,\text{Mpc}^{-1}$$

we obtain

$$T = 1.8 \times 10^{10}\ \text{years}$$

The assumption that the rate of expansion has always been the same is almost certainly an unreasonable one, and the figure is no more than an estimate. But it provides at least a first estimate which we can seek to refine. One way to refine the figure is to reduce the uncertainty in the value of the Hubble constant.

21 Estimates of the value of the Hubble constant vary from 40 to $100\,\text{km}\,\text{s}^{-1}\,\text{Mpc}^{-1}$. Taking each of these outside estimates, and then assuming that its value has never changed, what different estimates for the age of the Universe are obtained?

Attempts to increase precision of measurement of the Hubble constant

The Hubble constant is the gradient of the graph of velocity of recession against distance from us. We can measure velocity with which a galaxy is moving away from us from its red shift. But to plot a reliable graph, and obtain a reliable value for the Hubble constant, we also need reliable values of distance.

Study of Cepheid variables only provides reliable measurements for relatively short distances of a few Mpc. The only objects that lie within such distances are the members of the Local Supercluster. For longer distances we can use other effects.

One method uses radio wavelengths to determine the nature of the plasma content of a distant cluster. Plasma, ionised gas, associated with very distant clusters absorbs radio waves that form part of the cosmic background radiation. This is called the Sunyaev–Zel'dovich effect, and by measuring the intensity of the microwave background radiation in the direction of the cluster of galaxies, we can tell how hot the plasma is. This, in turn, predicts the rate at which we would expect the plasma to emit X-rays. It allows us to calculate its luminosity as an X-ray source. We can then compare this with the X-ray intensity we receive here on Earth, and so work out the distance to the source. However, there are still large uncertainties in the measurements.

An additional method for determining distance to very remote objects relies on **gravitational lensing** (Figures 8.23 and 8.24). This occurs when light from a distant source must pass a very large object on its journey to us. We see more than one image of the source, because light has taken different routes past the massive obstacle. The pathways of the light appear to be bent, as if they have passed through a lens. The effect is caused by the gravity of the large object. A **quasar** is a very intense flickering source, and because it is usually a very distant source its light may have to pass close to other objects before it reaches us. The light from quasars quite often shows gravitational lensing effects. A particular flicker of intensity may take longer to reach us by one route around a closer massive object than by another route, and from this difference we can obtain a value for the distance of the source.

Figure 8.23
The Einstein ring. It was Einstein who first suggested that gravity affected light. Here a very large galaxy (centre) is effectively bending the pathways of light from a source that is behind it from our viewpoint, so that the light appears as a ring. It is an example of gravitational lensing.

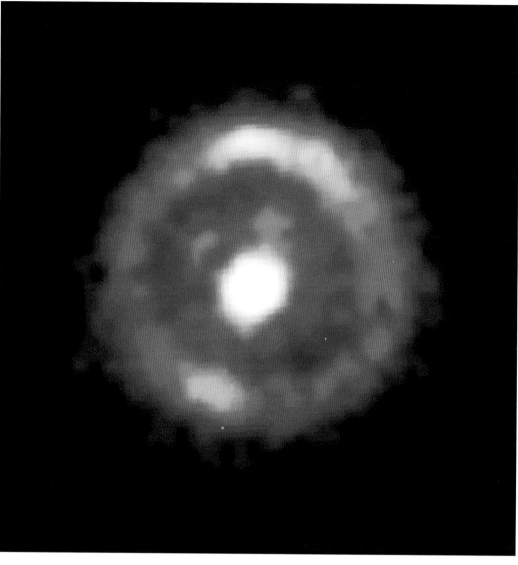

Figure 8.24
Gravitational lensing of light from a distant quasar.

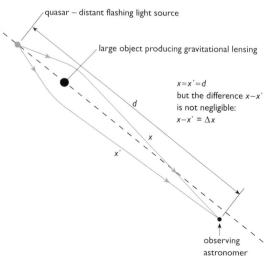

quasar – distant flashing light source

large object producing gravitational lensing

$x \approx x' \approx d$
but the difference $x-x'$
is not negligible:
$x-x' = \Delta x$

d

x

x'

observing astronomer

Light passing on opposite sides of the very large object must travel different path lengths. There is a path difference, Δx. Light following the two pathways takes different times to reach Earth. This is detectable when light is emitted in flashes, as by a quasar. The same emitted flash is received as two flashes, with a time difference Δt. The value of Δx can be worked out from Δt. From the geometry, distance d from quasar to Earth can be calculated (though still with significant uncertainty).

22 a What are the similarities in the principles of using Cepheid variables and the Sunyaev–Zel'dovich effect for measuring distances to objects in space?
b Why are the Sunyaev–Zel'dovich effect and gravitational lensing useful in providing an estimate of the age of the Universe?

These methods give us values for distance which we can plot against velocity of recession to obtain a somewhat more reliable value of gradient of the graph, the Hubble constant. Gravitational lensing observations suggest that the Hubble constant is less than $70\,\mathrm{km\,s^{-1}\,Mpc^{-1}}$.

Early, present and future Universe

The Big Bang and the early Universe

The Universe has changed. Its history is certainly not fully known, and many physicists and others are working to develop their ideas. But as it is now understood, the story is briefly as follows. It seems possible that the Universe has changed from a state in which there was no time, no space, no matter. This is sometimes called a singularity – suggesting a vanishing point. Time and space would have come into existence, but not much energy, and therefore not much matter (though what there was had a very high temperature). The Universe began to expand and cool. It very soon entered a new phase of extremely rapid expansion called **inflation**. What happened before inflation, whether or not the Universe began from a singularity, is a matter for debate amongst cosmologists. But most agree that the concept of inflation is a necessary part of a coherent theory of the origins of the Universe. Further detail of ideas about the early Universe can be found in books that are entirely about this subject, but here we provide a description of some interesting features.

During inflation, very quickly (in about 5×10^{-33} s), what is now the observable Universe expanded by a factor of 10^{50}. During this period the creation of a large amount of matter in the form of particles took place. An interesting feature of this is that the amount of matter created must have been bigger than the amount of antimatter created. Energy is required to create either kind, but it seems that the Universe had a bias in one direction. Though huge amounts of matter and of antimatter were created, and nearly all of this mutually annihilated, the excess of matter over antimatter led to the kinds of material we see in the Universe now.

Scientists suggest that before inflation forces were 'unified' – that is, the behaviour of interactions between particles was of only one kind. A little before, during and after inflation, these interactions developed different characteristics that now have the labels 'gravitational', 'strong', 'weak' and 'electric'.

We can now study the particles from which everyday matter is made, but particle physicists try to go further and recreate particles that existed in the post-inflationary period of the Universe, when particle interactions were extremely energetic and very massive particles could come into being (as predicted by *energy available = mass that can be created* $\times c^2$, or $E = mc^2$ – see *Introduction to Advanced Physics*, page 136, and Chapter 7 of this book). By using accelerators to create high-energy interactions, the particle researchers can investigate fundamental issues about the Universe now, and about the history of the early Universe. They are recreating the conditions of the early Universe.

Expansion continued at a reduced rate following inflation, and the temperature fell. At one second after the Big Bang, very massive particles had become extinct, and antimatter had become rare. Quarks began to gather in clusters that we now call the hadrons, including protons and neutrons. Still further expansion and cooling allowed neutrons and protons to gather together – creating nuclei of isotopes of hydrogen, helium and, eventually, in small quantities, lithium.

Figure 8.25
A representation of the whole sky, comparing microwave background radiation in all directions and showing that it is not uniform but 'lumpy'. The radiation was created in the early Universe, so this image tells a meaningful story of the developing Universe soon after the Big Bang.

The Big Bang theory predicts the relative amounts of hydrogen, helium and lithium that would have been created in the early phases. The figures work out as 77% hydrogen, 23% helium and 10^{-7}% lithium. We cannot look into relatively nearby space and expect to see these figures, because heavier elements – such as nitrogen, oxygen, calcium and iron – were later created from hydrogen and helium by nuclear fusion within stars, and these have altered the original proportions. But if we look very far into space we are also looking back in time – the light that reaches us from very distant parts of space has been travelling since the Universe was young, so that we can

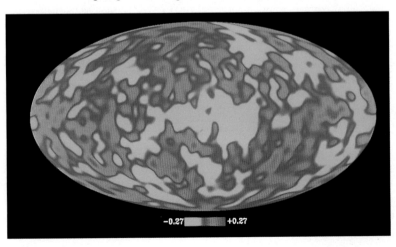

−0.27   +0.27

examine this light and find out what elements it is associated with. Such light has a spectrum that shows it has reached us from a source that is 77% hydrogen and 23% helium, providing support for our Big Bang ideas.

In the very early Universe, the high frequency of interactions of photons with other particles, such as free electrons, meant that a photon could never travel very far or last very long before being absorbed. Photons tended to have short lifetimes, and we do not detect photons from the very early Universe in any of our detectors. But as temperature continued to fall the plasma of surviving particles began to form into atoms and the rate of photon collisions fell. We can now detect such photons in the form of the microwave background radiation (see page 181).

Lumpiness of the early Universe (Figure 8.25) resulted in non-uniformities in the distribution of atoms of hydrogen and helium, and gravitational force made any regions of above-average density – clouds of gas – further attract, leading to collapse of the clouds and ultimately formation of stars and galaxies, with vast spaces between them.

When we look at very distant objects in space we are receiving light that left its source a very long time ago (Figure 8.26). If the Universe is about 10^{13} years old, then we can never see light from sources that are more than 10^{13} light-years away, whether or not such sources exist. There is a limit to how far we can see.

Figure 8.26
When we look at light from far distant sources we are looking at light that was emitted when the Universe was very young.

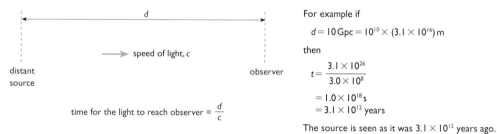

time for the light to reach observer $= \dfrac{d}{c}$

For example if
$$d = 10\,\text{Gpc} = 10^{10} \times (3.1 \times 10^{16})\,\text{m}$$
then
$$t = \frac{3.1 \times 10^{26}}{3.0 \times 10^{8}}$$
$$= 1.0 \times 10^{18}\,\text{s}$$
$$= 3.1 \times 10^{12}\,\text{years}$$

The source is seen as it was 3.1×10^{12} years ago.

New models of the Universe

People once believed the world to be geocentric (Earth at centre), then heliocentric (Sun at centre), and now having no centre. We see the Universe not only as having no special centre, but also as having no defined edges. Visual models have to relate to the familiar to be very useful, and our familiar experience is of three dimensions. Useful visual models are hard to develop, though mathematical models exist. Mathematicians can invent equations that consider large numbers of dimensions. If we want to fully understand the mathematical models, we have to accept that we have to do the maths. However, the mathematics of shape, **topology**, does provide some interesting visual representations (Figure 8.27).

Figure 8.27
Topological representations.

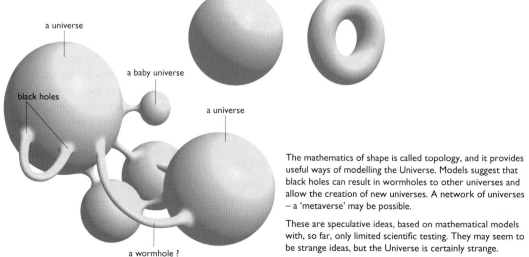

a universe

a baby universe

black holes

a universe

a wormhole ?

The mathematics of shape is called topology, and it provides useful ways of modelling the Universe. Models suggest that black holes can result in wormholes to other universes and allow the creation of new universes. A network of universes – a 'metaverse' may be possible.

These are speculative ideas, based on mathematical models with, so far, only limited scientific testing. They may seem to be strange ideas, but the Universe is certainly strange.

Predicting the far future

Will the Universe continue to expand, decreasing and decreasing in density, energy becoming more and more evenly distributed, until the Universe is a homogeneous soup of particles where nothing happens except endless expansion – a state of 'heat death'? That's the **open Universe** scenario (Figure 8.28). Will it turn out to be a **closed Universe** and contract, reversing the processes of expansion, into a **Big Crunch**? Or will expansion simply slow down and stop without reversing, creating static space? Such a state of balance is called a **critical Universe**. It depends on how much matter there is, and on the effect of gravity. We seem to live in a Universe that is very, very close to the critical condition. Is that mere chance, or are there reasons for this? At present, nobody knows. It is one of the questions in the continuing adventure of cosmology.

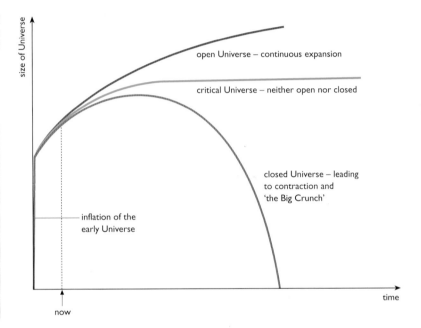

Figure 8.28
What kind of Universe are we part of?

23 Outline a prediction made by the Big Bang theory's account of the early Universe that is supported by quantitative observation.
24 a Why are there no surviving photons from the very early Universe?
   b What does this suggest about the time of creation of what is now the microwave background radiation?
25 Is it possible that the early Universe contained equal amounts of matter and antimatter?
26 How did nuclei bigger than lithium come into existence? (See Chapter 7.)
27 Why did the 1992 observation of non-uniformity in the cosmic background radiation cause so much excitement?

● **Comprehension and application**

SETI – the Search for Extra-Terrestrial Intelligence

Is there life out there? Well, if other stars have planetary systems, and some do, and if some of those planets have suitable temperature ranges, which they probably do, then it seems likely that chemical complexity has developed, aided by a supply of energy from a star. Will we ever get to shake hands with aliens? Leaving aside the fact that different evolutionary conditions are unlikely to have resulted in the same body structures, such as hands, the answer is 'probably not'. If there are planetary systems with the right combinations of temperature, chemical raw materials (minerals) and stability over long time periods, then there are almost certainly organisms like bacteria (Figure 8.29). It may even be that there are multicellular organisms, perhaps even with cell specialisation similar to that which developed on Earth some time in the last billion years (Figure 8.30). But the development of consciousness – of organisms that look at the Universe and contemplate its wondrous complexity – may or may not be a one-off. Just how special are we (Figure 8.31)?

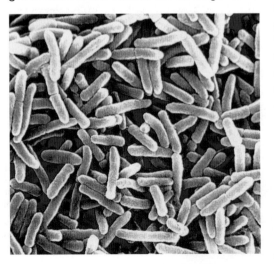

Figure 8.29
A colony of bacteria. Do such lifeforms exist on other planets? We really don't know, but a few scientists have even speculated that very simple organisms can exist, and first developed, in interstellar space.

Figure 8.30
Multicelled lifeforms are found in sedimentary rocks up to 600 million years old. On Earth, that seems to be when life got complicated.

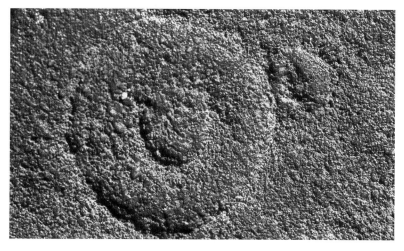

Figure 8.31
Intelligent life on Earth!

28 a Given that your body is 75–80% water, approximately what proportion of its atoms are hydrogen atoms?

b According to the Big Bang theory, at what stage did the nuclei of these atoms come into existence, and from what?

c At what stage and under what conditions did the remaining atoms come into existence?

d Lifetimes of stars range from about one billion (big, high-luminosity stars) to 50000 billion years (small, low-luminosity stars). Explain why complex life could not have existed for at least a billion years after the Big Bang.

● **Extra skills task** Application of Number, Information Technology and Communication

1 There are 10^{11} stars in the Milky Way, and at least 10^9 galaxies in the observable Universe. In recent years, astronomers have been developing better and better techniques for detecting the presence of planets, and it seems that many stars have planetary systems.

    a If one star in every thousand has a planetary system, how many such systems are there in the Universe?

    b If one planet in each thousand planetary systems has the right conditions of temperature and long-term stability that would allow the development of cellular life, how many such planets are there in the Universe?

2 Carry out an Internet search on SETI. You should try to find sites that relate to scientific research, as well as sites that are not scientific but are based on science fiction. Look at sites that make claims about contact with 'aliens'. Write a short report that provides an examination of the scientific validity of the claims that these sites make.

3 Use your work on **1** and **2** to contribute to a discussion on the following points:

    a Do you believe that there are lifeforms elsewhere in the Universe?

    b Do you believe that there are intelligent lifeforms elsewhere in the Universe?

Examination questions

1   In the laboratory, the so called hydrogen line has a normal wavelength of 656.285 nm, but in the spectrum emitted by the star Vega this line is located at 656.255 nm. What is Vega's velocity with respect to the Earth?   (3)

IB, Higher level 3, Paper 430, Specimen (part)

2   **a**   Explain how Hubble's law supports the Big Bang model of the universe.   (2)
  b   Outline *one* other piece of evidence for the model, saying how it supports the Big Bang.   (3)
  c   The Andromeda galaxy is a relatively close galaxy, about 700 kpc from the Milky Way, whereas the Virgo nebula is 2.3 Mpc away. If Virgo is moving away at 1200 km s^{-1}, show that Hubble's law predicts that Andromeda should be moving away at roughly 400 km s^{-1}.   (1)
  d   Andromeda is in fact moving *towards* the Milky Way, with a speed of about 100 km s^{-1}. How can this discrepancy from the prediction, in both magnitude and direction, be explained?   (3)

IB, Standard level 3, Paper 430, November 1998

3   This question is about investigating the properties of a *nearby star in our own galaxy*.
  a   With the aid of a diagram, explain how the distance of such a star from the earth could be determined.   (3)
  b   Could the method be used for more distant stars? Explain briefly.   (1)
  c   Explain how the surface temperature of the star could be estimated.   (3)
  d   Explain briefly how the chemical composition of the star could be determined.   (4)

IB, Standard level 3, Paper 430, November 1998

4   **a**   Describe briefly the principle of one method for measuring the surface temperature of a star.
      Describe briefly how the component of a star's velocity towards or away from the Earth may be found.   (6)
  b   State Hubble's law and illustrate your statement with a simple graph. Use your graph to explain the meaning of the Hubble constant.
      Assuming a value of 3×10^{-18} s^{-1} for the Hubble constant, find the distance from Earth of a galaxy for which the red shift for a particular spectral line is a tenth of the wavelength of the same spectral line observed from a stationary source.   (6)

London, A level, Module Test PH3, January 2000 (part)

5   **a**   Astronomers have discovered from their observations of the star Capella that:
    • its surface temperature T is 5200 K
    • its distance from the Earth is 4.3×10^{17} m
    • at the Earth's surface the intensity of the radiation received from Capella is 1.2×10^{-8} W m^{-2}

    i   Explain briefly how the surface temperature is determined.   (3)
    ii   Describe, in outline only, the parallax method for finding the distance of the star from the Earth.   (4)
    iii   Calculate the radius r of Capella, given that its luminosity L can be found by using

$$L = 4\pi r^2 \sigma T^4$$

    where σ is the Stefan–Boltzmann constant which is 5.7×10^{-8} W m^{-2} K^{-4}.   (3)
  b   **i**   Sketch and label an HR diagram showing the *main sequence* and the regions occupied by *red giants* and *white dwarfs*.
      Explain why main sequence stars of large mass have higher luminosities and shorter lives than main sequence stars of low mass.   (5)
    ii   Describe the processes which occur within a star similar in mass to the Sun when it leaves the main sequence.   (5)

London, A level, Module Test PH3, January 1997

6   **a**   The distance of Sirius A from the Sun, calculated from the measured value of its annual parallax, is 8.7 light years. One light year equals 9.46×10^{15} m.
      Explain the term *annual parallax* by sketching and labelling a suitable diagram.
      How would the annual parallax of Sirius A be measured if the distance of the Earth from the Sun is already known? On what assumption does the method depend?
      The distance of the Earth from the Sun is 1.50×10^{11} m. Calculate the annual parallax for Sirius A.   (8)
  b   Explain what is meant by the *luminosity* of a star.
      The luminosity of Sirius A is 8.17×10^{27} W. Calculate the *intensity* of Sirius A measured at the Earth.   (4)

London, A level, Module Test PH3, June 1999 (part)

7   **a**   Hubble's law relates the speed of recession of a galaxy to its distance from Earth. Show how the maximum distance to the edge of the Universe may be obtained from this law.   (2)
  b   The Hubble constant is assumed to be between 50 km s^{-1} Mpc^{-1} and 100 km s^{-1} Mpc^{-1}. Hence calculate the maximum age of the Universe.   (3)

NEAB, AS/A level, Module Test PH04, March 1998

8 a The figure shows a Hertzsprung–Russell diagram for stars in our neighbourhood of the Milky Way.

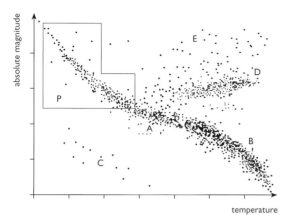

i Insert approximate values on each axis.

ii Identify the type of stars found in each of the following regions (ignoring the box marked P at this stage).

region A to B
region C
region D
region E (4)

b A globular cluster is a group of many thousands of stars, with a range of masses, that were all formed at the same time. The Hertzsprung–Russell diagram for an old globular cluster would be similar to that shown in the figure *except* that the stars in box P would be missing.

i With the aid of the Hertzsprung–Russell diagram, give *two* characteristics of the stars that would have occupied box P.

ii Suggest a reason why these stars are missing and hence deduce, with reasons, what they would have become when they left the main sequence. (6)

NEAB, AS/A level, Module Test PH04, June 1998 (part)

9 a The table below gives values of the masses of objects in the Universe. The masses are quoted in terms of $M_\odot$, the mass of the Sun.

Mass/$M_\odot$	Identity
1×10^{-10}	
1×10^{-5}	
1	the Sun
20	

Complete the table by suggesting what the objects might be. (3)

b The table below gives distances from Earth, or diameters, of objects in the Universe.

Distance (or diameter)	Identity
1.5 light-seconds	
8 light-minutes	distance between Earth and Sun
6×10^4 light-years	
1×10^7 light-years	

Complete the table by suggesting what the distances or the diameters might be. (3)

UCLES, A level, 4837, June 1998

10 a i Describe what is meant by *microwave background radiation*.

ii What do the characteristics of this radiation tell us about the Universe? (7)

b In recent years, red- and blue-shifts have been detected in the microwave background radiation.

i Explain what is meant by *red-* and *blue-shifts*.

ii State what information about the Earth can be gained from these shifts in the microwave background radiation. (4)

c i Explain whether the microwave background radiation is detectable at the Earth's surface.

ii Suggest why the shifts in the microwave background radiation were not detected at the Earth's surface. (3)

UCLES, A level, 4837, June 1998

11 A spectrum of light from a distant galaxy shows the spectral lines of a certain element. The spectral lines are compared with those produced in the laboratory from the same element. The diagram shows the two sets of lines. (The dark lines represent the galaxy lines; the dotted lines represent the laboratory lines.)

a i Name the effect which causes the shift in the position of the spectral lines. (1)

ii State which end (left or right) in the diagram represents the long-wavelength end of the spectrum. Explain your answer. (2)

b Explain how the shifts of spectral lines, such as those shown in the diagram, have been used to support a theory about the evolution of the Universe. (2)

OCR (Oxford), A level, 6844, June 1999

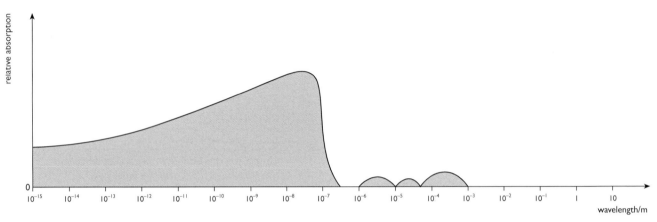

12 The diagram above shows how the relative absorption of electromagnetic radiation by the Earth's atmosphere varies with wavelength across the electromagnetic spectrum. The atmosphere has a high transparency where the relative absorption is low.

    a On the wavelength axis, identify a wavelength in each of the following regions of the spectrum.
      i infra-red (label this wavelength I)
      ii visible (label this wavelength L)
      iii ultraviolet (label this wavelength U)   (2)
    b i Observations are made from above the Earth's atmosphere even for wavelengths for which the atmosphere is very transparent. State two reasons why this is done.

    ii Suggest, in view of your answer to **b i**, two reasons why observations from the Earth's surface are still common.   (4)

OCR (Cambridge), A level, 4837, June 1999

13 a Explain what is meant by the *Hertzsprung–Russell diagram*, what observations and measurements are needed to plot it, how it leads to a classification of stars into different types and what the relationship is between these types of star.   (14)
    b State two types of information about a star, other than its temperature, that may be obtained from observation of its spectrum. Refer to the relevant physical principles.   (7)

OCR (Cambridge), A level, 4837, June 1999

PART B

THEMES AND APPLICATIONS

9 · Using logarithms

Base 10

Some natural phenomena are measurable over a huge range of values. Take human hearing, for example. We can hear sounds over a fairly large range of frequencies, from about 20 Hz to about 20 kHz. That is a range of, very nearly, 20 kHz. The range can also be expressed as a ratio. The highest frequency that we can hear is 10^3 times higher than the lowest.

Since our counting system is based on the number 10, for no good reason except that it is convenient and manageable, it makes sense to work with ratios or multiples of 10. We set up a number system whereby

$$
\begin{aligned}
1 &&&= 10^0 \\
10 &= 10 &&= 10^1 \\
100 &= 10 \times 10 &&= 10^2 \\
1000 &= 10 \times 10 \times 10 &&= 10^3
\end{aligned}
$$

and so on. The power is a shorthand way of writing out multiples. It is also called a logarithm. The logarithms here relate to the base number of 10. The idea can be expressed as a single sentence that is worth remembering:

> The logarithm of a number is the power to which the base must be raised to give the number.

It's a good idea to read the definition again more than once. There are three key words involved:

- number — which can be any number (above we have shown multiples of 10 only);
- logarithm — which is also the power;
- base — we could choose any base we like, but our numbering system uses a base of 10 (the binary numbering system uses a base of 2, and we will come back to that).

Figure 9.1

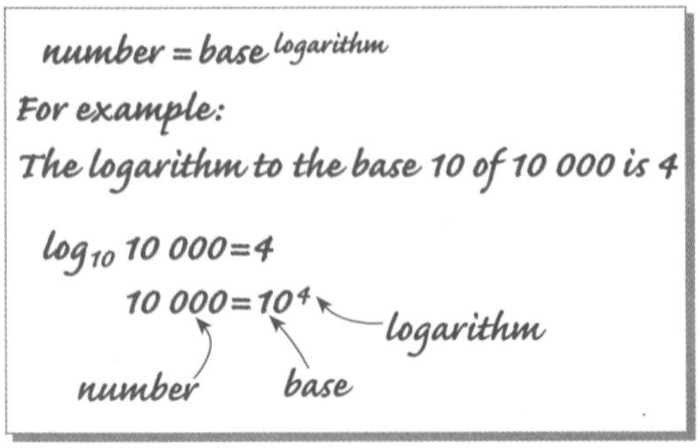

$number = base^{logarithm}$

For example:

The logarithm to the base 10 of 10 000 is 4

$\log_{10} 10\,000 = 4$

$10\,000 = 10^4$ — *logarithm*

number   *base*

Sticking for the moment with our base 10 system, what about a number that is not a multiple of 10, such as, say, 89? A scientific calculator will tell you the value of a logarithm of a number to the base 10 at the push of a button – the log button.

We know that the logarithm of a number is the power to which the base must be raised to give the number. That is, for the number 89,

$$\log_{10} 89 = 1.95$$

and so

$$89 = 10^{1.95}$$

We might have been able to guess that the logarithm would lie between 1.00 and 2.00, since 89 is more than 10 (or 10^1) but less than 100 (or 10^2). To give another example, say 36,

$$\log_{10} 36 = 1.55$$

and so

$$36 = 10^{1.55}$$

If the number is more than 100 but less than 1000, then its logarithm to the base 10 lies between 2 and 3. For example, for 258,

$$\log_{10} 258 = 2.41$$

and so

$$258 = 10^{2.41}$$

So what happens for much bigger numbers, like one-and-a-half million, 1500000? We can tell straight away that the logarithm lies between 6 and 7. How? Because the number lies between one million (10^6) and ten million (10^7). In fact,

$$\log_{10} 1500000 = 6.18$$

and so

$$1500000 = 10^{6.18}$$

Note that sometimes log to the base ten is abbreviated further to lg.

1 Evaluate the following, using a calculator where necessary:
a $\log_{10} 1$
b $\log_{10} 10000$
c $\log_{10} 10^9$
d $\log_{10} 3458$
e $\log_{10} 0.1$
f $\log_{10} 0.654$

Logs and human hearing

Logarithmic expression of numbers – such as the use of $10^{6.18}$ to represent 1500000 – presents a solution to a problem. It is impossible to draw a bar showing the simple (arithmetic) frequency range of human hearing without using a very large sheet or screen. We cannot fit in a full range of 20 kHz and at the same time show that the lower end of the range is a few Hz and not zero (Figure 9.2).

Figure 9.2
Range of hearing frequencies, on linear scales.

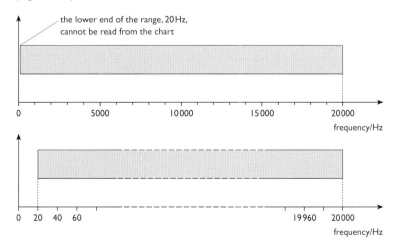

The wide range of frequencies that we can hear cannot be fully presented on a useful linear scale.

A broken scale allows the two limits of the range to be shown, but not all of the values in between.

Logarithmic scales, or log scales, provide a useful solution. Supposing that the lower end of the frequency range is 20 Hz, then we have

$$\log_{10} 20 = 1.30$$

and

$$20 = 10^{1.30}$$

and for the upper end of the range, 20 kHz or 20 000 Hz, we have

$$\log_{10} 20\,000 = 4.30$$

and

$$20\,000 = 10^{4.30}$$

Now we can plot our bar as shown in Figure 9.3. The scale is not linear but logarithmic. On a familiar linear scale, equal distances on the paper represent equal differences in values. On a logarithmic scale, equal distances on the paper represent equal multiples.

Figure 9.3
Range of hearing frequencies, on a log scale.

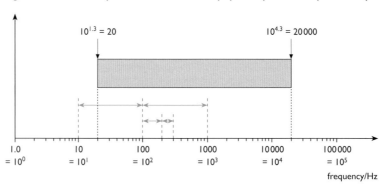

Equal increments of a logarithmic scale represent equal increases in logarithm.

This is shown, for example, by the equal lengths from 10 to 100, and from 100 to 1000 (indicated by the green arrows).

Equal increases in number (value of frequency in this particular case) are *not* represented by equal increments of the scale.

This is shown, for example, by the different lengths from 100 to 200, and from 200 to 300 (indicated by the orange arrows).

It is worth noting how a particular range of values, say 100 Hz to 1000 Hz, can be plotted on linear and logarithmic scales. On the linear scale, we can choose a distance on the paper to represent each 100 Hz. But for the logarithmic scale we choose a distance on the paper to represent each multiple of 10. Frequencies in the range 100 to 1000 Hz, and the logarithms of the values, are given in Table 9.1. Figure 9.4 opposite shows the frequencies plotted on linear, or lin, and logarithmic, or log, scales.

Table 9.1

Number (frequency/hertz)	Logarithm
100	2.00
200	2.30
300	2.48
400	2.60
500	2.70
600	2.78
700	2.85
800	2.90
900	2.95

Figure 9.4
Plotting values on linear
and logarithmic scales.

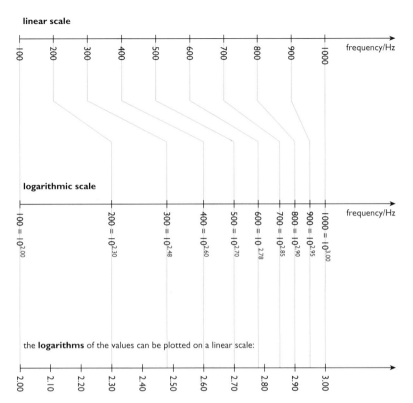

The decibel scale

The range of frequencies that are audible to the human ear is large, but the range of sound *intensities* that we can hear is much bigger. The lowest intensity, I_0, that an average person can hear at a frequency of 2 to 3 kHz is $10^{-12}\,\mathrm{W\,m^{-2}}$ and is called the threshold of hearing. At the same frequency and an intensity in the region of $1\,\mathrm{W\,m^{-2}}$, a sound is so loud that an average person begins to feel pain, and so this is called the threshold of pain.

We can consider the ratio of intensity, I, of a particular sound to that of the threshold of hearing. This ratio is I/I_0. For the speech of a person with whom we are having a conversation, the intensity is in the region of $10^{-5}\,\mathrm{W\,m^{-2}}$. So,

$$\frac{I}{I_0} = \frac{10^{-5}}{10^{-12}} = 10^7$$

If the person talks more loudly and the intensity rises, say, to $10^{-4}\,\mathrm{W\,m^{-2}}$, then

$$\frac{I}{I_0} = \frac{10^{-4}}{10^{-12}} = 10^8$$

On a logarithmic scale we can use the numbers 7 and 8 to compare the loudnesses. This scale is called the bel scale of sound levels, and is given by

$$\text{sound level in bels, B} = \log_{10}\left(\frac{I}{10^{-12}}\right)$$

where I is in $\mathrm{W\,m^{-2}}$. Since the quantity is calculated from a ratio without introduction of other physical values that have dimensions, the sound level is itself dimensionless, and it is not strictly necessary to assign a unit to it. However, it is convenient to use a unit, and the bel is the fundamental unit then used.

• For the threshold of hearing:

$$\text{sound level in bels} = \log_{10}\left(\frac{10^{-12}}{10^{-12}}\right)$$

$$= \log_{10} 1 = 0\,\text{B}$$

• For the threshold of pain:

$$\text{sound level in bels} = \log_{10}\left(\frac{1}{10^{-12}}\right)$$

$$= \log_{10} 10^{12} = 12\,\text{B}$$

As a matter of working habit, the decibel is used more often than the bel. The decibel is simply one-tenth of a bel. That is,

$$1\,\text{dB} = 0.1\,\text{B}$$

So on the decibel scale, sound levels range from 0 dB at the threshold of hearing to 120 dB at the threshold of pain.

A graph of threshold of hearing against frequency can be plotted, either in terms of sound intensity ratio, I/I_0, with both axes having logarithmic scales, or in terms of the decibel scale (Figure 9.5).

Figure 9.5
Graphs of
a minimum audible intensity (W m^{-2}) and
b minimum audible intensity level (dB) plotted against frequency.

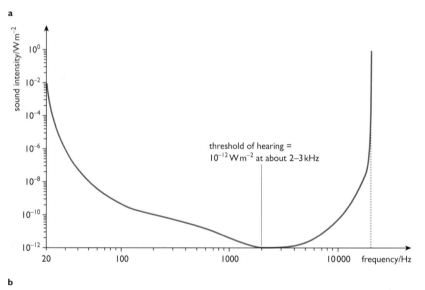

A graph with two logarithmic scales is called a log–log graph.

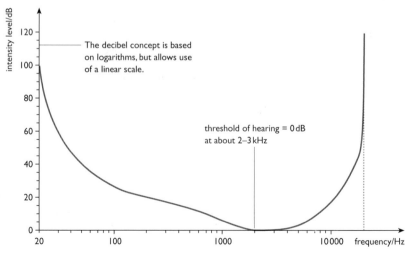

A graph with one logarithmic scale and one linear scale is called a log–linear or log–lin graph.

Log scale representation of the electromagnetic spectrum

Figure 9.6
A log scale used to show the range of electromagnetic wavelengths.

The electromagnetic spectrum includes a very wide range of wavelengths, from as much as 10^5 m for very long-wave radio waves to 10^{-16} m for high-energy gamma radiation. Textbook illustrations of the spectrum use logarithmic scales (Figure 9.6), except where only relatively small portions of the spectrum are shown.

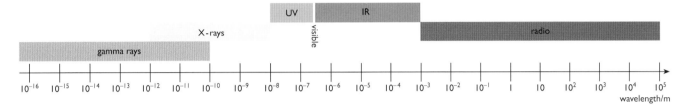

2 a What are, approximately, the lowest and highest radio wavelengths?
b What are, approximately, the lowest and highest gamma-ray wavelengths?
c Does the use of a logarithmic scale give a misleading representation of these wavelength ranges?
d i Sketch a diagram of the electromagnetic spectrum using lg (wavelength) for your numbered scale.
ii Repeat using lg (freqency) for the scale.
3 Use a logarithmic scale to show photon energies, using $c = \nu\lambda$ and $E = h\nu$ as appropriate, for the full range of electromagnetic frequencies, ν.

Log–linear resonance curves

The amplitude of a vibrating system – whether a guitar string, a ruler on the edge of a table, or a footbridge – tends not to have a very large range of values, and so can be plotted on an ordinary linear axis. We may be interested in the relationship between amplitude and driving or forcing frequency, and plotting a graph of these variables yields 'resonance curves'. The frequency may then have a wide range of values, requiring use of a log scale (Figure 9.7).

Figure 9.7
A resonance curve. This graph has one logarithmic scale and one linear scale; it is a log–lin graph.

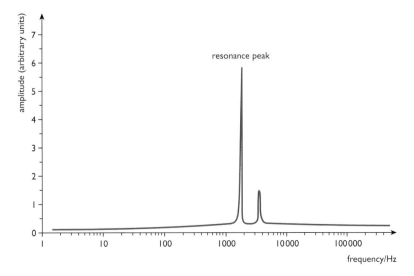

Use of log scales to represent wide resistance range for thermistors

Thermistors are very sensitive to temperature changes. Over relatively small temperature ranges, thermistor resistance can vary considerably. It is common to plot their resistance on log scales (Figure 9.8).

Figure 9.8
Resistance of a typical thermistor over the temperature range 20 to 120 °C.

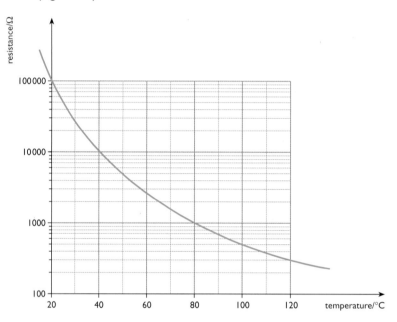

4 For this question, use Figure 9.8, showing variation of thermistor resistance with temperature.
  a Why is a logarithmic scale used here while it would be less useful for plotting resistance of a metal wire against temperature, for the same temperature range?
  b i What is the resistance at a temperature of 20 °C?
  ii What is the resistance at a temperature of 120 °C?
  iii What is the temperature when resistance is 500 Ω?
  iv What is the temperature when resistance is 1000 Ω?
  c Find the logarithms to the base 10 of the numbers 20, 40, 60, 80, 100 and 120, and hence sketch the shape of the graph that would be obtained when both axes are logarithmic.
  d Sketch the general shape of the graph that would be obtained with both axes being linear.
5 **a** What is the simple ratio of intensities of sounds of sound level 8 B and 7 B?
  b What is the simple ratio of intensities of sounds of sound level 90 dB and 20 dB?
6 A decibel scale is used to measure signal-to-noise ratio, for example in the telecommunications industry. Clarity of transmitted data depends on the rate of arrival of energy at the detector due to the data, and the rate of arrival of energy from other assorted sources. The first of these is the signal power, and the second is the noise power. They can be compared by a simple ratio or by using a decibel scale, which is similar to that used for sound intensity ratios, but using power P in place of intensity I. We then speak of 'signal-to-noise ratio on the decibel scale'.
  a In general, what is the difference between power and intensity?
  b Using the definitions of the bel and the decibel given in the text as a guide, write down an equation for signal-to-noise ratio using
  i bels
  ii decibels.
  c What is the signal-to-noise ratio on a decibel scale when
  i signal power = 10^{-3} W, noise power = 10^{-6} W
  ii signal power = 10^{-3} W, noise power = 10^{-3} W
  iii signal power = 5×10^{-3} W, noise power = 2×10^{-3} W
  iv signal power = 25 mW, noise power = 2 mW?

Base 2

Usually, in physics and other areas, we are counting using 10 as the base, and so logarithms to the base 10 are the most useful. When we simply say log 9 or log 200, for example, then these are to base 10. But it is possible to have other bases. The binary system, for example, uses 2 as its base.

Remember the definition of a logarithm: the logarithm of a number is the power to which the base must be raised to give the number. So if the base is 2 and the number is 7, then

$$\log_2 7 = x$$

and

$$7 = 2^x$$

x is the logarithm to base 2, and is equal to 2.81 in this example.

Note that using base 2, we get

$$\log_2 2 = 1$$
$$2 = 2^1$$

$$\log_2 4 = 2$$
$$4 = 2^2$$

$$\log_2 8 = 3$$
$$8 = 2^3$$

and so on.

7 What is the value of
  a $\log_{10} 2$
  b $\log_2 10$?
8 Evaluate the following, using a calculator where necessary:
  a $\log_2 16$
  b $\log_2 32$
  c $\log_2 20$
  d $\log_2 0.5$
  e $\log_2 0.25$
  f $\log_2 1$

Natural logarithms

It is worth mentioning one special number, e. This has a value of 2.718. We can use this number as a base and, following the usual rules for logarithms, we have

$$\log_e N = x$$

Note that $\log_e N$ can also be written $\ln N$, so

$$\ln N = x$$

Whichever way we write it, the accompanying statement is this:

$$N = e^x$$

$\ln N$ (or $\log_e N$) is called the natural logarithm of N.

One property of a natural logarithm is that the linear relationship

$$\frac{dN}{dx} = kN$$

leads always to

$$\ln\left(\frac{N}{N_0}\right) = kx$$

and so to the exponential relationship

$$N = N_0 e^{kx}$$

Because it is quite common for the rate of change of a quantity with time to be proportional to its own value, that is for $dN/dt \propto N$, the exponential relationship is an extremely useful equation in mathematical modelling of several physical and other systems.

9 Find the following, using a calculator as necessary:
  a $\ln 2.718$
  b $\ln 10$
  c $\ln 1$
10 Sketch straight line graphs to show the linear relationships on which each of the following are based:
  a $N = N_0 e^{-\lambda t}$ (N is the population of a radioactive isotope)
  b $P = P_0 e^{kt}$ (P is an unspecified quantity)
  c $Q = Q_0 e^{-t/CR}$ (Q is the charge on a capacitor)
  d $a = a_0 e^{-kt}$ (a is the amplitude of a vibration).
  e Suggest a quantity that might behave as given by part **b**.

Examination questions

I The activity A of a radioactive isotope was measured. The graph shows how $\ln A$ varied with time.

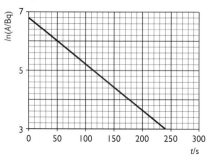

a Use the graph to determine
  i the initial activity of the isotope (1)
  ii the decay constant of the isotope. (2)
b Calculate the half-life of the isotope. (1)

London, AS/A level, Module Test PH2, June 1999

2 The Earth's atmosphere consists of a mixture of gases. Assuming that each behaves like an ideal gas at constant temperature, it is possible to use the kinetic theory to calculate how atmospheric pressure varies with height.

At a height h above sea level the contribution to the pressure p produced by a gas whose molecules are each of mass m is given by

$$p = p_0 e^{-\frac{mgh}{kT}}$$

where p_0 is the pressure exerted by this gas at sea level, T is the absolute (kelvin) temperature and g and k have their usual meanings.

a Show that the joule is the unit of both mgh and kT.
b Show that the height at which the pressure produced by a gas is half the pressure the gas exerts at sea level is given by

$$h = \frac{kT \ln 2}{mg}$$

c Estimate the value of h for nitrogen. The mass of a nitrogen molecule is 4.7×10^{-26} kg. (7)

Edexcel, A level, Synoptic Paper PH6, June 2000 (part)

3 The diagram shows, for a person with normal hearing, the variation with frequency of intensity level at the threshold of hearing.

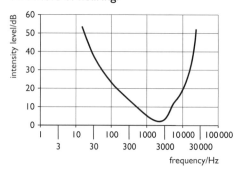

a State the threshold intensity of hearing I_0 [in $W\,m^{-2}$]. (1)
b **i** Use the diagram to find the threshold intensity level [in dB] at a frequency of 50 Hz.
  ii Calculate the intensity [in $W\,m^{-2}$] which corresponds to your answer to **i**. (4)
c Sensitivity s is defined by the equation

$$s = \lg\left(\frac{I}{\Delta I}\right)$$

where $\dfrac{\Delta I}{I}$ is the fractional change in intensity

which the ear can detect at a given intensity.
  At a frequency of 3.0 kHz, the ear can detect a change in intensity level of 0.40 dB when the intensity is $1.0 \times 10^{-6}\,W\,m^{-2}$.
i Show that, for this initial intensity of $1.0 \times 10^{-6}\,W\,m^{-2}$, an increase in intensity level of 0.40 dB would result in a new intensity of $1.1 \times 10^{-6}\,W\,m^{-2}$. (2)
ii Hence calculate
  1 the change in intensity ΔI [in $W\,m^{-2}$] corresponding to the 0.40 dB change in intensity level,
  2 the sensitivity s of the ear at 3.0 kHz and $1.0 \times 10^{-6}\,W\,m^{-2}$. (3)
iii By reference to the diagram, suggest how the sensitivity of the ear changes with frequency. (2)

OCR (Cambridge), A level, 4835, March 1999

4 In the transmission of an electrical analogue signal along a coaxial copper cable, the signal source can simply be connected to the cable and the signal recovered at the other end. When the same signal is transmitted as an analogue signal by optic figure, energy transformations must be performed at both ends of the fibre. This is illustrated in the diagram.

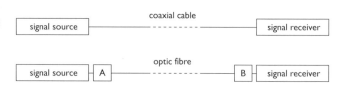

a Name the two transducers A and B which can perform these energy transformations,
  i transducer A at transmitting end
  ii transducer B at receiving end. (2)
b The maximum unbroken length of cable or fibre is governed by the minimum value of the signal-to-noise ratio which is acceptable at the signal receiver.
  i Explain what is meant by *noise*.

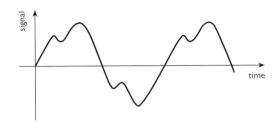

ii The graph shows a signal from the signal source. On the graph, show the effect of noise on this signal.
iii Explain why the signal-to-noise ratio does not remain constant along the cable or fibre. (3)
c In the systems shown in the first diagram, the output power of both signal sources is 480 mW. The minimum acceptable signal power at both receivers is 0.12 nW. The efficiency of each of the energy transducers A and B is 10%.

The ratio of two powers P_1 and P_2 is expressed as a number of decibels (dB) according to

$$\text{number of dB} = 10\lg\left(\frac{P_1}{P_2}\right).$$

i For the coaxial cable, calculate the reduction in signal power [in dB].
ii The coaxial cable has a loss per unit length of 8.0 dB km^{-1}. Calculate the maximum unbroken length of this cable.
iii For the optic fibre, calculate the reduction in signal power [in dB].
iv The optic fibre has a loss per unit length of 0.76 dB km^{-1}. Calculate the maximum unbroken length of this fibre. (5)

UCLES, A level, 4838, March 1998

5 The radiant power loss, P, of a filament bulb is investigated by immersing it in cool water, varying the current through the bulb, and calculating its power output and temperature. The filament temperature, T, and radiated power are recorded in the following table.

Filament temperature, T /10^2 K ±5K	Power radiated, P /W ±0.2W	log T /K	log P /W
5.00	1.7	?	?
6.00	3.3	?	?
7.00	7.1	?	?
8.00	8.9	?	?
9.00	16.0	?	?

The expected relationship between radiated power P, the filament temperature T and water temperature T_w is

$$P = \epsilon A\sigma(T^n - T_w^{\,n})$$

where A is the area of the filament, ϵ is its emissivity, and σ is Stefan's constant. The temperature of the water remains nearly constant at 20 °C throughout the experiment.

In this experiment the final term in the equation, $T_w^{\,n}$, is considered negligible. Taking logarithms, the expected relationship can therefore be written as

$$\log P = \log k + n \log T$$

where $k = \epsilon A\sigma$.
a *Copy* and complete the table. Include uncertainties in the endpoint values for the logarithm of radiated power only. (3)
b Plot a graph of $\log P$ against $\log T$. Include uncertainties in the endpoint values for radiated power only. (6)
c Do these data support the expected relationship? Explain. (2)
d Determine the value of the exponent n. (2)
e n is expected to be a whole number; what is its most likely value? (1)
f If the emissivity is 0.8 and the area of the filament is 1.1×10^{-3} m^2, determine the value of Stefan's constant. (4)
g Is the assumption that the temperature of the water T_w is negligible justified? Explain. (2)

IB, Higher level 2, Paper 430, November 1997

6 Three slide transparencies, each of different but uniform density, are placed in turn in a projector and the images produced on a screen are viewed. The *intensity* of the image produced by the second transparency is ten times as great as that produced by the first, and the *intensity* of the third image is one hundred times as great as that produced by the first.

Given that

response of the eye is proportional to $\log_{10}$ (*intensity*)

complete the table. (2)

Relative intensity	$\log_{10}$ (relative intensity)	Response of the eye
10	1	initial response R
100		
1000		

NEAB, AS/A level, Module Test (PH04), March 1997 (part)

10 Change and rate of change

THEMES AND TOPICS	● Change and rate of change
	● Gradients
	● Graphical representation
	● Calculus representation
	● Mathematical relationships and patterns of behaviour
	● Dimensional analysis

A world of change

We are surrounded by still images. This book is full of them. Such 'snapshots' of moments in time are very useful indeed. But they can quietly mislead us into thinking that the world is like that. It isn't. The camera lies.

We live in a world of change. A world without change is no world at all. A central requirement of physics is that it matches the world. In physics, we have to deal with change.

We often use arrows to show changes in time, such as flows and transfers. Many graphs have 'time' as their axes.

We need mathematics that deals with change. Unfortunately, simple addition and subtraction, multiplication and division, don't do the job. Newton realised that, and so did Gottfried Leibnitz a little later. We now use Leibnitz's way of writing down the ideas, but the principles of their ideas are the same. They developed calculus.

Calculus provides tools for the physicist, and for anyone else who wishes to deal quantitatively with processes of change – chemists, biologists, economists, sociologists, and so on. It provides a way to calculate the gradient of a graph (differentiation), and a way to calculate the area under the graph (integration). In considering rates of change, we use differentiation. For a graph with an *x*-axis that represents time, the gradient of the graph is a measure of rate of change with time.

Analysis of motion

For analysis of motion we can plot graphs of displacement against time, velocity against time, acceleration against time, and so on. In each case the gradient tells us the rate of change. The gradient of a displacement–time graph, for example, is rate of change of displacement, which is also called velocity. The relationships between the quantities can be written as follows:

$$\text{displacement} = s$$
$$\text{velocity} = v$$
$$\text{acceleration} = a$$

$$\text{rate of change of displacement} = \frac{ds}{dt} = v$$

$$\text{rate of change of velocity} = \frac{dv}{dt} = \frac{d^2s}{dt^2} = a$$

We can match these mathematical representations with visual representations (Figure 10.1).

Figure 10.1
Each group of three graphs, **a**, **b** and **c**, shows three representations of the same motion.

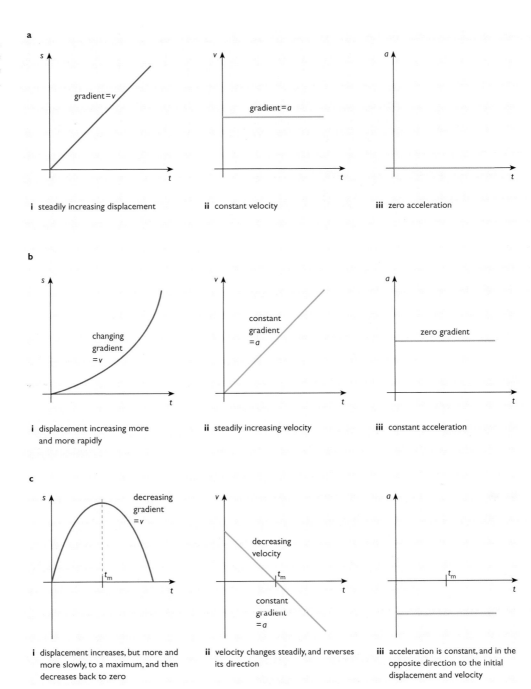

a

i steadily increasing displacement

ii constant velocity

iii zero acceleration

b

i displacement increasing more and more rapidly

ii steadily increasing velocity

iii constant acceleration

c

i displacement increases, but more and more slowly, to a maximum, and then decreases back to zero

ii velocity changes steadily, and reverses its direction

iii acceleration is constant, and in the opposite direction to the initial displacement and velocity

1 What is the significance of the sign (positive or negative) of a value of displacement, velocity or acceleration?

2 Give examples of real motions that would follow, either exactly or closely, the behaviour shown by each set of graphs **a**, **b** and **c** in Figure 10.1.

3 **a** On a single set of axes, sketch displacement–time graphs for motions with different constant velocities.
   b Sketch corresponding velocity–time and acceleration–time graphs.

4 For an acceleration–time graph that is a straight line through the origin, showing a steadily increasing acceleration, sketch a corresponding velocity–time graph.

5 **KEY SKILL** – INFORMATION TECHNOLOGY
   a Collect data on accelerated motion along an air track, in a form that can be entered directly or manually into a computer.
   b Use the data and a computer to plot displacement–time, velocity–time and acceleration–time graphs for the motion.
   c Compare your graphs with those obtained by other students.
   d Use the Internet or CD ROMs to research the work of Galileo in the study of motion.
   e Prepare a presentation suitable for AS Physics students on how your findings relate to Galileo's work.

Energy and power

There are other changes with which we wish to deal. We find it useful, for example, to calculate the rate at which energy is transferred. In fact it is so useful that the rate of energy transfer has its own name – power. The gradient of a graph of energy transferred against time is equal to power (Figure 10.2). If energy is transferred steadily, power is constant, if, however, more energy is transferred in one infinitesimally short period of time than the next, then the gradient of the energy–time graph varies, and power varies.

Figure 10.2
Energy–time graphs, for constant and non-constant power.

The gradient of an energy–time graph is equal to the power:

$$\text{gradient} = \frac{dQ}{dt} = \text{power}$$

where Q is the energy transferred.

a constant power

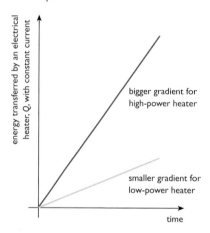

bigger gradient for high-power heater

smaller gradient for low-power heater

The area under a power–time graph is equal to the energy transferred.

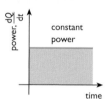

constant power

b decreasing power

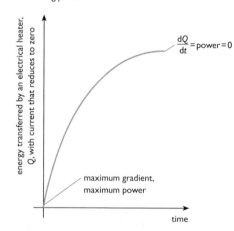

$\dfrac{dQ}{dt} = \text{power} = 0$

maximum gradient, maximum power

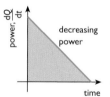

decreasing power

Note that we can also plot power against time (Figure 10.2, inset). Rate of change of power is of interest in some situations, but is not the most important point about such graphs. What is often useful is the fact that area under a power–time graph is equal to energy. Finding the area under a graph can usefully be thought of as the reverse process of finding the gradient; that is, integration can be thought of as the reverse process of differentiation.

Flow processes

Thermal conduction can be thought of as a flow process, with energy as the quantity that flows. Similarly, charge can flow along a wire, and liquid can flow along a pipe. Each of these flows is the result of a physical difference between one point and another. These systems behave in such a way as to tend to minimise difference.

Consider water in two containers, connected at their bases by a pipe with a tap. If the containers are filled to unequal levels, then as soon as the tap is opened flow takes place, and the flow eliminates the depth difference between the ends of the pipe. In doing so it eliminates pressure difference. (Pressure due to a body of liquid is related to depth by the formula $P = \rho g h$.) The rate of flow depends not only on the pressure difference, ΔP, but also on the length of the pipe, Δx. The ratio of pressure difference to length of pipe is the pressure gradient, $\Delta P/\Delta x$ (Figure 10.3).

Figure 10.3
Pressure gradients.

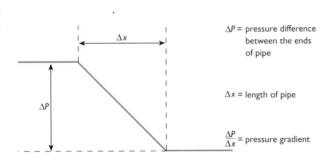

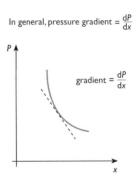

$\Delta P/\Delta x$ is the pressure gradient at any point along the pipe provided that the pressure gradient is the same at all points. In more general terms – that is, whether or not the pressure gradient is the same all along the pipe – pressure gradient is given by dP/dx (Figure 10.3, inset).

Rate of flow is found to be proportional to the pressure gradient. It also depends on the cross-sectional area, A, of the pipe and on physical factors such as the viscosity of the water. The relationship can be written as:

$$\text{rate of flow of water} = \frac{dm}{dt} = kA\frac{dP}{dx}$$

Here k is a constant whose value depends on physical factors of the system, m is mass and we are expressing rate of flow as dm/dt. We could deal with the flow in terms of volume rather than mass:

$$\text{rate of flow of water} = \frac{dV}{dt} = k'A\frac{dP}{dx}$$

Consideration of dimensions and/or units tells us straight away that the constants k and k' must be different.

In electricity it is charge that flows, and the rate of flow is current. The rate of flow depends not on pressure difference but on potential difference in a wire of given material and given proportions:

$$\frac{dQ}{dt} \propto \Delta V \qquad \text{for a given wire}$$

(Note that in dealing with circuits we usually write potential difference ΔV simply as V.)

We can also seek a more generalised relationship – one that applies to any wire of any size and any material. For this we find

$$\frac{dQ}{dt} = \frac{1}{\rho}A\frac{dV}{dx}$$

Here ρ is the resistivity of the material of the wire, A is its cross-sectional area, and dV/dx is the potential gradient (Figure 10.4). The formula is identical in form to that which relates rate of flow of liquid to pressure gradient.

Figure 10.4
Potential gradients.

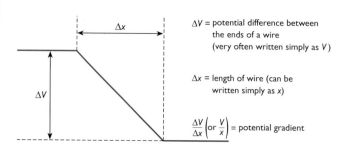

ΔV = potential difference between the ends of a wire (very often written simply as V)

Δx = length of wire (can be written simply as x)

$\dfrac{\Delta V}{\Delta x}\left(\text{or } \dfrac{V}{x}\right)$ = potential gradient

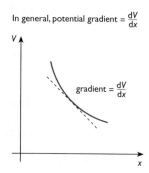

In general, potential gradient = $\dfrac{dV}{dx}$

gradient = $\dfrac{dV}{dx}$

We can also use the expression dQ/dt to represent rate of thermal transfer of energy through a conductor (Figure 10.5). A temperature gradient across the conductor is the condition that leads to the flow, and other factors concerned are the area normal to the flow (through which the flow takes place) and the conducting property of the material, which we quantify in terms of its thermal conductivity. Thus we can say:

$$\frac{dQ}{dt} = kA\,\frac{dT}{dx}$$

where k in this case is the thermal conductivity of the material, A is the cross-sectional area and dT/dx is the temperature gradient.

Figure 10.5
Temperature gradients.

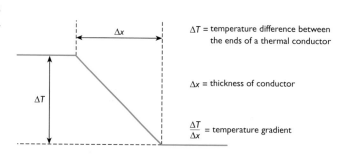

ΔT = temperature difference between the ends of a thermal conductor

Δx = thickness of conductor

$\dfrac{\Delta T}{\Delta x}$ = temperature gradient

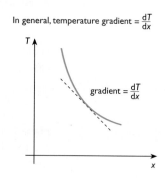

In general, temperature gradient = $\dfrac{dT}{dx}$

gradient = $\dfrac{dT}{dx}$

6 Use dimensional analysis or another method to determine, for liquid flow, how the ratio k/k' is related to the density of the liquid.

7 a Compare $\dfrac{dQ}{dt} = \dfrac{1}{\rho}A\,\dfrac{dV}{dx}$ with $I = \dfrac{V}{R}$

where Q is charge, ρ is resistivity, A is conductor cross-sectional area, dV/dx is potential gradient, I is current and R is resistance.

b It is possible to define a quantity, thermal resistance, which is analogous to electrical resistance. Suggest such a definition.

c Explain how thermal resistance may be useful for considering rates of energy transfer through layers of material (materials 'in series'), such as the layers of a double glazing system.

8 Electrical conductivity is the inverse of electrical resistivity.
a Write an equation relating rate of flow of charge to potential gradient, in terms of conductivity rather than resistivity.
b Write an equation relating electrical conductivity of a material to electrical resistance of the particular wire from which it is made.

9 The temperature difference between the core of a human body and its surroundings is normally constant from one moment to the next. Explain, with reference to types of clothing that can be worn and using sketch diagrams where appropriate, what other factors influence rate of thermal transfer of energy between body and surroundings.

Sinusoidal relationships

The relationship $x = x_0 \sin \omega t$ turns out to be a useful one for the creation of mathematical descriptions of periodic phenomena. The quantity ωt is an angle, which can be called a phase angle, and so it is measured in radians. From this we can see that ω must be measured in radians per second and, in fact, is an angular velocity.

This seems strange because, for a mass on a spring or an electron in an a.c. circuit, motion is in a straight line – there seem to be no angles involved. Yet their motion can be considered to be the straight line projection of circular motion – the shadow on a flat surface of a body moving in circular motion, and lit by a distant source of light, moves in a way that is the same as that of the mass on the spring. It is to this 'matching' circular motion that the angular velocity applies (Figure 10.6).

Figure 10.6
Circular motion and linear projection.

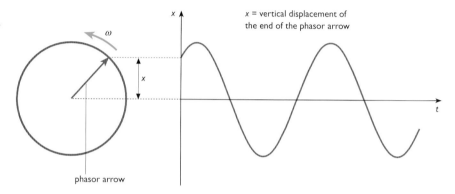

Now we know that velocity v is related to displacement x by $v = \mathrm{d}x/\mathrm{d}t$. Differentiation of the relationship between x and t yields $\mathrm{d}x/\mathrm{d}t$. That is, if

$$x = x_0 \sin \omega t$$

then

$$\frac{\mathrm{d}x}{\mathrm{d}t} = x_0 \omega \cos \omega t$$

So the relationship relating displacement with time leads to a relationship relating velocity with time.

We also know, of course, that the value of the velocity at any time can be found from the gradient of the displacement–time graph. We can plot the graphs to see this (Figure 10.7).

Figure 10.7
Displacement, velocity and acceleration versus time.

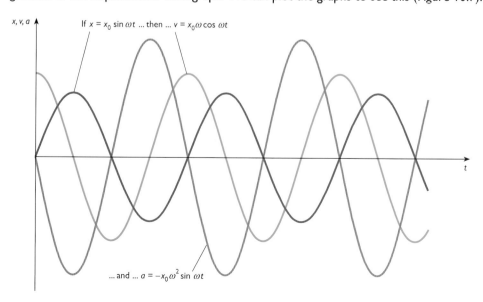

Acceleration is equal to d^2x/dt^2, and is also equal to the gradient of a velocity–time graph (Figure 10.7). Differentiation of the relationship

$$\frac{dx}{dt} = x_0\omega\cos\omega t$$

gives

$$\frac{d^2x}{dt^2} = -x_0\omega^2\sin\omega t$$

10 a Explain how the quantity ω used to describe oscillating motion is related to period and frequency.
   b Give the following formulae in terms of frequency, f, instead of ω:
   i $x = x_0\sin\omega t$
   ii $v = x_0\omega\cos\omega t$
   iii $a = -x_0\omega^2\sin\omega t$
   c In words, describe the frequency dependence of amplitude and peak acceleration for a body in simple harmonic motion.
11 a Sketch graphs to illustrate the following relationships:
   i $x = kt$
   ii $x = mt - c$

   iii $\dfrac{dx}{dt} = kt$ (sketch graphs of dx/dt against t and x against t)

   iv $\dfrac{dx}{dt} = kx$ (sketch graphs of dx/dt against x, and x against t).

   b For each graph, give an example of a physical situation to which it might apply.
12 a Sketch graphs to show variation of displacement of a mass on a spring and current in an a.c. circuit with time. Annotate each graph to show how it relates to frequency.
   b What are the relationships between

   i $\dfrac{dx}{dt}$

   ii $\dfrac{dI}{dt}$
   and frequency?

Exponential changes

13 What is the relationship involving rate of change of a quantity that leads to an exponential decay curve?

A quite simple relationship, proportionality, between a quantity and its rate of change with time leads to an exponential relationship. You can read about these relationships in Chapters 9 and 16.

Examination questions

1 The graph shows the variation of temperature with time for the contents of a freezer during a power cut. The contents warm up and start to defrost at time T.

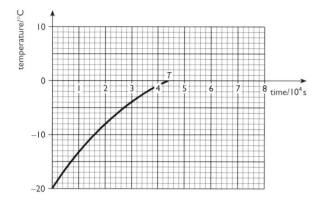

a Heat transfer through the walls of the freezer happens by thermal conduction. State *two* ways in which the rate of heat transfer can be reduced. (2)

b Show that the rate of change of temperature for the curved part of the graph at the time T ($= 4.4 \times 10^4$ s) is approximately 2.3×10^{-4} K s^{-1}. (3)

c Calculate the rate of transfer of energy into the freezer at time T. (3)

  Mass of freezer contents $= 30$ kg
  Specific heat capacity of
    freezer contents $= 2.1$ kJ kg^{-1} K^{-1}

d Estimate the time taken, from the time T, for the contents of the freezer to defrost. (3)

  Specific latent heat of freezer contents $= 330$ kJ kg^{-1}

AEB, A level, Paper 2, Summer 1997 (part)

2 Throughout, neglect air resistance and take the value of the acceleration of free fall, g, as 10 m s^{-2}.

a A ball is dropped from rest from a height of 80 m. Calculate the time taken to reach the ground.

b A ball is thrown horizontally at 20 m s^{-1} from the top of a cliff 80 m high. It falls into the sea. Showing appropriate numerical values, draw graphs of

  A the horizontal component of velocity, v_H, against time, t,

  B the vertical component of velocity, v_V, against time, t,

  C the height of the ball above the sea, h, against time, t. (6)

NEAB, A level, Paper 1, Section B, June 1997 (part)

3 The activity (disintegration rate) of a sample of a radioactive isotope of iodine is 3.7×10^4 Bq.

  48 hours later the activity is found to be 3.1×10^4 Bq.

a Show that the decay constant is about 1.0×10^{-6} s^{-1}.

b Calculate the half-life of the iodine isotope. (4)

NEAB, AS/A level, Module Test (PH02), June 1998

4 An astronaut conducts a dynamics experiment on the surface of the Moon where there is no air resistance. In the experiment, a ball is dropped from a height h on to a flat horizontal surface. A graph of the variation with time t of the vertical velocity v of the ball, from its time of release until it strikes the horizontal surface, is shown in the graph.

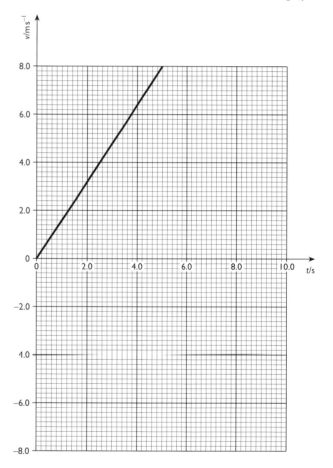

a Define *velocity*. (2)

b Use the graph to

  i show that the height above the surface from which the ball is dropped is 20 m,

  ii calculate the acceleration of free fall near the surface of the Moon. (4)

c i Use your answer in **b** to calculate the change in potential energy of the ball [in J], of mass 120 g, during the whole of its descent.
ii State the value of the kinetic energy of the ball [in J] just before impact with the horizontal surface. (3)
d The ball rebounds from the horizontal surface and, during the process, loses 25% of its kinetic energy.
i Calculate the speed [in $m\,s^{-1}$] at which the ball leaves the surface.
ii On the graph, show the variation with time of the velocity of the ball as it rebounds from the surface and rises to its new maximum height. You may assume that the time of impact of the ball with the surface is negligible. (5)

UCLES, AS/A level, 4830, March 1998

5 An athlete of mass 80 kg completes a 1500 m race in a time of 4.0 minutes. For each stride during the race, the centre of mass of the athlete rises and falls through a vertical distance of 15 cm. The average length of each stride is 1.8 m.
a Show that the average rate of working against gravity during the race is about 400 W. (3)
b Suggest why the total useful power output of the athlete's muscles during the race may be different from that in **a**. (3)
c The total output power for this athlete is 1600 W. The muscles are working with an efficiency of 25% and the remaining 75% is converted into thermal energy.
i Calculate the rate of production of thermal energy [in W] for this athlete during the race.
ii Perspiration and its evaporation remove 40% of the thermal energy generated in the muscles during the race. 2300 J of thermal energy are required for the evaporation of 1 g of sweat. Calculate the mass of sweat [in g] evaporated from the athlete as a result of the race. (5)
d Explain why, at the end of marathon races, athletes are sometimes wrapped in foil. (2)

OCR (Cambridge), A level, 4835, June 1999

6 For each of the statements below indicate whether the statement is true or false.

The gradient of a displacement/time curve is the acceleration at that instant.

The acceleration of a ball which is thrown upwards is a maximum at the instant the ball changes direction.

For a body to move at constant speed along a circular path, the centripetal force must be constant.

An equation which is homogeneous must be correct. (4)

London, AS/A level, Module Test PH1, January 1996

7 Complete each of the following statements with a single word:

The rate of change of displacement is called…
The rate of flow of charge is called…
The rate of doing work is called…
The rate of change of momentum is called… (4)

London, AS/A level, Module Test PH1, June 1998

8 a The half-life of bismuth-211 is 130 seconds. Using the axes below, *sketch* a graph representing the decay of a sample of bismuth-211 with an initial activity of A_0 during a period of 400 seconds. (2)

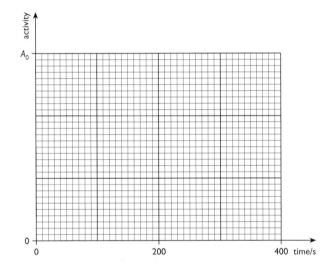

b As the sample of bismuth-211 decays, thallium-207 is produced. Thallium-207 is unstable and decays with a half-life of 290 seconds to form lead-207. Using another set of axes like those above, *sketch* a graph to represent the activity owing to thallium-207 present during the 400 second period. Assume that the sample initially contains only bismuth-211 atoms. (2)

UODLE, A level, Paper S2, March 1998

Principles of imaging

THEMES AND
TOPICS
● Visual representation of data
● Vision
● Light–matter interaction
● Technologies – applications of static electricity, semiconductor behaviour, photoelectric effect, electron microscopes

Visual representation of data

The business of making images raises some deep questions about the world and about ourselves. First of all, it says a lot about our species that we like images so much. We much prefer a picture to a string of numbers – a stream of data – that carries the same information. Often, huge amounts of information can be expressed as a picture, providing an almost instant understanding of what is going on. The picture might simply be a sketch diagram or a graph, or it could be, for example, an EEG (electroencephalograph) image of the brain. Here, the electrodes pick up electrical activity in different parts of the brain. The voltages themselves are just numbers. A list of numbers is not just uninteresting – it is almost impossible to extract meaning from it. As a picture – either as a single trace or as a colourful reconstruction of activity across the brain – the data come alive to us. The data do not change but they mean more.

Figure 11.1
An EEG line trace. A graph has the benefit of relative simplicity, and is an ideal way of making a visual record of the change of one variable over time.

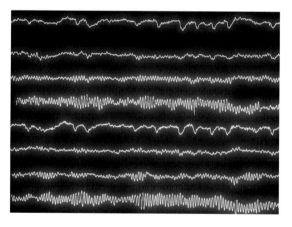

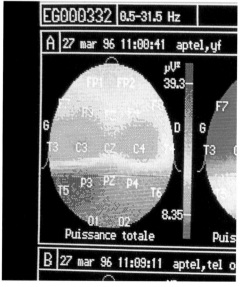

Figure 11.2 (right)
A false-colour EEG of the whole brain, showing spatial variation in activity. Colours can represent quantities – usually in a crude way, with one colour representing a range of values. But the information is easily and quickly understood.

1 A map of the world is a visual representation of reality.
  A seismograph is produced by sensing ground vibrations, usually recording these with a pen past which a roll of paper continuously moves.
  a Compare these with different types of EEG images.
  b In what ways do the map and the seismograph complement each other?
2 A local map carries information. What are the similarities and differences between a local map and a graph?

There are two principal types of EEG image (Figures 11.1 and 11.2). One is a line trace, which is produced by a pen responding to voltages as it moves steadily across a strip of paper. It is an excellent guide to what is happening moment by moment. The other is a representation of the whole brain area, showing which locations are particularly active during a limited period of time. Here, voltage levels can be represented by colours. The colours are easy for the human mind to interpret.

The first of these types of visual representation, or image, provides temporal information (what is happening from one time to the next) and the second provides spatial information (relative intensities of activity from place to place).

Some light–matter interactions

Emission of light

Electromagnetic radiation emerges from matter in a number of ways – such as by radioactive (gamma) emission, by atomic excitation and electron transitions, and by electrical interaction that results from vibration either randomly, as in thermal emission, or in a prescribed way, as in radio emission.

Absorption of light by the retina

Vision involves processes of interaction of light and matter. Light strikes chemicals in the cells that coat the inner surfaces of the eyes, and it induces reversible changes. Enough of these changes happening together create the pulses of in–out ion flow that travel along the fibres of nerve cells to the back of the brain.

The chemicals in the cells of the retina have limited sensitivities, especially in terms of the ranges of frequencies (or wavelengths) to which they respond (Figure 11.3). They are sensitive to the light that travels most easily through the earthly atmosphere, which is the environment in which vision evolved. Different types of chemical response take place in different kinds of cells – rods and cones. Three types of cone cells respond to three different wavelength ranges. As a result we see colour, which, as is shown by the EEG scan (Figure 11.2), is useful.

Figure 11.3
Frequency sensitivities of
a rods and **b** cones.

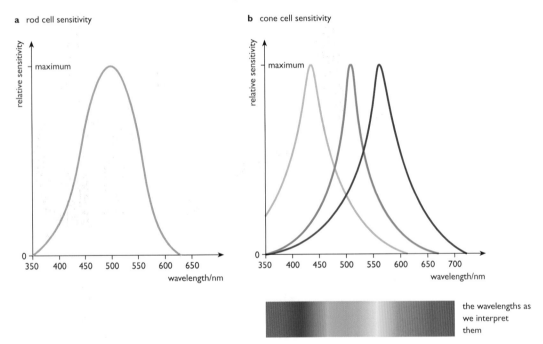

Reflections and absorptions by other surfaces

Light–matter interactions do not happen only at our retinas, of course. Light is not capable of standing still, and having emerged from a source then, unless it is in a vacuum, further interactions take place. For example, it is by reflection that surfaces become visible, and we tell one kind of surface from another. At each of the reflecting surfaces that surround us there is usually significant change in observable characteristics of the light that hits and then leaves it. Very often, the reflections and absorptions are frequency dependent. Then we describe the surface of the matter as having colour.

Light that enters our eyes from each direction tells us something about the interactions it has experienced on its journey from its source, and so it tells us about the matter that is there. Our sense of touch usually confirms our sense of vision – we can experience surfaces not only by their reflection of light but by electrical interactions between particles in the surface and in the nerve endings in our skin.

3 When looking at a building, such as an office block, how can you tell whether the windows are closed or open?

Imaging technologies using light–matter interactions

Some light–matter interactions considerably alter the material as well as the light. Light-sensitive materials are chemicals that are changed, usually permanently, by the light that falls on them. Many surfaces are light-sensitive. Paper is made yellow, slowly, by exposure to light. Many pigments similarly change, making the care of paintings, whether in caves or in art galleries, very challenging. Some chemicals change quite rapidly, so that a layer (or 'film') of chemicals in a camera can be sufficiently changed by an exposure time of, say, a millisecond to produce an image that our eyes accept as a clear (if static) representation of the world. Photocopiers, fax machines and TV cameras work rather differently.

A photocopier (Figure 11.4) relies on static electric charge. Bright light is reflected from a sheet of paper to be copied, and falls on the main drum inside the machine. The main drum carries an electric charge, but is coated with photoconductor material – one that only conducts electricity when exposed to light. It is a semiconductor material, and the light provides the energy needed for electrons to transfer across the 'energy gap' and become free. Areas exposed to light reflected from the paper lose their charge – leaving charge only on the areas from which a low intensity of light was reflected – the areas covered in ink. These areas attract the toner powder, just as a rubbed rule attracts scraps of paper, and transfer it to a new sheet of paper. The toner contains both the black 'ink' and a plastic material, and when the sheet is then heated the plastic softens and adheres to the paper.

Figure 11.4
How a photocopier works.

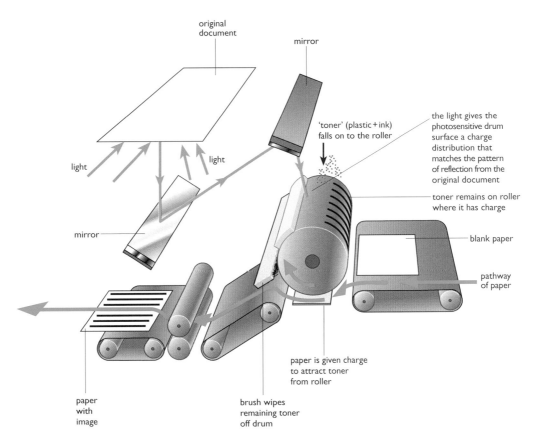

In a fax machine, light reflected by a sheet of paper is detected by a row of about 2000 semiconductor sensors – rather like tiny light-dependent resistors. Each of these is connected into a system which provides outputs that depend on whether the small area of paper that reflects light to each sensor is dark (that is, low reflection) or bright (high reflection). Thus the machine, in reading the level of reflection at each small area in turn, creates a stream of digital signals.

TV cameras use CCDs, or charge-coupled devices. These are arrays of small squares or pixels of semiconductor material from which electrons are liberated by photoelectric emission, and collected on electrodes. Periodically the charge gathered on each small electrode is attracted from it by a burst of applied potential difference. The current that flows during this burst is a measure of the number of electrons set free by photoelectric emission and hence of the light that fell on the pixel since the charge was previously removed from it.

A photomultiplier tube (Figure 11.5) also uses the photoelectric effect to begin a process of creating a detectable signal from incident light. Low-intensity light causes liberation of an electron from a photocathode, and the electron is then accelerated to a metal plate or dynode where the collision sets more electrons free. These newly freed electrons are accelerated towards a second dynode, where the number of free electrons increases again. The end result is a current that is big enough to be detected and used as a signal.

Figure 11.5
How a photomultiplier tube works.

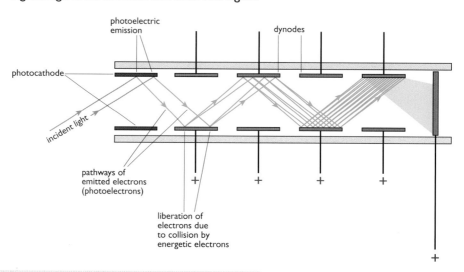

4 Which processes of emission of light are involved in:
   a a radio transmitter
   b a TV screen
   c a discharge tube
   d a hot filament
   e the Sun?
5 Name three uses of photographic film in physics. In each case, what benefits over the human eye does photographic film provide?
6 A photomultiplier tube and a microphone can both be considered to be transducers. Compare them in terms of their inputs and outputs.
7 When using photographic film, the duration of exposure of the film to the radiation is very important in determining the image quality.
   a What happens if the film receives too much radiation?
   b What happens if it is not exposed enough?
   c In what ways, other than exposure time, can image brightness be controlled?
   d What is the relevance of spherical aberration (see *Introduction to Advanced Physics*, page 224) to this?

Imaging technologies using matter–matter interactions

An electron microscope relies on the interaction of matter (the electrons) with matter (the object that is being imaged). Light only then becomes important in detecting the electrons and in making the interaction interpretable to human eyes. Electrons that have interacted with the object hit a fluorescent screen, as in a TV, but unlike in a TV the flashes in the electron microscope do not have sufficient intensity to produce a directly visible image. An image is produced with the help of photomultipliers, which generate pulses of electrical current corresponding to the flashes of light on the screen. It is these pulses that are used to induce emission of light by a computer screen. Figures 11.6 and 11.7 show the kinds of images that may be obtained.

Figure 11.6
A transmission electron microscope produces images of slices of material.

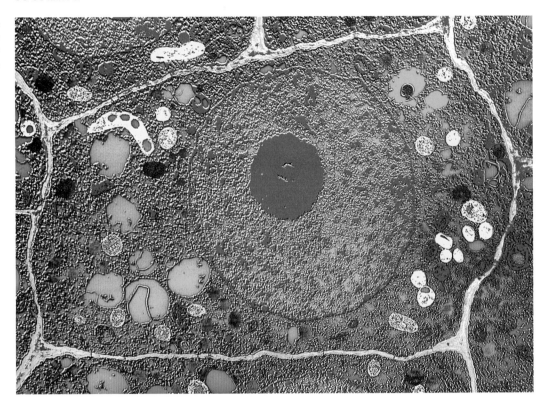

Figure 11.7
Once gathered and stored, data from an electron microscope can be further manipulated. Here, many electron micrographs have been made of slices through two nerve cells, and these separate images have been put together by computer to produce this particularly informative image, with false colour added to highlight features. The light blue structure is part of one cell and the yellow is another.

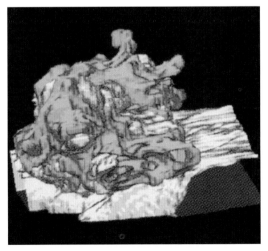

The two nerve cells (neurons) have been imaged at the points at which there is a very small gap (synapse) between them, in order to study how nerve cells pass signals from one to another. Chemicals pass across the gap from cell to cell.

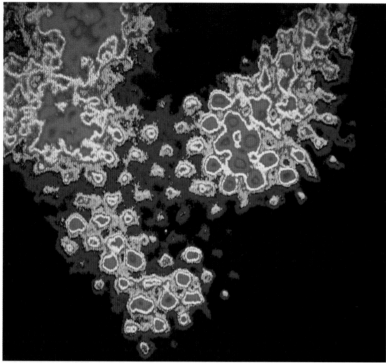

Figure 11.8
A tunnelling microscope image showing the pattern of individual surface atoms.

All interactions of matter with matter are due to fields – electric fields and gravitational fields at the macroscopic or human scale, and strong and weak interactions at the nuclear scale. Tunnelling microscopes illustrate interaction by the electric field. A probe passes over a surface, and an applied potential difference tends to encourage electron transfer between them. But electrical forces also hold the electrons in their atoms, and these forces are dominant. However, it is possible for electrons to escape from an atom of one material, into an atom of the other, by the process of quantum tunnelling. The principle of this is similar to the tunnelling effect that allows some nuclei to emit alpha particles. The movement of electrons is detectable and can be used to generate an image, as shown in Figure 11.8.

Matter–matter interaction can also be due to gravitational interactions. It is possible to measure gravitational field strength to sufficient precision to be able to detect local variations due to underlying rock structures. The data can then be entered into a computer, which can produce an image of the landscape – a gravitational map (Figure 11.9).

Figure 11.9
A gravitational map showing variations in the local value of gravitational field strength at the Earth's surface.

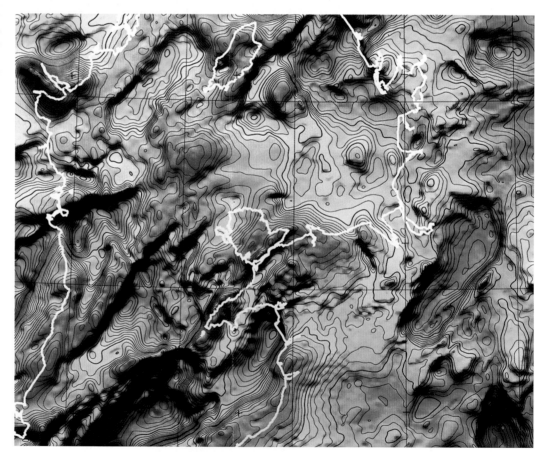

A summary of the principles of producing images is shown as a flowchart in Figure 11.10.

Figure 11.10
There is more to imaging than meets the eye!

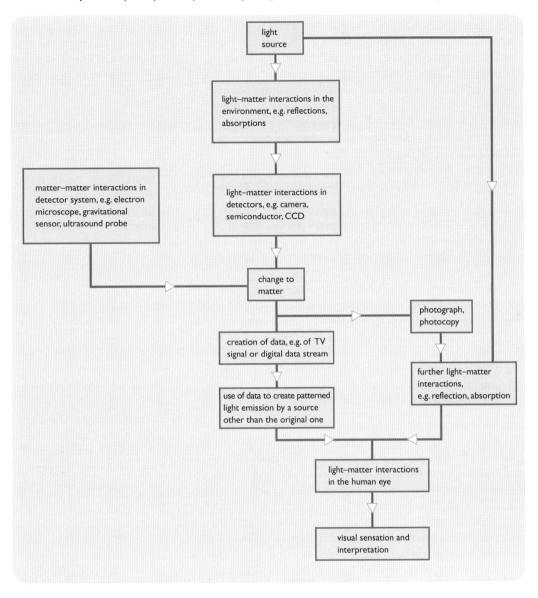

8 **a** Examine an enlarged photograph and a CCD image, such as the image produced by the Hubble Space Telescope, page 183, with a magnifying glass. Both of the images you look at will have been affected by the printing process as well as by the degree of enlargement, so direct comparison has only limited meaning. Nevertheless, compare in general terms the different image recording processes.
   b Why do astronomers usually prefer CCDs to photographic film?
9 Describe the interactions between light and matter and between matter and matter that occur in the production of:
   a photocopier images
   b transmission electron microscope images
   c CCD images.

10 **DISCUSS**
   a Suggest how, for people living in pre-agricultural communities (more than about 6000 years ago) colour vision might have provided an aid to survival.
   b Suggest how our present-day understanding of the world is influenced by our natural ability to distinguish light frequencies.
11 **a** What processes initiate the creation of images in high-energy particle physics?
   b Describe one way in which a permanent visual record of particle interaction events is made.
12 **a** To what extent can the tunnelling microscope image in Figure 11.8 be considered to be a picture of atoms?
   b Does the image make particle theory of matter significantly more acceptable than it would be without the image?
13 In what ways can use of images help you to maximise your personal physics examination performance?

Examination questions

1 Late one night, a student was observing a car approaching from a long distance away. She noticed that when she first observed the headlights of the car, they appeared to be one point of light. Later, when the car was closer, she became able to see two separate points of light. If the wavelength of light can be taken as 500 nm and the diameter of her pupil is approximately 4 mm, calculate how far away the car was when she could first distinguish two points of light. Take the distance between the headlights to be 1.8 m. (5)

IB, Higher level 3, Paper 430, Specimen Paper

2 In order for the unaided eye to detect a distant source, the rate of energy arriving at the eye from this source must be 1.0×10^{-16} W. The energy of each photon arriving at the eye is 3.6×10^{-19} J and the light emitted by the source has wavelength 550 nm.

a Calculate the number of photons arriving at the eye per second if the eye can just detect the source.

b Estimate the number of photons which are detected by the retina per second in order for the source to be observed.

c Give *one* reason why fewer photons reach the retina than are incident on the pupil of the eye. (3)

NEAB, AS/A level, Module Test (PH04), March 1998 (part)

3 a State the *two* types of photoreceptors found in the eye.

b State how the two photoreceptors differ from each other in their response to the intensity of light falling on them.

c Hence, explain why stars of low apparent brightness, when observed by the unaided eye, appear to have no colour, but can be seen to be of different colours when viewed through a telescope. (4)

NEAB, AS/A level, Module Test (PH04), June 1998 (part)

4 Optical microscopy, electron microscopy and X-ray diffraction are methods commonly used to investigate the microstructure of materials. For each of the techniques, state *one* property of the microstructure that may be investigated and *one* limitation of the technique.

a Optical microscopy
b Electron microscopy
c X-ray diffraction (6)

NEAB, AS/A level, Module Test (PH05), June 1998

5 In the course of diagnosis and treatment several images are required of a child's broken arm. Similarly, to check the progress of a woman's pregnancy, several images or 'scans' of the foetus are required. *In each case*, state which imaging technique would probably be used and give *two* reasons for the choice.

a Child's broken arm
b Foetus (4)

NEAB, A level, Paper 2 Section B (PH07), June 1997

6 The figure shows the tip of the probe of a scanning tunnelling microscope a few nanometres above a horizontal metal surface. The tip is at a constant potential of -1.0 V relative to the surface.

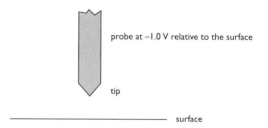

a Explain why electrons pass from the tip to the surface even though there is a gap between the two. (3)

b Electrons move across the gap, creating a current which is amplified by an electronic amplifier. The tip was moved a small distance horizontally above the surface at constant height. The graph shows how the current changed with the horizontal displacement of the tip.

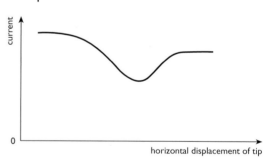

Explain why the current was not constant and describe how the surface varies along the line covered by the tip. (3)

NEAB, A level, Paper 2 Section B (PH09), June 1997

Visual models of energy

THEMES AND TOPICS

● Energy and potential
● Energy–separation graphs in different fields
● Energy levels in atoms and nuclei
● Energy transfers
● Laws of thermodynamics

● Energy–distance graphs

We know from Chapter 3 that the electrical potential energy of a two-body system is given by

$$W = \frac{1}{4\pi\epsilon_0} \frac{qq'}{r}$$

A plot of potential energy against separation of the bodies, r, has a familiar shape, as shown in Figure 12.1.

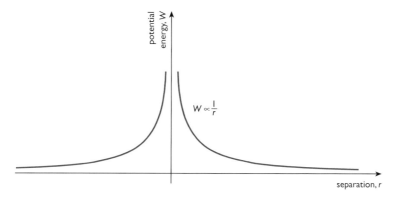

Figure 12.1
PE–separation graph for two bodies of like charge.

$W \propto \frac{1}{r}$

Where only one body is present the system does not have potential energy, but every point in the field of a charge q has 'potential', V:

$$V = \frac{1}{4\pi\epsilon_0} \frac{q}{r}$$

A graph of potential against separation r has the same general shape as the graph for potential energy (Figure 12.1).

The graphs have particular use because they link to everyday experience. Work must be done to roll a heavy object – a log, or a barrel, for example – up a hill. The steeper the hill, the 'harder' the work. Likewise, two charges repel each other, so that work must be done to push them closer together – and the closer together they get, the steeper the graph, and the 'harder' we must work. The shape of the energy–separation curve is sometimes called an energy hill. (For a potential–separation graph we use the terms 'potential hill' or 'potential barrier'.)

Also from Chapter 3, we know that

$$\frac{dV}{dr} = -E, \text{ where } E \text{ is field strength} \quad \text{and} \quad \frac{dW}{dr} = -F, \text{ where } W \text{ is potential energy and } F \text{ is force}$$

That is, force is equal to the gradient of the potential energy–separation graph.

If one of the charges, q or q', is negative, then the force is attractive. The graph of potential energy against distance is rather different (Figure 12.2).

VISUAL MODELS OF ENERGY

Figure 12.2
PE–separation graph for two bodies of opposite charge.

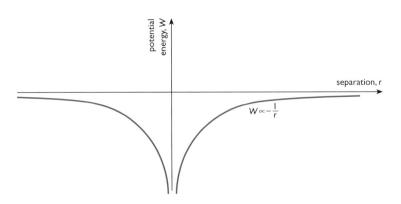

The same ideas can be applied to gravitational fields, and they can be related to the effects of the strong nuclear force that influences nucleon behaviour. Consider the examples shown in Figures 12.3 to 12.6.

Figure 12.3
PE–separation graph for a pair of atoms.

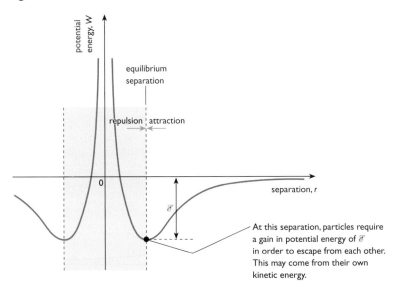

If separation increases above the equilibrium value, then the particles attract; if it is less than the equilibrium value, then they repel. The result is oscillation (unless the particles have enough energy to escape from each other). Note that the curve is symmetrical only for separations close to the equilibrium value. For particles oscillating with large amplitude, the oscillation is asymmetrical and their *mean* separation increases. For collections of large numbers of atoms, increase in amplitude of oscillation results in expansion of the material.

Figure 12.4
PE and alpha emission from a nucleus.

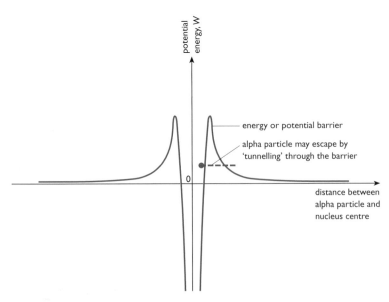

Even an alpha particle that is already at some distance from the nucleus (such as the one shown) must gain considerable extra energy in order to climb over the energy (or potential) barrier and escape. This is very unlikely. Still unlikely, but less so, and allowed by the Heisenberg Uncertainty Principle of quantum mechanics, is that the alpha particle disobeys the principle of conservation of energy for a short time – long enough for it to travel 'through' the barrier. The phenomenon is called 'tunnelling', and without it there would be little alpha emission.

Figure 12.5
PE and orbits in a planet's gravitational field – two representations of the same reality.

a potential energy versus separation

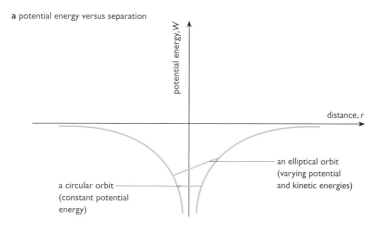

a circular orbit (constant potential energy)

an elliptical orbit (varying potential and kinetic energies)

b potential energy versus distance in two directions

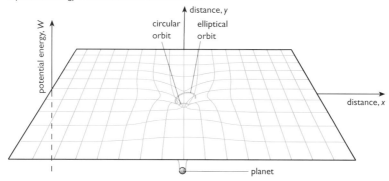

circular orbit

elliptical orbit

planet

Figure 12.6
The potential energy of a ball rolled into this structure behaves very much as an object in space does, because the curvature of the surface is the shape of a PE–distance graph.

1 Sketching diagrams and graphs is an important communication and examination skill, and one that improves with practice. Refer back to Chapter 3 for help in sketching a force–separation graph for a pair of atoms. Also sketch a potential energy–separation graph, using the same scale for the separation. At first sight the graphs have similar shapes. Highlight the differences between them.

2 **a** Sketch a graph to show how the potential energy of an astronaut varies along a line that joins the surface of the Earth to the surface of the Moon.
   b On your graph, show the point at which the astronaut experiences zero net force.
   c Explain why the astronaut's potential energy is not zero at this point.

3 **a** A comet, such as Hale-Bopp or Halley's, has a very elliptical orbit. Show such an orbit on a potential energy against distance graph.
   b Use the graph to describe the changes in kinetic energy experienced by a comet during its orbit of the Sun.

4 Explain why many more alpha particles can escape by tunnelling than by 'climbing' over the top of the energy barrier.

5 **a** Nuclei can absorb neutrons much more easily than they can absorb protons. Sketch graphs and annotate them to explain why.
   b Nucleon emission is a rare form of radioactive change. Predict whether you would expect more nuclides to be natural neutron emitters than natural proton emitters.

Modelling energy levels

Figure 12.7
An energy level diagram, showing a typical set of atomic electron energy levels.

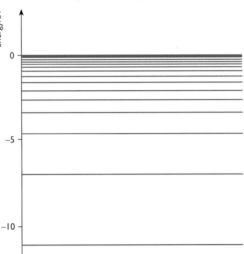

A bar chart is a familiar type of graph which represents the value of a single variable. For some data we don't even have to draw bars with any thickness, but can represent quantities as levels on a single axis. This becomes useful for considering electrons in atoms (Figure 12.7).

We use similar diagrams to show energy changes experienced by a mass of material in a chemical reaction (Figure 12.8). Also, at a very different scale, energy levels provide a description of internal states of nuclei and how these relate to gamma emission by nuclei following emission of alpha or beta radiation (Figure 12.9).

Figure 12.8
Representations of total potential energy changes in chemical reactions.

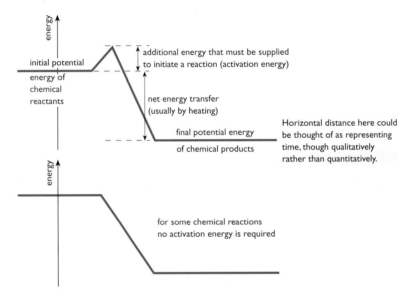

Figure 12.9
The energetics of the decay of iodine-133.

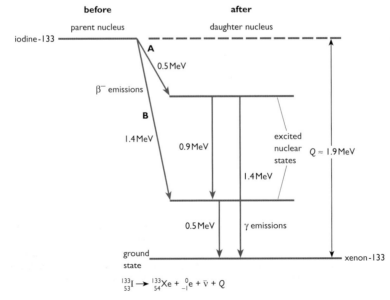

Iodine-133 can decay by emitting beta⁻ particles of two energy ranges:
A up to 0.5 MeV;
B up to 1.4 MeV.
In either case the xenon nucleus that results is in an excited state. It 'falls' to its ground state by emitting gamma radiation, the photon energies of which are as follows:
A 0.9 MeV followed by 0.5 MeV; *or* 1.4 MeV;
B 0.5 MeV.

Some pathways of electrons in atoms can be described as circular, although many cannot. By plotting the energy–separation curve and superimposing on this the atomic energy level diagram for those orbits that *are* circular, we can see how the radii correspond to energy levels (Figure 12.10).

Figure 12.10
The circular orbits of the electron in a hydrogen atom related to the three lowest energy levels.

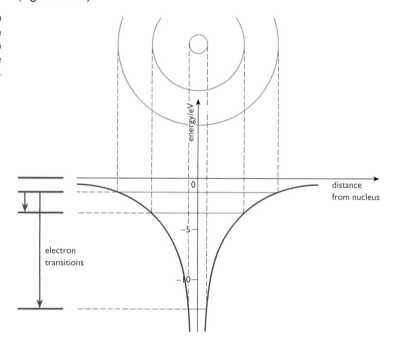

6 Figure 12.11 shows the energy levels for electrons that are available in a many-atom system as the separation of the atoms decreases.

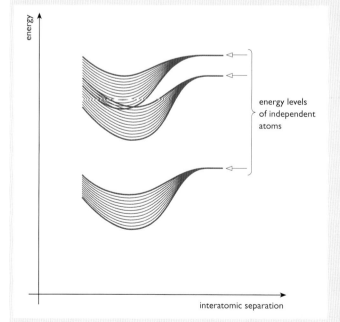

Figure 12.11

Explain how the 'splitting' of energy levels is related to the conducting (electrical and thermal) abilities of the element.

7 Figure 12.12 is an example of a decay process that results in emission of gamma radiation of several different possible photon energies.

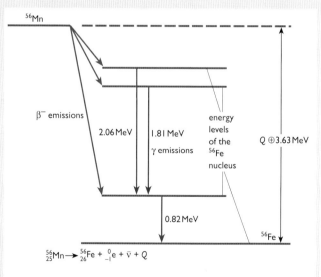

Figure 12.12

a Use the information to sketch a 'line spectrum' showing the frequencies of the photons, similar to an emission spectrum that can be produced for light emitted by atoms due to electron transitions.
b What are the principal differences between this line spectrum and that due to atomic transitions of a typical element?

Modelling energy transfers

Diagrams can be used to show energy that is transferred for a required purpose, and energy that is, as part of the same processes, transferred in a way that has no use to us. We can do this, for example, for an electric fire or an electric light bulb (Figures 12.13a and b). We can also show the relative amounts of energy that are transferred thermally (by heating) and mechanically (by doing work), as in a power station turbine (Figure 12.13c).

Figure 12.13
Energy transfer diagrams.

a

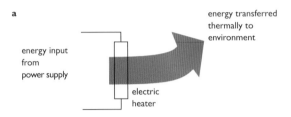

energy input from power supply

energy transferred thermally to environment

electric heater

Since the requirement here is for thermal transfer, we can think of all of the energy transfer as useful.

b

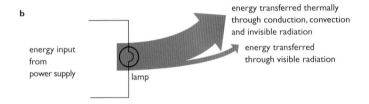

energy input from power supply

lamp

energy transferred thermally through conduction, convection and invisible radiation

energy transferred through visible radiation

Here, only a small proportion of the energy is transferred usefully.

c

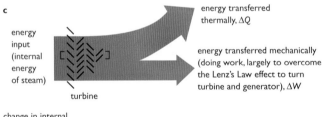

energy input (internal energy of steam)

turbine

energy transferred thermally, ΔQ

energy transferred mechanically (doing work, largely to overcome the Lenz's Law effect to turn turbine and generator), ΔW

change in internal energy, ΔU

$\Delta U = \Delta Q + \Delta W$

Since
- the steam *loses* internal energy,
- energy is transferred *from* it thermally, and
- energy is transferred *from* it mechanically,

ΔU, ΔQ and ΔW will here all be numbers with negative values.

For the power station turbine, the diagram shows total energy transfer, which is the change in internal energy of the steam, and also shows energy transferred thermally and mechanically. It can provide a useful visual representation of the First Law of Thermodynamics as it applies to the steam.

A turbine is a heat engine. It uses a hot energy source and a cold energy sink. It then partially hijacks the thermal energy transfer between the two and transfers some of the available energy mechanically. This can be represented as in Figure 12.14.

For a refrigerator, thermal transfer of energy takes place from a cooler to a hotter body, which is the reverse of the more usual process. It is only made possible by doing work on the system. A visual representation of the energy transfers again provides the best foundation for understanding the system (Figure 12.15).

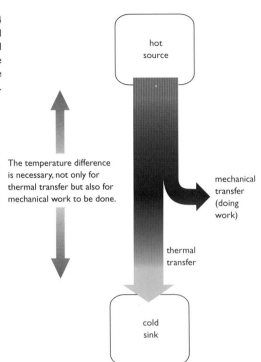

Figure 12.14
A heat engine: the natural tendency for thermal transfer of energy to take place also allows some mechanical transfer.

The temperature difference is necessary, not only for thermal transfer but also for mechanical work to be done.

hot source

mechanical transfer (doing work)

thermal transfer

cold sink

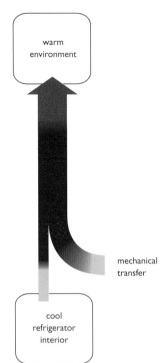

Figure 12.15
A heat pump: some thermal transfer can take place in opposition to the natural tendency by mechanical transfer of energy to the system.

warm environment

mechanical transfer

cool refrigerator interior

---

8 a Draw a diagram showing energy transferred by a gas fire that heats a room but also heats air in a chimney. Show the relative amounts of energy transferred usefully and dissipated immediately in a non-useful way.
b Do the same for a car engine cylinder.
c In each case, to what extent do these relative amounts correspond to the quantities ΔW and ΔQ?

9 KEY SKILL – COMMUNICATION
Read the following statements A–D:

A For a body to maintain a constant mean temperature without any work being done, energy must transfer to it at the same mean rate as energy transfers away from it.
B For the Earth, the rate at which energy transfers to it is fixed (by the nature of the Sun, the Earth–Sun distance, and so on). However, changes to the atmosphere have the potential to change the rate at which energy transfers back out to space at a given temperature.
C An increase in temperature of the Earth (relative to surrounding space) can cause energy to transfer outwards more quickly.
D Equilibrium between rates of energy transfer to and from the Earth can be re-established once atmospheric change ceases, but at a changed temperature.

a Sketch diagrams to illustrate statements A, B, C and D. Use your diagrams to explain the following:

i why a small one-off change in the atmosphere may result in the Earth's mean temperature stabilising at a new, higher, temperature
ii why continuing change to the atmosphere may result in continuing temperature instability.
b Use the Internet to research mean global temperature and climate stability.
c Take part in a group discussion about the physics of climate change. Points that might be covered include:

- evidence that climate change is or is not happening
- who originated the evidence, and whether this has any significance
- what is and is not known
- what scientists are working to find out
- the scientific problems to be solved
- who funds the scientists
- the relationship between scientists and politicians at national level
- public demand for low fuel taxation
- different viewpoints of politicians from different nations
- problems of achieving international agreement
- whether the score for length of time, in millions of years, surviving on Earth will be: dinosaurs 200 – humanoids 1 (nearly).

d Prepare a well-illustrated presentation on 'The Science Behind Climate Change', suitable for peer-group students who are not studying physics.

Examination questions

1 a The working part of a freezer is a heat pump which pumps energy from the inside of a freezer to the outside. The diagram shows the energy flows for one day of operation.

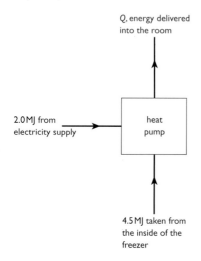

Q, energy delivered into the room

2.0 MJ from electricity supply

heat pump

4.5 MJ taken from the inside of the freezer

 i What is the value of Q, the energy flow out of the freezer? (1)
 ii State the physical law you used to calculate your answer. (1)
 iii Suggest a reason why you need an energy source to pump energy from the inside of the freezer to the outside. (1)
 iv What is the power flow through the walls of the freezer? (1)
b The freezer has an internal temperature of $-16\,^{\circ}\text{C}$ and is in a room whose temperature is $20\,^{\circ}\text{C}$. The walls of the freezer are insulating foam, 5 cm thick, and have an area of $4.1\,\text{m}^2$.
   Calculate the thermal conductivity of the walls. (3)
c Four containers of liquid milk, each having mass of 2.3 kg and initially at $0\,^{\circ}\text{C}$, are placed in the freezer. It takes 24 hours before the milk freezes. (The specific latent heat of fusion of milk is $334\,\text{kJ}\,\text{kg}^{-1}$.) What is the extra energy that the heat pump must pump out of the freezer as the milk freezes? (3)

d Inside the freezer there are no cooling fins at the bottom, but there are a large number of them towards the top. Explain how these fins cool the freezer and why there are none at the bottom. (2)

London, A level, Module Test PH3, January 1997

2 The diagram shows some of the energy levels for atomic hydrogen.

————————————————— 0 eV
————————————————— −1.51 eV
————————————————— −3.39 eV

————————————————— −13.6 eV

a Add arrows to the diagram showing all the single transitions which could ionise the atom. (2)
b Why is the level labelled $-13.6\,\text{eV}$ called the ground state? (1)
c Identify the transition which would result in the emission of light of wavelength 660 nm. (4)

Edexcel, AS/A level, Module Test PH2, June 2000

3 Draw a fully labelled energy band diagram for a typical intrinsic semiconductor.
   In what way does the energy band diagram of a typical insulator differ from that of a typical intrinsic semiconductor? (4)

NEAB, AS/A level, Module Test PH05, March 1997 (part)

13 Particle accelerators

THEMES AND TOPICS	• Particle energies
	• Acceleration by electric and magnetic fields
	• Nuclear changes
	• Circular motion

Searches for knowledge, and technological spin-offs

Fast particles are useful. Electron beams can weld and drill metals, as well as generate X-rays or create visible images in TVs and electron microscopes (Figure 13.1). Fast protons and neutrons can interact with nuclei of target material to create radioisotopes for medical and industrial use. Applications of ion beams include bombardment of semiconductor material as part of the process of developing computer microchips.

Figure 13.1
An electron beam in a scanning electron microscope (SEM) was used to produce this image. It shows, at a magnification of about ×10 000, crystals (purple) and 'nanotubes' (green) of carbon. In an electron microscope, as in a TV, acceleration of electrons is produced by a potential difference between a cathode and an anode system.

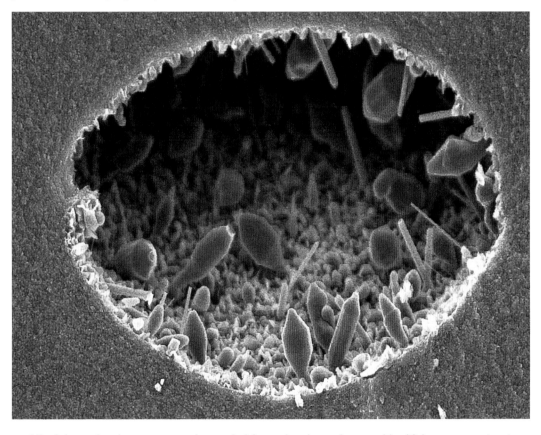

All of these developments are descended from the glass tubes used by 19th-century scientists in their explorations of the structure of matter. The explorations led to J. J. Thomson's identification of cathode rays as electrons, in 1895. We still use essentially the same method as Thomson used, thermionic emission, to provide sources of electrons.

The tradition of those 19th-century scientists continues at places such as CERN in Geneva, where researchers from several dozen countries probe matter at the deepest possible level. Their work is paid for by their governments, who know that, while the scientists are motivated by fundamental questions about the nature of the world, their work develops skills and new technologies. The World Wide Web is one such spin-off from work at CERN.

Today's big accelerators

A very large accelerator at CERN is called LEP, which stands for Large Electron Positron Collider. Beams of electrons and their antiparticle, positrons, are accelerated in opposite directions around a huge circular tunnel, 27 km in circumference (see Figure 13.6, page 231). The beams are allowed to collide within detectors, and because of the very high energies of the particles – up to nearly 100 GeV (10^{11} eV) each – the mutual annihilations allow not just photons to be created but other exotic particles, many of much larger mass than the electron or the positron.

The new Large Hadron Collider, LHC, is also being constructed at CERN, using the same tunnel as LEP. The intention here is to collide hadrons – which are particles with quark structures – with other hadrons. For example, protons can be collided with antiprotons. It was by colliding protons and antiprotons that experimental evidence was found in 1994, in the USA, in support of the predicted existence of the sixth and heaviest kind of quark – called the 'top' quark. (Other quarks include the 'up' and 'down' of which most of ordinary matter – you and your surroundings – is made. More exotic forms of material contain 'strange' quarks, 'charmed' quarks and 'bottom' quarks.)

1 A TV incorporates a particle accelerator.
a With what other kinds of particles do the accelerated electrons collide?
b What kinds of 'particles' are produced? Comment on the use of the word 'particles' here.
c Why do the collisions result in the production of no relatively heavy particles (such as protons)?
2 a Define the electronvolt.
b What is the total mass that can be created by the annihilation of two 100 GeV particles (electron and positron)? Use $E = mc^2$.

The cyclotron

All charged particles – electrons, positrons, protons, antiprotons, ions and so on – can be accelerated by an electric field. In a TV, the applied potential difference is between 1 and 10 kV, so each electron gains an energy of between 1 and 10 keV. Experimental particle accelerators like LEP and LHC provide particles with up to about a billion times that amount of energy. This cannot be done simply by using potential differences a billion times bigger. Even if such voltages could be generated, they could not be maintained without discharge. So the answer is to arrange for the accelerating particles to experience an applied voltage many times over, receiving a boost in energy each time.

The first type of accelerator that achieved this was the cyclotron – first developed in the 1930s. A cyclotron is relatively small. The first one to be built was a bench-top affair (Figure 13.2), and modern machines are just a bit bigger. Cyclotrons are still used for accelerating protons and deuterons. A deuteron is a proton together with a neutron (Figure 13.3). It is identical to a nucleus of deuterium, an isotope of hydrogen. Many larger hospitals, for example, use a cyclotron (Figure 13.4) to accelerate deuterons at target materials in order to create a supply of useful radioisotopes. An example of these interactions is:

$$^{2}_{1}H + ^{24}_{12}Mg \rightarrow ^{22}_{11}Na + ^{4}_{2}\alpha$$

The isotope sodium-22 is a useful radioactive tracer – providing a source of gamma radiation.

Figure 13.2
The world's first cyclotron, built by E. O. Lawrence in California.

Figure 13.3
Proton and deuteron.

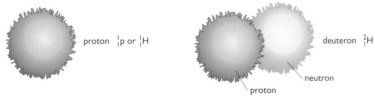

proton ^1_1p or ^1_1H

deuteron ^2_1H

neutron

proton

Figure 13.4
Many hospital physics departments have their own cyclotrons for production of radioactive tracers.

A cyclotron has two half-cylinders, or Ds, between which the voltage is applied (Figure 13.5). A proton or deuteron source is placed at the centre of the circle made by the Ds. As a particle is released, it is attracted to one of the Ds and repelled by the other. A magnetic field surrounding the Ds causes the particle to tend to follow a circular motion, so that it is forced to travel from D to D. The voltage alternates, so that the D that a particle is leaving always repels it, and the one that it is approaching always attracts it. On each journey across the gap between the Ds, the particle gains energy and speed.

Figure 13.5
A cyclotron uses a potential difference between two Ds to accelerate protons or deuterons, while a magnetic field holds them in a 'circular' path whose radius increases as the particles become faster – so that they 'spiral' outwards.

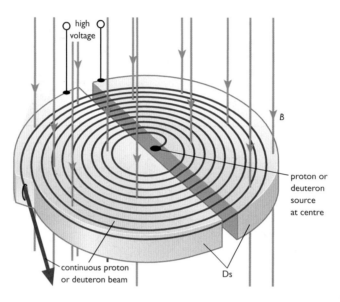

Magnetic force on a particle of charge e travelling at speed v perpendicularly to a field of flux density B is given by

$$F = Bev$$

and this provides a centripetal force for which

$$F = \frac{mv^2}{r}$$

So,

$$Bev = \frac{mv^2}{r}$$

and

$$r = \frac{mv}{Be}$$

So as speed increases, radius of orbit increases and hence the spiral pathway.

The cyclotron provides a continuous beam of particles, which originate from continuous emission at the centre. For each particle in a cyclotron, its first energy boost comes as it is emitted into the centre of the apparatus. Then as it travels within one of the Ds the applied voltage reverses, so that when it again reaches the gap between the Ds it is repelled by the one that it is leaving – the one that at first attracted it – and it is pulled towards the other. At some stage during the next part of its journey inside a D the voltage again reverses.

But synchronisation of D to D motion and the alternating voltage is eventually lost because of a relativistic effect. The rest mass of a proton – its mass as measured by a person in the same frame of reference – is 1.67×10^{-27} kg. But when at high speed relative to an observer or detecting apparatus, its mass is seen to increase. This increase in mass tends to increase the radius of the path, and hence the time spent in each D. The period between movements from D to D increases, while the period of the alternating voltage remains the same. This effect places an upper limit on proton speed, and hence on proton energy, that can be provided by a cyclotron.

3 A beam of deuterons collides with a target of beryllium-9, producing boron-10 and a beam of neutrons.
a Write the process as a nuclear change.
b Explain why the neutrons are produced as a fast beam.
c Suggest some uses for a beam of fast neutrons.

The synchrotron

Synchrotrons, by comparison with cyclotrons, are huge. The LEP at CERN (Figure 13.6) can be described as a synchrotron, and it has a circumference of 27 km. The large circumference, and hence radius, is necessary because otherwise the very high speeds would require unfeasibly strong magnetic fields to contain the particle beam.

Figure 13.6
The LEP and LHC tunnel is 27 km in circumference, and lies under the border between France and Switzerland.

The acceleration of particles in a synchrotron is achieved by applying the electric field by means of sectors of the tunnel, or cavities (Figure 13.7). The particles pass, not continuously but in short pulses, from cavity to cavity so rapidly that the manner in which the field is applied to the cavities can be considered to be as a wave. By using short bursts, the alternating voltage applied to the cavities can be kept in a state of synchronisation with the speed of each pulse of particles.

Figure 13.7
The huge radius of a synchrotron allows very fast particles to be contained within a beam by magnetic fields, while voltages are applied between cavities. The voltage alternates, and is synchronised with a burst of particles.

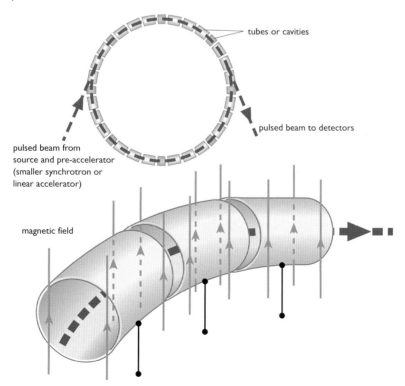

The magnetic fields needed to contain and steer the beam (Figure 13.8) are provided by superconducting coils that can carry large current without heating effect, but which, of course, require initial cooling and a high standard of thermal insulation. It is this kind of challenge that forces the designers of particle physics experiments to create new technologies. The skills and technologies developed are applicable outside the field of particle physics.

Figure 13.8
Inside the tunnel, thin pulsed beams of particles are steered around the huge circle by magnetic fields.

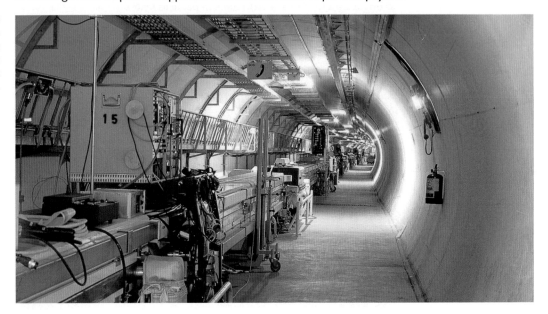

The linear accelerator

A linear accelerator (Figure 13.9) manages without using magnetic fields for steering the particles' pathways, by allowing the beam to travel in a straight line. As in the synchrotron, an alternating voltage is applied to a series of tubes through which the particles travel. As particles move from one tube to the next, the voltage applied to the tube ahead results in attraction, and the tube behind repels. While the particle pulse is inside each tube, the voltage switches, so that the same pattern of attraction and repulsion is experienced as the pulse moves to the next tube ahead. The frequency of the applied voltage stays the same, and because of the increasing particle speed, the pulse of particles must travel through longer and longer tubes.

Note, however, that, as particle speed increases, particle mass also increases, as predicted by Einstein's special relativity. So energy supplied to particles is required partly to increase speed and partly to increase mass. If a particle could reach the speed of light, its mass would be measured as infinite. The more nearly the particle speed approaches the speed of light, the more rapidly mass increases.

Figure 13.9
The principle of a linear accelerator.

The beam collides with a fixed target at the end of the accelerator. Collisions of the accelerated electrons with neutrons and protons in the target may be inelastic, transferring energy to (or from) the internal structure of the neutrons and protons, and so providing evidence for this internal structure – which we believe to be based on quarks.

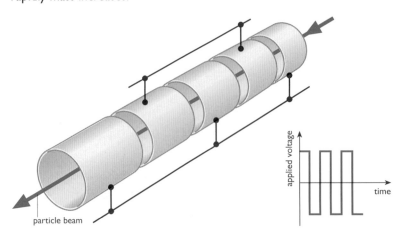

particle beam

Tubes or cavities become longer to compensate for increasing particle speed, until speed is so high that the effect of providing more energy to particles is to increase their measured mass rather than increase their speed significantly.

Where will fast particles lead us?

A summary of accelerator types and energies (in increasing order) is shown in Figure 13.10. But where is all of this leading? The curiosity of the 19th-century scientists led to vacuum tubes – and so to the development of electronics and communications, including both the radio and the TV. It also led to quantum mechanics and a view of the world which is very different to that which existed before, and with which we are still coming to terms. Who knows how we will change our views when we have a better understanding of the fundamental nature of matter, light and energy? Who knows what technologies might emerge? But the evidence of the past suggests that it *is* leading somewhere.

Figure 13.10
Summary of accelerator types.

a electron gun

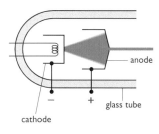

potential difference up to 25 kV,
particle energies up to 25 keV

b Van de Graaff generator

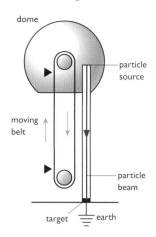

dome

moving belt

particle source

particle beam

target ‖ earth

potential difference up to 20 MV,
particle energies up to 20 MeV

The action of the moving belt results in accumulation of charge on the dome, and so in a high potential difference between dome and earth.

c cyclotron

potential difference between Ds ~ 10 kV
particle energies up to 25 MeV

d linear accelerator

particle energies up to 50 GeV

e synchrotron

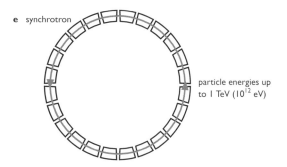

particle energies up to 1 TeV (10^{12} eV)

4 a Describe the different ways in which a cyclotron, a synchrotron and a linear accelerator subject a particle to a relatively low voltage over and over again.
b Why is this necessary in order to achieve very high particle energies?

5 a Calculate the magnetic field strength (flux density) that is needed to maintain an electron in circular orbit at speed $2 \times 10^8 \, \text{m s}^{-1}$ with radius
i 0.1 m
ii 1.0 km
Ignore relativistic effects.

electron charge $= 1.6 \times 10^{-19} \, \text{C}$
electron mass $= 9.1 \times 10^{-31} \, \text{kg}$

b What is the relevance of your calculation to accelerator design?

6 a For particles with constant mass and charge in a cyclotron that lies in a fixed and uniform flux density,

sketch a graph of particle speed against radius (assuming that there are no relativistic effects).
b A particle gains the same amount of energy each time it passes from D to D. Sketch a graph of number of journeys from D to D (x-axis) against particle energy.
c The energy gain takes the form of kinetic energy. Sketch a graph of number of D to D journeys against particle speed.
d Suggest what a graph of number of D to D journeys against pathway radius will look like, and sketch a matching spiral shape.

7 Make a list of specialist vocabulary (with definitions) used in this chapter, and spread the list across a large piece of paper. Include words that are new in this chapter and words that you have already come across, from all parts of the physics specification. Lay out the page so that associated words are close together.

Examination questions

1 In a cyclotron, a positive charge, q, travels at right angles to a uniform magnetic field of flux density B.

a If the mass of the charge is m and its speed v, show that the radius, r, of the circle in which it travels is given by

$$r = \frac{mv}{Bq}$$

b Show that the time taken for the charge to complete a half circle is independent of v.

c Explain how this principle is used in the cyclotron. (7)

NEAB, A level, Paper 1 Section B, June 1997 (part)

2 a Define *magnetic flux density*. (2)

b In the diagram, the arrowed circle is the path of a proton in an evacuated circular tube at a nuclear research centre. The circular path is maintained by a uniform magnetic field acting in a direction perpendicular to the plane of the circle.

i At a certain instant, the speed of the proton is $2.5 \times 10^7 \, \text{m s}^{-1}$ and the radius of its path is $120 \, \text{m}$. The mass and charge of a proton are $1.7 \times 10^{-27} \, \text{kg}$ and

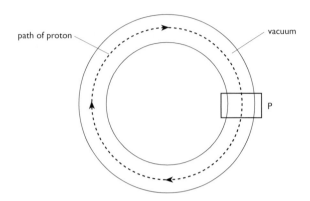

$1.6 \times 10^{-19} \, \text{C}$ respectively. Calculate the flux density [in T] of the uniform magnetic field. (3)

ii Each time the proton passes through the region labelled P in the tube, its speed is boosted by the application of an appropriate electric field. Without further calculation, describe and explain what action must be taken to maintain the proton in its original circular path. (2)

OCR (Oxford), A level, 6843, June 1999

Low temperatures and superconducting circuits

Useful low temperatures

Figure 14.1
Superfluid helium can creep up the sides of its container. It coats the container in a thin film and then flows over the brim and coats the outside too. It can drip off the bottom of the container, which looks as if it is leaking.

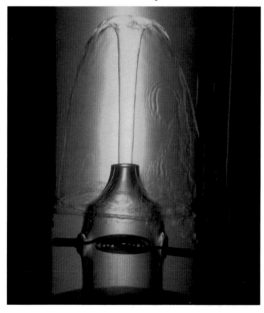

Figure 14.2
Energy transfers in normal and superconducting wires.

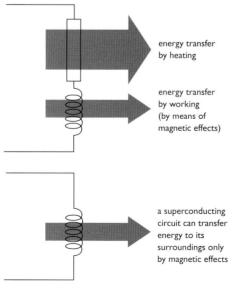

energy transfer by heating

energy transfer by working (by means of magnetic effects)

a superconducting circuit can transfer energy to its surroundings only by magnetic effects

Energy transfer by working is made possible by magnetic interactions, as in electric motors, loudspeakers, etc., or by transformer action. Lenz's Law applies in all such interactions.

Some very major changes in behaviour take place to many materials when their temperature is very low, usually less than about 10 K. At about 2.2 K or less, for example, helium is a superfluid (Figure 14.1). It has no viscosity and presents no resistance to motion through it. It also can flow in some very unexpected ways, such as through very narrow gaps that normal liquids cannot penetrate.

Superfluidity is not the only example of change in behaviour that happens at low temperature. Many metals lose their resistivity. Since there is no resistance of a wire made from such a metal, there is no electrical heating effect, whatever the size of current (Figure 14.2). This behaviour is superconductivity. It is extremely useful where large currents are needed but heating is not – as in creation of magnetic fields of large flux density.

It is only possible to take advantage of superconductivity by maintaining the conductor at very low temperature. A problem here is that the conductor is then inevitably much cooler than the general environment, and energy transfers into it, tending to raise its temperature. This can be minimised by surrounding the conductor by liquid helium and by good thermal insulation. So liquid helium is an essential requirement for applications of superconductivity using metal wires. At suitable pressure, liquid helium is useful for cooling other materials, which can then be studied at low temperature.

Liquefying gases

An ideal gas obeys Boyle's Law perfectly. For a fixed mass of ideal gas at a constant temperature, pressure is inversely proportional to volume, and so a graph of pressure against volume is a curve. Temperature can cause both pressure and volume to change. At a different temperature, the form of the relationship between pressure and volume still applies, though the curve is rather different. Each such curve, showing the relationship at a fixed temperature, is called an isotherm (Figure 14.3) – suggesting 'same temperature'.

Real gases do not differ too much from this, except that real gases can change state and become liquid (or solid). This can also be shown on a pressure–volume graph (Figure 14.3). Provided the temperature is below a certain critical value, the gas can become a liquid. It appears on the graph as a change in volume with no accompanying change in pressure.

Figure 14.3
Isotherms and critical temperature.

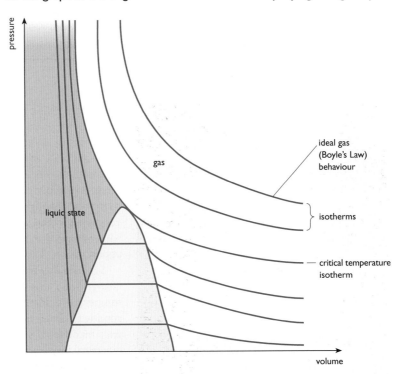

If temperature is too high, then the gas cannot condense into a liquid at all no matter what pressure is applied. The temperature at which condensation just becomes possible is called the critical temperature for change of state (Figure 14.3), which is a constant for each gas. For carbon dioxide, for example, this critical temperature for change of state is quite low – only 31 °C. Carbon dioxide can exist as a solid at higher temperatures provided pressure is high, but not as a liquid. At room temperature we very rarely see carbon dioxide in liquid form – it usually sublimes directly from the solid into the gas state.

The critical temperature for change of state of water is significantly higher, 647 °C. At temperatures higher than this, water cannot exist as a liquid but only as a solid or a gas.

1 Why is it possible for carbon dioxide to exist as a liquid at room temperature only for a narrow range of pressures? Sketch a graph to help you to answer this.

2 a Sketch isotherms for water at 100 °C and 647 °C.
   b Sketch a diagram showing the triple point of water.
   c Describe any differences and similarities between the two representations in a and b.

3 In this question, 'critical temperature' is that relating to change of state.
   a Helium can only exist as a liquid below 5 K. What is its critical temperature?

b The critical temperature of butane is 152 °C. Sketch isotherms for butane at 20 °C, 152 °C and 200 °C.
c At 20 °C, what happens to butane gas as pressure increases? Explain this by referring to your sketch in b.
d The critical temperature of methane is 191 K. Why is it not practicable to transport methane as a liquid in pressurised containers?

4 Suggest how temperatures of less than 5 K can be measured.

Cooling by adiabatic expansion

Adiabatic expansion of a gas is a change that involves no net exchange of energy by thermal processes between the gas and its surroundings. From the First Law of Thermodynamics, we have

$$\Delta U = \Delta Q + \Delta W$$

where ΔU is change in internal energy, ΔQ is energy transferred by thermal processes to the gas, and ΔW is the work done on the gas. For an adiabatic change, ΔQ is zero, and so

$$\Delta U = \Delta W$$

Work is not done *on* the gas when it expands but *by* the gas. It must push the surrounding atmosphere out of the way, and it must do work against internal forces of attraction. This means that the value of ΔW is negative, and the value of ΔU must therefore also be negative. The internal energy of the gas decreases. Its temperature falls.

In practice, gases such as oxygen and helium are cooled by expansion, doing work against their own intermolecular forces of attraction, by allowing them to travel through a nozzle from a region of higher to lower pressure (Figure 14.4).

Figure 14.4
Cooling (and liquefaction) occur by adiabatic expansion.

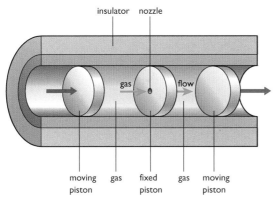

insulator nozzle

gas flow

A constant pressure difference is maintained between the gas on either side of the fixed piston by the moving pistons. Gas passes through the nozzle and expands. The energy required to overcome intermolecular attraction is taken from the internal energy of the gas, lowering its temperature.

moving piston | gas | fixed piston | gas | moving piston

Superconductivity

Figure 14.5
MRI scanners use magnetic flux densities of as much as 10 T. The coil providing this field carries a large current, and must be cooled to less than 4 K with liquid helium.

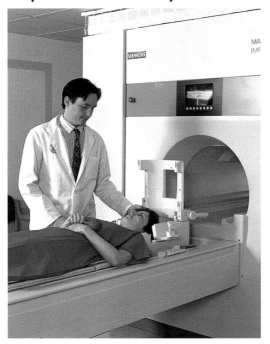

In a normal conductor, a large current causes powerful heating. But since superconducting metals have no resistivity, they can sustain a large electric current with no heating effect and no thermal transfer of energy to the surroundings.

Magnetic resonance brain scanners (Figure 14.5) require very strong magnetic fields. Anyone who has had a scan will tell you how noisy the experience is. It's rather like being inside a huge loudspeaker, since interactions of magnetic fields cause vibration. For such strong magnetic fields, scanners require large currents. The current flows in a coil that is cooled to a very low temperature, so that the wires are superconducting and thus heating effects and associated energy loss are eliminated. The cooling system must be well insulated to minimise thermal transfer of energy into the system, which would result in heating of the helium and the coil, and loss of superconductivity. There is a limitation on the size of the magnetic flux density attainable with the superconducting coil, as very strong magnetic fields can actually destroy the superconductivity, but nevertheless strong fields are possible.

The need for cooling is problematical – although a very high level of insulation can reduce rate of transfer of energy, it cannot completely eliminate a net thermal transfer from surroundings to the jacket of liquid helium and the superconducting wires. So continuous cooling is essential. Materials that become superconductors at higher temperatures reduce the need for cooling and so save money. Superconductivity would be a much more useful phenomenon if such materials could be developed. The temperature at which a material becomes superconducting is called its superconductivity critical temperature, T_c (Figure 14.6). Considerable funding is being given to research laboratories around the world to develop materials with relatively high superconductivity critical temperatures. So far the highest that has been achieved is 125 K (Table 14.1).

Table 14.1
Superconductivity critical temperatures.

Material	Superconductivity critical temperature/K
aluminium	1.20
gallium	1.08
tin	3.72
niobium/aluminium, Nb_3Al	18.7
$YBa_2Cu_3O_7$	80
$Tl_2Ba_2Ca_2Cu_3O_{10}$	125

Figure 14.6
Resistivity versus temperature graphs for materials becoming superconducting.

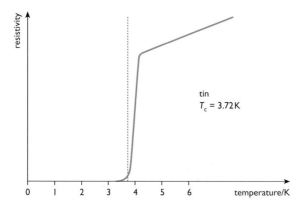

tin
$T_c = 3.72$ K

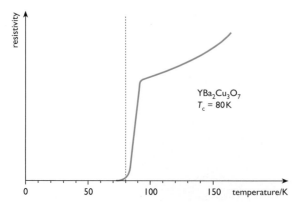

$YBa_2Cu_3O_7$
$T_c = 80$ K

5 a In a superconducting coil, why is no power source needed provided that the coil does no work through magnetic effects?

b Suggest why metal structures in MRI scanners vibrate energetically.

c What is the effect of such vibration on energy required to sustain current in the coils? Refer to Lenz's Law.

6 a Calculate the rate of change of temperature of 1 kg of aluminium cable if energy transfers into it at a constant rate of 100 W. The specific heat capacity of aluminium at 273 K is 899 J kg^{-1}K^{-1}. (Assume this to be constant.)

b The rate of thermal transfer of energy through any material is given by

$$\frac{dQ}{dt} = kA\frac{dT}{dx}$$

where k is the thermal conductivity of the material, A is its cross-sectional area and dT/dx is temperature gradient.

Explain why

i choice and arrangement of the insulating material around superconducting coils are important,

ii a superconducting material with a critical temperature of 125 K is, in principle, a better choice than aluminium for use in MRI coils. (In practice there are problems in use of high critical temperature superconducting materials that have yet to be overcome.)

7 a What current is required to create a flux density of 10 T at the centre of a long air-filled solenoid of 2000 turns per metre? ($\mu_0 = 4\pi \times 10^{-7}$H m^{-1})

b Why can such a flux density only be sustained by using superconducting technology?

8 Large magnetic fields can destroy superconducting behaviour. If this happens suddenly, then a superconducting coil may become very hot very quickly, and energy must be transferred away from it. This can be done by allowing the helium to boil.

a What is the flux density at a distance of 0.4 m from a straight conductor, in air, carrying a current of 800 A? ($\mu_0 = 4\pi \times 10^{-7}$H m^{-1})

b What would be the flux density at the centre of a solenoid of 500 turns per metre, in air, carrying the same current?

c Explain how the presence of an iron core would considerably increase the flux density as calculated in **b**.

d If the resistivity of a wire is $2.5 \times 10^{-8}\,\Omega$ m when not superconducting, what would be the power at which thermal transfer takes place in each metre of wire of cross-sectional area 1 mm^2, at a current of 800 A?

e At what rate must helium boil, in kilograms per second, to transfer energy away from the wire at the rate required to balance the rate of heating in the same wire if, having been used as a superconductor, it suddenly loses it superconductivity? (At 4 K, the effective latent heat of vaporisation of helium is 2.2×10^4 J kg^{-1}.)

15 Fluids in motion

Principles of flow

All everyday motion that we experience involves movement of fluids – gases or liquids. The movement may be within the fluid, or it may be movement of the fluid relative to solid surfaces.

A small but continuous squirt of coloured dye into the middle of a flow of fluid will, until it diffuses, trace a line of colour. It follows a pathway, called a streamline (Figure 15.1), that would be taken by a very small volume of the fluid in the flow.

Within a body of flowing fluid it is possible to trace any number of streamlines. Where the flow of liquid is such that layers do not mix then the streamlines remain stable and smooth (approximately parallel to each other from one length to the next). Such flow is called laminar flow.

Laminar flow can happen in the middle of a stream and even at its edges where the flow meets a solid boundary. But here there is an effective friction with the solid. At the boundary itself where particles of the fluid and solid are next to each other, the velocity of the flow, *v*, is zero. It increases with distance, *x*, into the liquid (Figure 15.1). The rate of change of velocity with distance, d*v*/d*x*, is a velocity gradient.

Figure 15.1
Representations of flow in a pipe – velocity vectors and streamlines.

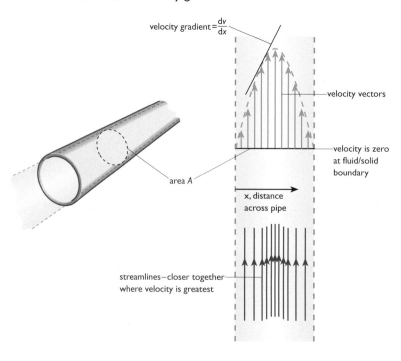

velocity gradient $=\dfrac{\mathrm{d}v}{\mathrm{d}x}$

velocity vectors

velocity is zero at fluid/solid boundary

area *A*

x, distance across pipe

streamlines—closer together where velocity is greatest

As the speed of a fluid across a surface increases, random motions may begin to develop on a larger and larger scale. The flow ceases to be laminar, streamlines cease to be stable and smooth, and the flow becomes turbulent (Figures 15.2 and 15.3).

Figure 15.2
Laminar flow and
turbulent flow.

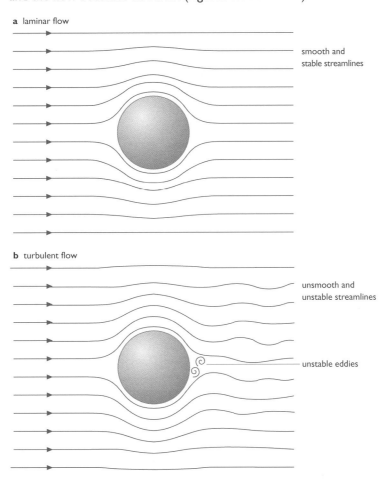

a laminar flow

smooth and
stable streamlines

b turbulent flow

unsmooth and
unstable streamlines

unstable eddies

Figure 15.3
An example of turbulent
flow in a laboratory test,
modelling flow of air
around tall buildings.

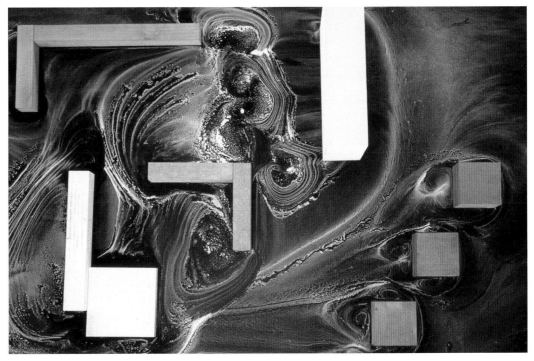

Viscous resistance to motion – viscosity

Figure 15.4
The laminar flow of layers of fluid, and its relationship to resistive force.

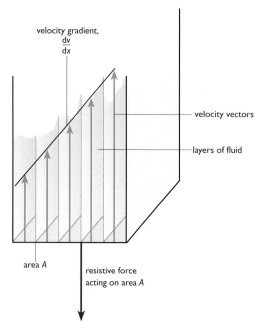

velocity gradient, $\dfrac{dv}{dx}$

velocity vectors

layers of fluid

area A

resistive force acting on area A

A fluid resists motion through it, and resists motion between neighbouring layers of the fluid itself. The level of this 'viscous' resistance depends on the nature of the fluid – its density and the forces between its particles.

The viscosity – also called the coefficient of viscosity and given the symbol η – of a fluid is a physical constant. Like many physical 'constants' of materials, however, it is dependent on temperature. Warm oil is much less viscous than cool oil, for example.

Viscosity is defined in terms of the resistive force, F, acting on laminar flow of the fluid, the cross-sectional area normal to the flow, A, and the velocity gradient. It is simplest to consider flow of parallel sheets of the fluid, each sheet with a velocity slightly different from that of its neighbours (Figure 15.4).

We get:

$$\eta = \frac{F}{A\,dv/dx}$$

The units of quantities on the right-hand side are:

F	N
A	m^2
v	$m\,s^{-1}$
x	m

The unit of viscosity is therefore the $N\,s\,m^{-2}$ (sometimes written as $Pa\,s$). Some typical values are given in Table 15.1.

Table 15.1
Some values of viscosity.

State or phase	Material	Viscosity/$N\,s\,m^{-2}$
gases	air, at 273 K	1.71×10^{-5}
	methane, at 273 K	1.03×10^{-5}
liquids	water, at 298 K	8.91×10^{-4}
	silicone oil, at 298 K	4.95×10^{-4}
	glycerol, at 298 K	9.420×10^{-1}

1 a Make a list of at least five physical constants of materials that you have studied.
b Identify any that are *not* temperature-dependent.
c For one of the others, describe how particle theory provides an explanation for its temperature dependence.

2 a Sketch the pathway of a cannonball in an idealised flight for which there is no resistance to motion. Draw vectors to show the horizontal and vertical components of velocity at two different points in the flight (one for the rising cannonball, and one for its fall).
b Consider the effects of air resistance on each of the components. Make a second sketch to show the flight of the cannonball taking account of the air resistance. Again, show the horizontal and vertical components of velocity at two different points in the flight.

3 a Fluid in a pipe can be considered to have zero velocity at the surface of the pipe. As the velocity of the fluid at the centre of the pipe increases, what would you expect to happen to velocity gradient?
b Explain in terms of the appropriate mathematical formula why total resistance to flow increases, therefore, when maximum velocity increases.
c Name the factors that influence whether wider pipes offer more or less total viscous force than narrower pipes.

4 What are the dimensions of each quantity on the right-hand side of:

$$\eta = \frac{F}{A\,dv/dx}$$

and hence what are the dimensions of η?

5 a What is the ratio of the viscous forces acting on air and methane flowing with identical maximum velocities in pipes of identical cross-sectional areas at 273 K? (Use data from Table 15.1.)
b Repeat part a for water and glycerol at 298 K.

The drag force equation for motion through air

Figure 15.5 (right)
Streamlines around a moving cylinder.

Figure 15.6 (below)
Engineers look for laminar flow and turbulent flow, study streamlines and measure drag coefficient. Turbulent flow produces greater resistance to motion than laminar flow does, and is more difficult to analyse.

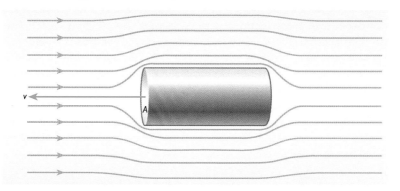

v is the relative velocity of cylinder and air, so though it may be the cylinder that is moving, we can draw streamlines just as we would for a stationary cylinder and flowing air.

Consider a solid cylinder moving through the air, in a direction parallel to its axis (Figure 15.5). The equation that describes the size of the viscous drag force on the cylinder is:

$$F = \tfrac{1}{2}C_D A \rho v^2$$

This equation applies not only to cylinders but also to bodies of other shapes. A is the frontal area of the body, C_D is the drag coefficient, ρ is the density of the air and v is the speed of the body through the air. The drag coefficient is a constant for a particular body. For a shape as complex as a car, the drag coefficient is difficult to calculate even with the most sophisticated computer models. It has to be measured in a wind tunnel test (Figure 15.6).

Note that the unit of drag coefficient can be determined from the equation, written as

$$C_D = \frac{2F}{A\rho v^2}$$

The dimensions of C_D are those of the right-hand side of the equation, which we can list:

2   a number, dimensionless

F defined by the formula $F = ma$, and so with dimensions $[M][L][T]^{-2}$

A with dimensions $[L]^2$

ρ defined by the formula $\rho = m/V$, and so with dimensions $[M][L]^{-3}$

v defined by the formula $v = dx/dt$, and so v^2 has dimensions $[L]^2[T]^{-2}$

The collected dimensions of the right-hand side of the equation are therefore:

$$\frac{[M][L][T]^{-2}}{[L]^2[M][L]^{-3}[L]^2[T]^{-2}}$$

or

$$[M][L][T]^{-2}[L]^{-2}[M]^{-1}[L]^3[L]^{-2}[T]^2$$

which can also be written

$$[M][M]^{-1}[L][L]^{-2}[L]^3[L]^{-2}[T]^{-2}[T]^2$$

which simplifies, such that C_D is a dimensionless quantity with no unit.

6 a For a car with drag coefficient 0.45 and cross-sectional area 2.4 m² moving at a speed of 25 m s⁻¹, calculate the viscous drag force in air of density 1.3 kg m⁻³.
b Calculate the rate at which the car must do work in order to balance this force. Give your answer in watts.
c If rolling resistance results in an additional and constant force of 250 N, what is the total output power of the car that is necessary for motion at a constant speed of 25 m s⁻¹?
7 a Use the equation for viscous drag force to help you to sketch a graph of the relationship between drag force acting on a car and its speed.
b Rolling resistance, which results in the resistive force between the car and the road, is much more nearly speed-independent than is the drag force. At a speed of about 20 m s⁻¹ the two forces contribute equally to opposition of the car's motion. On the same axes as used in part **a**, sketch a graph of rolling resistance against speed, assuming rolling resistance to be constant. Also sketch a graph of total resistive force against speed.
c Hence explain why fuel efficiency is strongly influenced by car speed, and particularly so at high speeds.

The Bernoulli effect in liquids

Imagine a flow of liquid along a pipe of cross-sectional area A_1, which at a point X becomes narrower, with cross-sectional area decreasing to A_2 (Figure 15.7a). The number of liquid particles approaching X every second must be the same as the number leaving. (Otherwise liquid particles are being created or destroyed at X.) So the masses of liquid approaching X and leaving X every second must also be identical. Since the liquid is incompressible and has constant density, the volumes of liquid approaching and leaving X each second must, likewise, be the same.

But if we draw equal volumes of liquid either side of X, we see that they must have different lengths. If the volumes we draw are the volumes approaching and leaving X per second, then the lengths that we see are the distances travelled by the liquid in each second. We see that the liquid has to travel faster in the narrow pipe.

Figure 15.7
Flow of liquid from a wide to a narrow pipe.

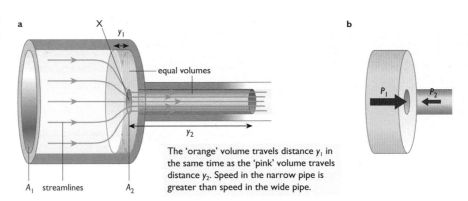

The 'orange' volume travels distance y_1 in the same time as the 'pink' volume travels distance y_2. Speed in the narrow pipe is greater than speed in the wide pipe.

The liquid accelerates as it passes through X. This requires a net force.

net force = force acting on area A_2 of liquid at X due to liquid in wide tube
  − force acting on the same area due to liquid in narrow tube

net force = $P_1 A_2 - P_2 A_2$

where P_1 is the pressure exerted by the slower liquid in the wide tube and P_2 is the pressure exerted by the faster liquid in the narrow tube (Figure 15.7b).

For net force to be positive, which clearly it is, then it is necessary that $P_1 > P_2$. Thus in the flowing liquid, the pressure is lower where the speed is greater. This is the Bernoulli principle or the Bernoulli effect.

Venturi meters (Figure 15.8) make use of the Bernoulli effect to measure speed of flow of liquid in pipes – for example in a kidney dialysis machine (Figure 15.9). The faster the speed of flow in the wide pipe, the more liquid is passing from wide to narrow sections per second, and the bigger the net force that is needed. That produces a bigger pressure difference between the sections.

Figure 15.8
Principle of a Venturi meter. Liquid levels in the vertical tubes allow comparison of the pressure, and hence speed, of the flowing liquid in the different sections of pipe.

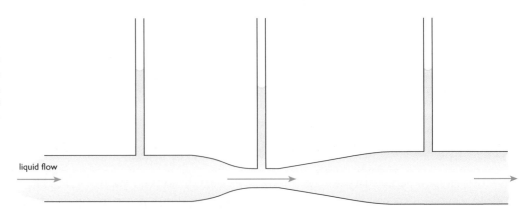

liquid flow

Figure 15.9
In a kidney dialysis machine the flowing liquid is the patient's blood. Measurement and control of the flow is of vital importance.

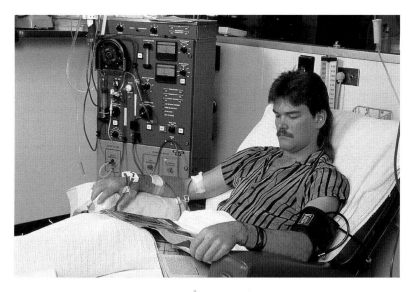

The Bernoulli effect in gases

A gas must also accelerate as it passes from a wider section of pipe to a narrower section (Figure 15.10). Unlike a liquid, a gas is compressible and its density changes. However, the number of particles arriving at an area X must still equal the number of particles leaving in the same period of time. The mass of gas arriving at area X is the same as the mass of gas leaving:

mass approaching area X per second = mass leaving area X per second

$$\rho_1 A_1 v_1 = \rho_2 A_2 v_2$$

where ρ_1 and ρ_2 are densities of the gas before and after passing area X, A_1 and A_2 are the corresponding cross-sectional areas on either side of X, and v_1 and v_2 are the corresponding velocities. (Note that a liquid is incompressible and $\rho_1 = \rho_2$, so for a liquid the same equation simplifies to $A_1 v_1 = A_2 v_2$.)

Figure 15.10
Flow of gas from a wide pipe to a narrow pipe.

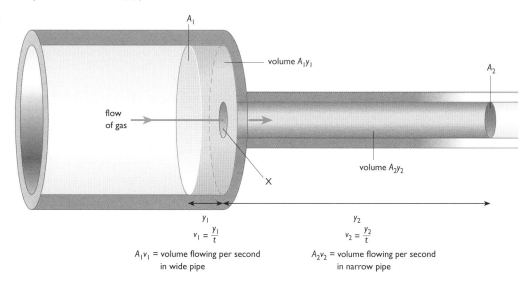

Similarly to the liquid case,

net force experienced by gas passing through area $X = P_1 A_1 - P_2 A_2$

and, again, we know that P_1 must be greater than P_2 because we know that the gas accelerates. Pressure exerted by the faster gas is less than that exerted by the slower gas. Gases obey the Bernoulli principle.

Lift force

Aircraft wings must deflect air downwards, so that the aircraft experiences an equal and opposite, upwards, force (Figure 15.11). There is more than one way to do this, but a very efficient way is to use the aerofoil effect (Figure 15.12). An aerofoil is shaped such that air has to travel further over the upper surface than over the lower surface, and so tends to travel faster. The difference in speed produces a difference in pressure between the upper and lower surfaces, and hence a net upwards force, or lift.

Figure 15.11
A metal tube full of people, defying gravity.

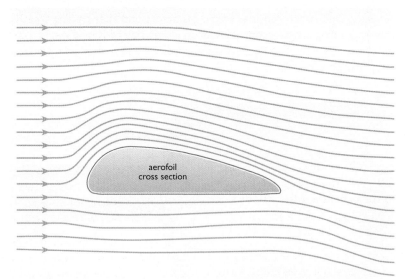

Figure 15.12
An aerofoil creates unequal speeds of air, and hence unequal pressures.

• Pressure on upper surface of wing is less than pressure on lower surface.
• Air is deflected downwards.

aerofoil
cross section

8 a Explain how we know that, for any volume of a pipe along which a liquid flows, particles enter and leave at the same rate at every instant. Why do we not know this for gases except when flow is laminar and steady?
b The formula

$$\rho_1 A_1 v_1 = \rho_2 A_2 v_2$$

applies to gases and liquids. Explain in what way, and why, it can be simplified for liquids.

9 What is the ratio of velocities for liquid flowing in a pipe of area $0.010\,\mathrm{m}^2$ and then flowing in a pipe of area $0.004\,\mathrm{m}^2$?

10 a If the mass of an aircraft is $100\,\mathrm{t}$ ($10^5\,\mathrm{kg}$) and the area of its wings as seen from above is $40\,\mathrm{m}^2$, what is the pressure difference between upper and lower wings?
b Suppose that the pressure on the lower surface of the wing is $1.0 \times 10^5\,\mathrm{Pa}$. By what percentage does the pressure acting on the upper wing surface differ from this?

Examination questions

1 The volume flow rate of a fluid through a pipe of cross-sectional area A and length l is given by

$$\frac{V}{t} = \frac{cA^2\Delta p}{\eta l}$$

where c is a constant having no unit.

a i What quantity does the term η represent? (1)
ii The SI unit of pressure in base units is $kg\,m^{-1}\,s^{-2}$. State the SI unit for each of the following quantities:

V, t, A, l (2)

iii Determine the unit of η in terms of SI base units. (2)

b The diagram shows a system used to provide a patient with a blood transfusion. The required fluid pressure to enable blood to flow through the hollow needle, and into the patient, can be achieved by raising or lowering the container.

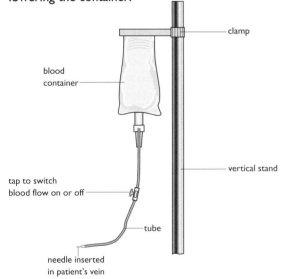

In one case a needle, 0.035 m long and of internal cross-sectional area $3.0 \times 10^{-7}\,m^2$, is used. The required volume flow rate into the patient is $2.5 \times 10^{-7}\,m^3\,s^{-1}$.

The magnitude of η in SI units $= 4.0 \times 10^{-3}$
The value of $c = 0.040$

i Use the equation given to determine the pressure difference between the ends of the needle that will produce the required blood flow rate. (2)
ii Use your value from **i** to determine the vertical distance between the top surface of the blood in the container and the end of the needle that will produce the required blood flow rate when the needle is inserted horizontally into a vein. (4)

The density of blood $= 1050\,kg\,m^{-3}$
The acceleration of free fall, $g = 9.8\,m\,s^{-2}$
Average blood pressure of a patient above atmospheric pressure $= 140\,Pa$

AEB, A level, Paper 2, Summer 1998

2 The diagram shows the cooling water from a power station joining a river. The areas given on the diagram are the cross-sectional areas of the respective flow regions.

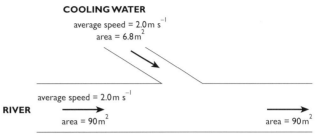

a i Calculate the volume flow rate of the water in the river before the cooling water joins it. (1)
ii Calculate the river velocity after the cooling water has joined it. (2)

b One end of a tube is lowered into the water. The other end of the tube is connected to a pressure gauge as shown.

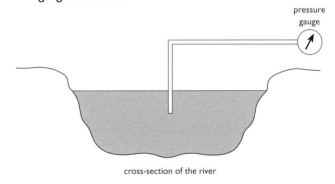

cross-section of the river

State what will happen to the reading on the pressure gauge when the velocity of the water in the river increases. Give a reason for your answer. (2)

c Suggest a suitable type of thermometer for use when monitoring the temperature of the cooling water, giving a reason for your choice. (1)

AEB, A level, Paper 2, Summer 1997

3 a Suggest why it is difficult to build a practical passenger car with a very low drag coefficient. (3)

b A car has a frontal area of $1.6\,m^2$ and a drag coefficient of 0.28. At the car's designed maximum speed of $45\,m\,s^{-1}$ the forces opposing motion, other than aerodynamic drag, amount to $220\,N$.

density of air $= 1.2\,kg\,m^{-3}$

Calculate:
i the total retarding force [in N] on the car at a speed of $45\,m\,s^{-1}$; (1)
ii the power [in W] needed to propel the car at $45\,m\,s^{-1}$. (1)

UODLE, A level, Paper 55, March 1998

4 The Venturi meter shown in the diagram is used to measure the flow rate of liquid through pipework in a distillery. Tubes A, B and C are open to the atmosphere at their top ends, and the liquid flows in the direction shown.

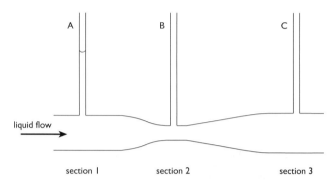

a i The level of liquid in tube A is shown in the diagram. Mark on the diagram the levels of liquid which might be expected in tubes B and C.
ii Explain why the levels in B and C are as you have drawn them. (5)
b Liquid of density $850\,\mathrm{kg\,m^{-3}}$ enters section 1 of the meter at a flow rate of $0.50\,\mathrm{kg\,s^{-1}}$. The diameter of the pipe at section 1 is 30 mm, and at section 2 it is 18 mm.
i Calculate the speed of the liquid in section 1 of the meter.
ii Calculate the speed of the liquid in section 2 of the meter. (4)

NEAB, A level, Paper 2 Section B (PH08), June 1997

5 a Explain, using sketches, the meaning of
i *laminar* flow,
ii *turbulent* flow
in relation to the motion of a fluid. (4)
b The pressure drag force *F* acting on a body as it moves at velocity *v* through a fluid is given by the equation

$$F = A\rho v^2$$

i Define the other symbols in this equation, and explain the conditions under which it applies.
ii Derive the equation. (8)
c An engineer designing a car uses the equation

$$F = \tfrac{1}{2}C_\mathrm{D}A\rho v^2$$

for the pressure drag force, where C_D is the drag coefficient.
   Explain why the engineer would not use the equation in **b**. Discuss the factors which determine the value of C_D. (5)
d A motorist wishes to buy a car which will be economical in its use of fuel. He is interested in two models each of which has a drag coefficient of 0.30. Will the two cars necessarily be equally economical (assuming that their engines have the same efficiency)? Briefly explain your answer. (4)

UCLES, A level, 4836, June 1998

Absorption and transmission of X-rays

THEMES AND TOPICS
● Light–matter interactions
● Energy and momentum conservation
● Mathematical modelling
● Collimation and absorption
● Inverse square law and exponential behaviours

Travel of X-rays with and without absorption

The intensity of any beam of radiation is the rate of flow of energy normal to a unit of cross-sectional area, and is measured in watt per square metre, $W\,m^{-2}$. Where the radiation spreads out from a point source into a vacuum, then its intensity obeys the inverse square law. In a parallel beam, however, there is no such spreading, and in a vacuum the beam will maintain a constant intensity. It shows no divergence or convergence, and can be represented by parallel rays. Absorption processes, however, can greatly influence beam intensity. These situations are summarised in Figure 16.1.

Figure 16.1
Point source beam and parallel beam, with and without absorption.

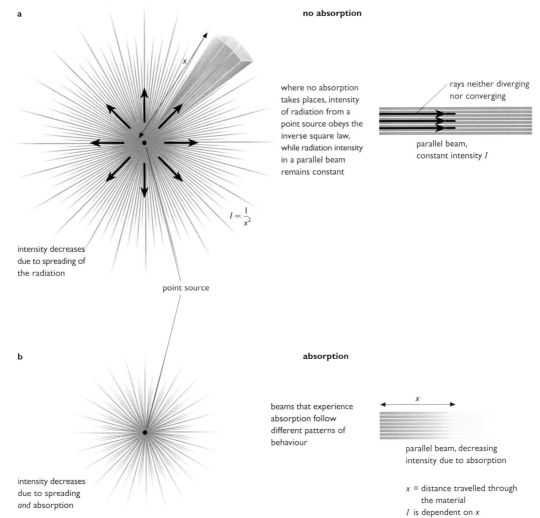

a

no absorption

where no absorption takes places, intensity of radiation from a point source obeys the inverse square law, while radiation intensity in a parallel beam remains constant

rays neither diverging nor converging

parallel beam, constant intensity I

$$I \propto \frac{1}{x^2}$$

intensity decreases due to spreading of the radiation

point source

b

absorption

beams that experience absorption follow different patterns of behaviour

parallel beam, decreasing intensity due to absorption

x = distance travelled through the material
I is dependent on x

intensity decreases due to spreading *and* absorption

Interactions of photons and atoms

Matter can take energy away from X-radiation, resulting in scattering or absorption. It can take some or all of the energy away from each photon. Human tissue is no different from other matter in this respect, and in fact human tissue behaves almost, though not quite, like a sack of water. The obvious difference is the presence of bone, which contains large proportions of calcium atoms and is denser than ordinary tissue. With an atomic number of 20, calcium atoms are significantly more massive than most of the atoms of the body, and scatter and absorb X-rays much more effectively. Other tissues also vary a little in their ability to transmit or absorb the radiation, so the human body is like a complex bag of water. An additional complexity, of course, is that ionisation within the body can cause damage at a chemical level.

There are several processes by which X-ray photons can be scattered or can lose energy, or both of these. A low-energy photon may not ionise or excite an atom but may interact with the complex electric field in the vicinity of the atom and be absorbed, and an identical photon emitted in another direction. We could think of this, with equal justification in terms of the end result, as a single photon experiencing a change in direction. This is simple scattering.

Alternatively, excitation or ionisation may take place, the latter resulting in the liberation of an electron. Where the electron escapes from the material, then the process is the same as photoelectric emission, and the electron can be called a photoelectron. Often, however, the electron will cause further ionisation on its journey within the material, and will lose its initially high level of kinetic energy and be unable to escape.

Photons can interact with electrons within a material by another process – called the Compton effect (Figure 16.2). During such an interaction, an electron is knocked into motion, gaining both momentum and kinetic energy. The photon continues, deflected in its path, and now having reduced energy and frequency. It makes little difference whether we think of the photon emerging from the process as a new photon, or the same photon with its properties changed. An important point about the Compton effect, however, is that since the electron gains momentum, the 'initial' photon (and also the 'emergent' photon) must have momentum – so the effect is evidence for 'particulate' behaviour of photons. The de Broglie relation can be applied to analysis of photon momentum.

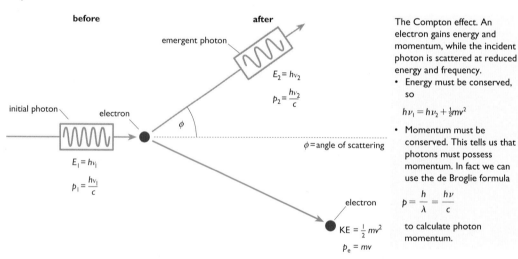

Figure 16.2 The Compton effect.

before | after

emergent photon

$E_2 = h\nu_2$

$p_2 = \dfrac{h\nu_2}{c}$

initial photon | electron

ϕ

$\phi =$ angle of scattering

$E_1 = h\nu_1$

$p_1 = \dfrac{h\nu_1}{c}$

electron

$KE = \frac{1}{2}mv^2$

$p_e = mv$

The Compton effect. An electron gains energy and momentum, while the incident photon is scattered at reduced energy and frequency.
- Energy must be conserved, so

$$h\nu_1 = h\nu_2 + \tfrac{1}{2}mv^2$$

- Momentum must be conserved. This tells us that photons must possess momentum. In fact we can use the de Broglie formula

$$p = \frac{h}{\lambda} = \frac{h\nu}{c}$$

to calculate photon momentum.

There is a fourth phenomenon, which only occurs with high-energy X-rays – sometimes called 'hard' X-rays. This is pair production – the simultaneous creation of an electron and a positron (Figure 16.3) that takes place most frequently in the strong electric fields of larger nuclei. The photon ceases to exist, all of its energy being carried away by the masses (according to $E = mc^2$) and kinetic energies of the electron and positron. A positron produced in such circumstances usually has a short lifetime. As a particle of antimatter surrounded by a world of matter, it does not take long for it to meet with one of its mirror image particles, an electron, and for mutual annihilation to take place. The annihilation results in a pair of photons, which each must have much less energy than the lone photon that began the process.

Figure 16.3
Pair production, followed by a mutual annihilation.

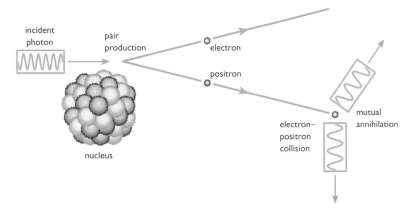

When receiving a diagnostic X-ray examination (Figure 16.4), a patient should receive the minimum possible dose of radiation that is required for creation of the image. A problem here is that many of the incident photons are scattered, so that they emerge from the patient in many different directions and do not help in creation of a clear 'shadow' image. Also, due to scattering, even if a perfectly parallel beam is incident on the patient's skin, X-rays will still reach parts of the body that are not in the line of the beam. This creates a problem in that the testes and ovaries, in which sperm and ova are generated, may receive exposure, with potential damage to DNA that is then transmitted to offspring. Such a risk is a very small but real one in any X-ray examination, so unless the patient has some special problem in which a high level of risk is judged to be acceptable, X-ray examinations should not be frequent. Scattering of X-rays is also the reason why the radiographer performing the examination does not stay in the same room but retreats behind a protective screen or window.

Figure 16.4
X-rays interact with matter by several different processes, most of which depend on the nucleon number of the atoms concerned. Where there are higher densities of larger atoms than in surrounding material, a higher degree of absorption and scattering takes place. Bone, tooth fillings and swallowed safety pins(!) form distinct shadows with X-rays.

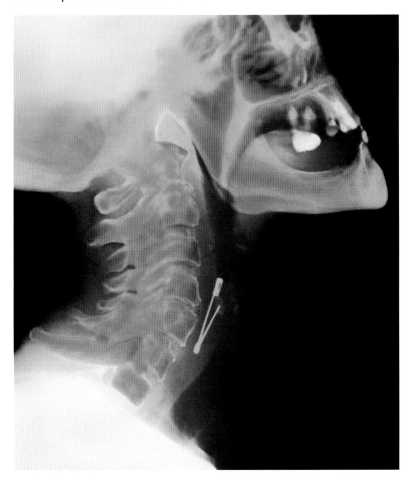

A summary of these interaction processes, showing also their general dependence on nucleon number, is given in Figure 16.5.

Figure 16.5
Summary of X-ray interactions with matter. The shaded bars show the photon energy ranges in which different processes take place.

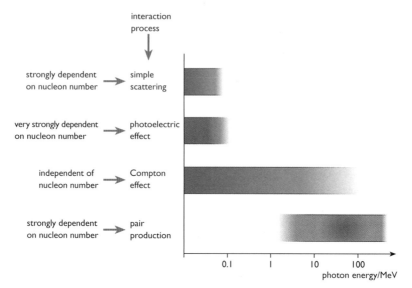

X-ray absorption is strongly influenced by:
- atomic mass number (nucleon number) of the material
- photon energy – different processes dominate at different photon energies.

I a Which interactions of X-rays and matter
i leave X-ray photon energy unchanged
ii always completely absorb energy of a photon
iii have little effect on photons of energy more than 100 keV
iv have little effect on photons of energy less than 1 MeV?
b Figure 16.6 represents X-rays incident on a bone. Sketch a similar diagram and use it to show
i scattering within the bone
ii how a shadow of the bone is created for photon energies of up to 100 keV.

2 a State a formula that shows how electron momentum is related to electron wavelength.
b In what way does the Compton effect provide evidence for the particle nature of electromagnetic radiation?
c For a Compton effect event as shown in Figure 16.2, develop expressions for the components of electron momentum parallel to and perpendicular to the initial photon direction.

3 a Use the information in Figure 16.5 to explain why the photoelectric effect results in a much higher level of X-ray absorption in bone than in soft tissue.
b For treating cancerous tumours, high exposure is required in order to kill the tumour cells. If X-rays that are absorbed predominantly in bone are used, then a tumour in soft tissue will absorb a low proportion of the incident X-rays. To give the tumour a sufficient dose, the bones must receive a very large dose indeed. Suggest therefore why X-rays with energies in the range 100 keV to 1 MeV are used for such therapeutic purposes.
c Why are X-rays in this energy range less suitable than softer (lower-energy) X-rays for producing images of shadows of bones?
d Even with soft X-rays, such X-ray images are rather cloudy. Why is this?

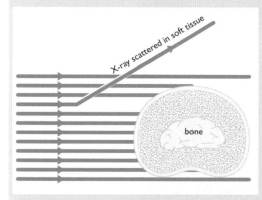

Figure 16.6

X-ray beams

X-rays cannot be refracted at boundaries in the same way as visible light can, and so they cannot be 'focused' in the same way. They can, however, be collimated – a beam can be made near-parallel by an absorbing grid or collimator (Figure 16.7).

Figure 16.7
A collimator fitted to an X-ray tube.

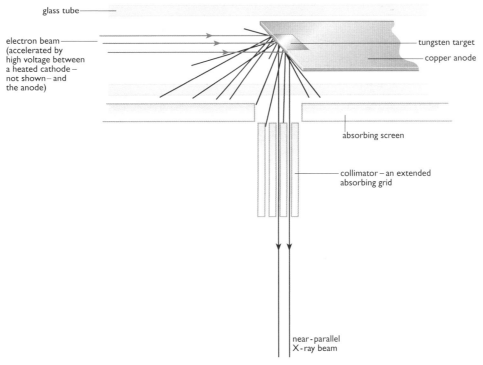

glass tube

electron beam
(accelerated by
high voltage between
a heated cathode –
not shown – and
the anode)

tungsten target

copper anode

absorbing screen

collimator – an extended
absorbing grid

near-parallel
X-ray beam

Since a parallel beam is relatively simple, the mathematics that can describe it is relatively simple. For a beam entering a body of material from a vacuum, or a human body from air, absorption processes begin immediately. The beam loses energy and the material gains energy. The intensity of the beam falls as distance x into the material increases. The more intense the beam, the more absorption takes place for each unit of distance travelled. This means that the rate of absorption is proportional to the intensity:

rate of absorption $\propto$ intensity

Therefore

rate of change of intensity $\propto$ intensity

$$\frac{dI}{dx} \propto -I$$

The minus sign is necessary because, while intensity can only ever be positive, it is decreasing and so rate of change of intensity is a negative quantity. We can introduce a constant of proportionality and write an equation:

$$\frac{dI}{dx} = -\mu I$$

where μ is called the linear attenuation coefficient. Its value depends on X-ray frequency and on the absorbing material.

Just as for radioactive emission and charge storage by capacitors, situations in which a rate of change of a quantity is proportional to its own value result in an exponential relationship:

$$I = I_0 e^{-\mu x}$$

where I_0 is the incident or initial intensity. Thus a graph of parallel beam intensity against depth of material is an exponential decay curve (see Figure 16.8).

We can define a quantity that is analogous to radioactive half-life. This is the distance over which the intensity of a parallel beam is halved (Figure 16.8), and it is called half-value thickness, $x_{1/2}$. After travelling for this distance, an initial intensity of I_0 is reduced to $I_0/2$:

$$\frac{I_0}{2} = I_0 e^{-\mu x_{1/2}}$$

Figure 16.8
Different X-ray photon energies result in different penetrating properties in the same material (copper).

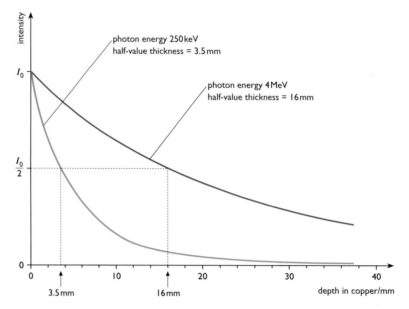

Practical medical X-ray beams contain X-rays with wide energy and wavelength spectra, and so such a beam has a half-value thickness that results from this complexity. (For some purposes, filters can be used to narrow the X-ray spectrum.)

4 a What type of beam, in what circumstances, obeys
i the inverse square law
ii exponential decrease in intensity?
b On the same axes, sketch graphs to show these two types of intensity variation with distance.
c Which type of behaviour is dominant for ultraviolet radiation travelling into space from the Sun?
d Would you expect a significant difference between UV intensity at sea level and that on a mountain top 10 000 m above sea level due to
i effects that follow the inverse square law
ii absorption?
5 Half-value thickness may be quoted from measurements on absorption by copper. X-rays with energy 250 keV, for example, have half-value thickness of 3.5 mm of copper, whereas for 4 MeV X-rays the half-value thickness in copper is 16 mm.
a What is the effective linear absorption coefficient in each case?
b For the same initial intensity, what is the ratio of the two rates of change of intensity?
c Which of the X-ray types delivers energy most rapidly to the copper for the same initial intensity? Explain as fully as you can.
6 a Sketch a graph of intensity of a parallel beam against distance travelled into a material and show how such a graph can be used to find values of half-value thickness and linear absorption coefficient.
b If a source were close to the surface of a material, without collimation, how, in general terms, would the graph be different?

Examination questions

1 a X-rays generated at voltages around 40 kV are attenuated principally by photoelectric absorption. X-rays generated at voltages around 11 MV are attenuated largely by Compton scattering. Describe the key event in each of these two processes. What is the difference between absorption and scattering? (5)

b Explain the term *half-value thickness*.

The intensity of a beam of 30 kV X-rays drops by 70% when a block of wood 27 mm thick is placed across its path. Calculate the half-value thickness of wood for 30 kV X-rays.

You may use the equations
$$I = I_0 e^{-\mu x}$$
$$\mu x_{1/2} = \ln 2$$
where the symbols have their usual meanings. (4)

London, A level, Module Test PH3, January 2000 (part)

2 The graph below shows the photon intensity/energy curve of a heterogeneous X-ray beam before and after passing through an aluminium filter of thickness 1.55 mm.

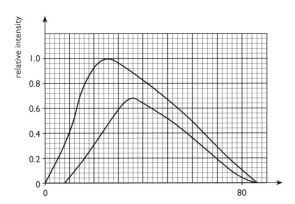

a Describe and explain the effect of filtration on
i the intensity of the X-ray beam,
ii the quality of the X-ray beam. (4)

b i What mechanism is mainly responsible for the attenuation of the X-rays by the filter? For what photon energy is the half value thickness in aluminium equal to 1.55 mm?
ii Calculate the linear absorption coefficient in aluminium for X-rays of photon energy 35 keV. (6)

London, A level, Module Test PH3, January 1996 (part)

3 a The intensity of an X-ray beam passing through matter is reduced by absorption and by scattering. Explain the difference between *absorption* and *scattering*. Name and describe briefly two interactions which involve either one or both of these processes. (6)

b Explain the term *half-value thickness* applied to a tissue irradiated with a monochromatic X-ray beam.

Why is it best to use high energy X-ray photons for therapy? (4)

London, A level, Module Test PH3, June 1998 (part)

4 The photoelectric effect and Compton scattering are two mechanisms by which X-rays are attenuated in matter.

a Describe *Compton scattering*. (4)

b State how the attenuation in tissue for each of the two mechanisms depends on a nuclear property of the tissue:
i photoelectric effect
ii Compton scattering. (2)

c State and explain which of the two attenuation mechanisms is irrelevant for use in
i the imaging of bone,
ii therapy for treating cancers of soft tissue. (4)

OCR (Cambridge), A level, 4835, March 1999

5 In different situations, there are two relationships between the intensity, I, of X-rays and the distance, namely:

1     $I = I_0 e^{-\mu x}$

and

2     $I = \dfrac{k}{r^2}$

where μ and k are constants and x and r are distances.

a In what circumstances does each relationship apply?
i Relationship 1
ii Relationship 2 (2)

b i Sketch the following logarithmic graphs. Note that the axes are different for the different relationships. (2)

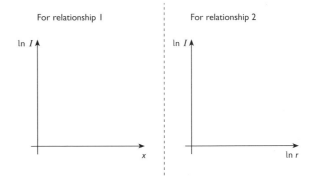

ii If either or both of your graphs is/are linear, express the gradient(s) in terms of the quantities in the equations. (2)

c Both of these equations are relevant when considering the protection of radiographers from X-rays. By referring to them, explain *two* ways in which these health workers protect themselves.
i Relationship 1
ii Relationship 2 (2)

d Identify and describe *two* biological effects that can be caused by different types of radiation. (4)

IB, Higher level 3, Paper 430, May 1998

Answers to numerical questions

● Chapter 1

Marginal questions

1 a 0.0175   **b** 0.000 175   **c** 0.0200   **d** 0.002 00

3 a 0.833 m s^{-2} **b** 0.167 rad s^{-1}

4 a 50 N, assuming mass of 60 kg
   b friction between shoe and track

5 a gravitational pull of Earth
   b reaction from water

7 7.1 m s^{-1}

11 a i 200 rad s^{-1}, 40 m s^{-1} **ii** 200 rad s^{-1}, 20 m s^{-1}
   b i reaction
      ii assuming mass of 20 g:
         weight $= mg = 0.020 \times 9.81 = 0.20$ N

         $$\text{centripetal force} = \frac{mv^2}{r}$$

         $$= \frac{0.020 \times 40^2}{0.2} = 160 \text{ N}$$

         $$\text{ratio} = \frac{\text{weight}}{\text{centripetal force}}$$

         $$= \frac{0.20}{160} = 0.001\,25 \text{ or 1 part in 800}$$

12 a 31.8 rev s^{-1} **b** 1910 rpm

13 20.9 m s^{-1}

14 a i 24 h   **ii** 86 400 s
   b 1.16 × 10^{-5} Hz   **c** 7.27 × 10^{-5} rad s^{-1}
   d 465 m s^{-1} **e** 60°

   f i $r^3 = \dfrac{k}{\omega^2}$ **ii** 4.23 × 10^7 m

15 a 365 days or 3.15 × 10^7 s

   b i 2.97 × 10^4 m s^{-1} **ii** 1.99 × 10^{-7} rad s^{-1}

Comprehension and application

16 $\omega = \dfrac{v}{r}$

20 12 m s^{-1}

21 a 0.12 m s^{-1} **b** towards centre of wheel

22 a 1070 N, 1920 N
   b i 1370 N down, weights add
      ii 1370 N down, weights add
      iii 1370 N down and 850 N horizontally
      iv 2220 N down, assuming 90 kg passenger at bottom

23 a 17.3 N, assuming head pointing towards centre
   b 21.4 N

25 friction between tyres and road

26 $\tan\phi = \dfrac{\omega^2 r}{g}$

27 b $860 \times 2 = 1720$ N

Examination questions

1 a direction changes
   c 19 ms^{-2}
   d friction between driver and seat

2 a i 52 rad s^{-1} **ii** 1.2 ms^{-1}

   b r increases by factor $\dfrac{115}{45} = 2.56$; ω increases by same factor to $\dfrac{500}{2.56} = 195$ rpm

3 a 8.7×10^{-4} rad s^{-1}

4 a 365 days or 3.15×10^7 s; 1.99×10^{-7} rad s^{-1}
   b 3.55×10^{22} N
   c gravitational pull of Sun

5 a 12.8 ms^{-1} **b** 205 ms^{-2} **c** 205 ms^{-2}
   d tension and weight in opposite direction, therefore tension higher than at top of circle

6 a i 2.9 ms^{-1} **ii** 1.8 rad s^{-1} **iii** 5.2 ms^{-2} towards centre

7 b 3.6×10^{22} N

8 a i 126 rad s^{-1} **ii** 950 ms^{-2} **iii** 0.095 N
   b i 645 rad s^{-1} **ii** 39 ms^{-1} tangentially to wheel

● Chapter 2   Marginal questions

3 a 325 km

15 $x = x_0 \cos \sqrt{\dfrac{k}{m}}\, t$

16 $x = x_0 \sin \theta$

18 a 5.0 Hz   **b** 490 N m^{-1} **c** -990 s^{-2}

19 a 0.05 m   **b** 2.6 Hz

20 a $2\pi \sqrt{m}$ **b** $\dfrac{1}{\sqrt{m}}$

25 a $4T$ **b** $\dfrac{8\pi}{\omega}$

27 decrease length of pendulum; g is smaller

29 a $x = \pm x_0$ **b** $x = 0, v = \omega x_0$ **d** 1.0 ms^{-1}

Comprehension and application

30 a i $3T, 3.5T, 10T, 10.5T$
   ii $x_0, -x_0, x_0, -x_0$
   c i $\phi, 2\pi - \phi, \phi, 2\pi - \phi$

36 a 1.3 ms^{-1}; 3200 ms^{-2} **b** 63 N

Examination questions

1 b   i 11 mm   **ii** 3.45 ms   **iii** 290 Hz   **iv** 20 m s^{-1}

3 a 0.5 Hz   **b** at centre; 0.31 m s^{-1}
c at maximum displacement; 0.99 m s^{-2}

4 d 6.15 pm approx.

6 a 0.023 m s^{-2}

7 a 4.0 s

8 b 37.1 N m^{-1} **c** 1.66 kg
d equal to, if resistive forces can be made negligible

9 a

	B	C	D	E	F
Displacement	+	0	−	0	+
Acceleration	−	0	+	0	−
Velocity	0	−	0	+	0

b 39.2 kJ   **d** 200 N m^{-1} **e** 4.0 s

● Chapter 3   Marginal questions

1 a 10 000 N C^{-1} **b** 0.0004 N

2 a −2000 MJ kg^{-1} **b** 2 MJ   **c** 2000 MJ

8 m^3 s^{-2} kg^{-1}

9 force against $\dfrac{1}{r^2}$

10 $\dfrac{G}{r^2}$; GM

11 a zero   **b** you would be weightless

13 $G = \dfrac{Fr^2}{Mm}$

14 $E = \dfrac{1}{4\pi\epsilon_0}\dfrac{q}{r^2}$

15 N m^2 kg^{-2}; N m^2 C^{-2}

16 a 58 N

17 a   i 0.062 N C^{-1} **ii** 0.187 N C^{-1}
b   i 0.118 N C^{-1} **ii** 0.102 N C^{-1}

18 a 9.58 × 10^7 C kg^{-1} **b** 1.35 × 10^{20} kg^2 C^{-2} **c** 1.2 × 10^{36} (no units)

21 $g = -\dfrac{V}{r}$

23 a 36 V   **b** 7.2 × 10^{-10} J   **c** 7.2 × 10^{-10} J

24 a   i 5000 V m^{-1} **ii** 5000 N C^{-1}
b 5.0 × 10^{-13} N   **c** 2.5 × 10^{12} m s^{-2} **d** 1.0 × 10^{-14} J

26 a acceleration is not constant

27 a $E = \dfrac{-qQ}{4\pi\epsilon_0\,(R+r)^2}$; $V = \dfrac{-qQ}{4\pi\epsilon_0\,(R+r)}$

Comprehension and application

31 $5.3 \times 10^4 \, \text{m s}^{-1}$

35 a $2 \times 10^{-4} \, \text{C}$, assuming mass of 60 kg
   b $200 \, \text{kV m}^{-1}$; $200 \, \text{kN C}^{-1}$

Examination questions

1 c $0.0060 \, \text{N}$

2 b i $-120 \, \text{kV}$ **ii** $-240 \, \text{kV}$; V is inversely proportional to r
   iv $-400 \, \text{N C}^{-1}$ **v** $6.4 \times 10^{-17} \, \text{N}$

5 a $\dfrac{m_s v_s^2}{(R_E + h)^2}$

   b $\dfrac{GM_E}{(R_E + h)^2}$; N kg^{-1}

   c v_s is inversely proportional to $\sqrt{(R_E + h)}$

6 a N kg^{-1} **e** $M =$ mass of Sun, $r =$ distance from Sun

7 c gravitational force *down* $= 9.8 \times 10^{-18} \, \text{N}$;
   electric force *up* $= 1.6 \times 10^{-17} \, \text{N}$;
   resultant force *up* $= 6.2 \times 10^{-18} \, \text{N}$

8 a ii $1.8 \, \text{m s}^{-2}$ **iv** $8.7 \times 10^{22} \, \text{kg}$
   b i $1.8 \, \text{N}$
     ii $0.20 \, \text{s}$; depends only on mass and spring constant
   c ii yes; direction of velocity changing; towards Jupiter
     iii $1.8 \times 10^{27} \, \text{kg}$

9 b v $2.2 \, \text{N C}^{-1}$ away from objects

Chapter 4   Marginal questions

2 $33 \, \mu\text{F}$

3 $10 \, \text{V}$

4 $600 \, \text{pF}$

5 $1.4 \times 10^{-6} \, \text{m}^2$ $(1.4 \, \text{mm}^2)$

6 a $11 \, \text{pF}$
   b i radius $0.009 \, \text{m}$ **ii** radius $9 \, \text{km}$

8 $1 \, \text{mm}$

9 $2 \, \mu\text{m}$

12 a $0.37 \, \mu\text{C}$ **b** $0.14 \, \mu\text{C}$ **c** $0.05 \, \mu\text{C}$

17 a $0.018 \, \text{s}$

18 a $1.2 \, \text{nF}$ **b** $14 \, \text{nF}$

21 capacitor stores $0.50 \, \text{J}$, spring stores $0.56 \, \text{J}$, so spring stores more energy

Comprehension and application

23 a $13 \, \text{MV m}^{-1}$ **b** $13 \, \text{MV m}^{-1}$

24 a $8.9 \times 10^{-7} \, \text{F}$ $(0.89 \, \mu\text{F})$ **b** $0.014 \, \text{F m}^{-2}$

25 $1.9 \, \text{nJ}$

26 b $3 \times 10^{-14} \, \text{J}$

Examination questions

1 b i 3300 μF   **ii** 1.0 J

2 a 1200 μC   **b** 3 V

4 a 5.6 mC
   b initial current $= \dfrac{12}{22 \times 10^3} = 0.55\,\text{mA}$

   current when $t = CR = 0.20\,\text{mA}$
   $CR = 10.3\,\text{s}$
   c 2.8 mC

5 a $9.0 \times 10^{-7}\,\text{F}$ (approx. 1 μF)   **b** 8.1 μC   **c** 4.1 μA; 2.0 s

6 a 54 μJ   **b** 8.6 μJ

7 b i $CR = 12\,\text{s}$ from graph, giving $C = 8\,\text{mF}$

8 a i 2.0 mC   **ii** 8.9 mJ
   b i 4.5 W   **ii** 1.0 A
   c 18 W

9 a i 66 pF   **ii** $1.7 \times 10^{-7}\,\text{C}$
   c 0.83 mm

10 a 1.6 MΩ
   b i 72 s   **ii** 18 MΩ
   c i 0.013 C

● Chapter 5 **Marginal questions**

1 1.025

3 a 0.25 mm

7 $2.0 \times 10^{-6}\,\text{T}$

10 b 750   **c** 375 mT

11 5.6 cm

13 radius of orbit $= 3.4\,\text{m}$, so deflection of beam will be negligible within the dimensions of a TV set

15 4.2 A

17 a $2.0 \times 10^{-4}\,\text{N m}$ **b** zero

18 a 5.7°   **b** $9.6 \times 10^{-7}\,\text{N m deg}^{-1}$

24 a increases by large factor   **b** doubles   **c** increases fourfold

Comprehension and application

27 a $3.2 \times 10^{-19}\,\text{N}$

Examination questions

1 c $4.0 \times 10^{-13}\,\text{N}$

2 a i 25 mA   **ii** 1.6 m s^{-1}
   b i $1.2 \times 10^{-20}\,\text{N}$ **v** 0.39 mV

3 a 220 m   **b** radius of path of helium nucleus is 440 m

4 b current flows from west to east; 410 A; wire will get very hot

5 b v $V_H = 64\,\mu\text{V}$, so suitable meter range is 100 μV

6 c i 16 kN C^{-1} **ii** $2.6 \times 10^{-15}\,\text{N}$
   e 730 V

9 b 0.9 mT (half the value at the centre)

● Chapter 6   **Marginal questions**

  4 a 76 µV
   b 1.2 mV (it is now the horizontal component of the flux that is being cut)

  8 a 2.0×10^{-4} Wb   **b** 1.0×10^{-2} Wb

  10 a $I_0 = \dfrac{BAN\omega}{R}$

  13 0.20 V

  17 a 25 µV

  22 a 50
   b power in secondary is less than power in primary, unless transormer is ideal

  23 a 15
   b i 48 W   **ii** 24 W
   c 0.10 A   **d** transformer not 100% efficient

  24 0.58 or 58%

  Comprehension and application

  28 a i 3.1 Ω   **ii** 32 kΩ
   b 5.0 kHz

  34 a 325 V   **b** $V = 325 \sin 100\pi t$

  35 a 300 A   **b** 3 kV (resistance of cable = 10 Ω)
   c 0.9 MW   **d** 4.3×10^4 A, 4.3×10^5 V, 18 GW

  Examination questions

  1 a i 8.8 V rms   **ii** 0.13 A rms
   c 3.2 µF (use any corresponding values of X_C and f)

  2 b 40 mH   **c** 20 Ω   **d** 2.3 mT

  3 a 7000 turns
   b ii 2800 V

  5 b ii 0.31 A (back e.m.f. = 9.375 V)

  6 b 310 V ($X_L = 157$ Ω)

  8 b 500 Hz

  9 b i primary current = 130 mA, assuming power in primary − power in secondary

  10 a 0.011 N m

  11 b 2.2 mm s^{-1}

  12 a i 1.6×10^{-6} Wb   **ii** 48×10^{-6} Wb
   b 6.5 mm s^{-1}

  13 a 5.0×10^{-4} T

● Chapter 7   **Marginal questions**

  5 use carbon-12 data: 1 u = 1.661×10^{-27} kg

  7 a 1070 u   **b** 1070 times

  11 b 0.0291 u   **c** 27.1 MeV

  15 b i 0.002 39 u   **ii** 2.22 MeV; 1.11 MeV per nucleon

16 b $p = h/\lambda$ **c** $p = mv$
     d $mv = h/\lambda$ in magnitude

22 zero in all cases

34 a between 0.8 and 0.9 MeV per nucleon
     b approximately 200 MeV per nucleus
     c 8.3×10^{13} J per kilogram   **d** 8.3×10^{4} s or 23 h
     e about 1 kg   **f** about 230 kg

Examination questions

3 b ii 186 MeV or 3.0×10^{-11} J; released as kinetic energy of fragments

5 a ii 7.58 MeV per nucleon

6 a ii 28.3 MeV

10 d i 13 600 MW   **ii** 4.3×10^{17} J   **iii** 1.3×10^{28} fissions   **iv** 5200 kg

11 b i 1.20×10^{57} protons   **ii** 1.0×10^{11} y

12 a i 8000 MW   **ii** 2.7×10^{20} s^{-1} **iii** 3300 kg

● Chapter 8   Marginal questions

3 a 2.5×10^{-7} rad, assuming $\lambda = 500$ nm   **b** 250 ly

8 a 4.8×10^{-6} pc   **b** 1.6×10^{-5} ly   **c** 8.2 light-minutes

9 9.3×10^{20} km

10 a 6.7×10^{-9} **b** 0.001

11 a 6×10^{6} **b** 1.0×10^{4} **c** 100

16 1.8×10^{-18} s^{-1}

18 4900 Mpc

21 2.4×10^{10} y, 9.8×10^{9} y

Comprehension and application

28 a approximately 50%

Extra skills task

1 a 10^{17} planetary systems   **b** 10^{14} planets

Examination questions

1 13.7 km s^{-1} towards Earth

4 b 1.0×10^{25} m

5 a iii 7.3×10^{7} m

6 a 1.8×10^{-6} rad or 0.38 seconds   **b** 9.6×10^{-8} W m^{-2}

7 b 2.0×10^{10} y

● Chapter 9   Marginal questions

1 a 0   **b** 4   **c** 9   **d** 3.54   **e** −1   **f** −0.184

4 b i 100 kΩ   **ii** 300 Ω   **iii** 100 °C   **iv** 80 °C

5 a 10   **b** 10^{7}

6 b i ratio in bels $= \log_{10} (P_{\text{signal}}/P_{\text{noise}})$
     ii ratio in dB $= 10 \log_{10} (P_{\text{signal}}/P_{\text{noise}})$
   c i 30 dB   **ii** 0 dB   **iii** 4 dB   **iv** 11 dB

7 a 0.301 **b** 3.32

8 a 4 **b** 5 **c** 4.32 **d** -1 **e** -2 **f** 0

9 a 1.00 **b** 2.30 **c** 0

Examination questions

1 a i 900 Bq **ii** $0.016 \, s^{-1}$
b 43 s

2 c 6000 m

3 a $1 \times 10^{12} \, W \, m^{-2}$
b i 30 dB **ii** $1 \times 10^{-9} \, W \, m^{-2}$
c ii $\Delta I = 0.1 \times 10^{-6} \, W \, m^{-2}$; $s = 1$

4 c i 96 dB **ii** 12 km **iii** 76 dB **iv** 100 km

5 d use gradient of straight line
e 4
f use intercept on $\log P$ axis

6

10	1	R
100	2	2R
1000	3	3R

Chapter 10

Marginal questions

6 $k/k' = \rho$

8 a $\dfrac{dQ}{dt} = \sigma A \dfrac{dV}{dx}$

b $\sigma = \dfrac{x}{RA}$

10 b i $x = x_0 \sin(2\pi f)t$
ii $v = x_0(2\pi f)\cos(2\pi f)t$
iii $a = -x_0(4\pi^2 f^2)\sin(2\pi f)t$

Examination questions

1 c 14 W **d** 6.8×10^5 s or 190 h

2 a 4.0 s

3 b 190 h

4 b ii $1.6 \, m \, s^{-2}$
c i 3.8 J **ii** 3.8 J
d i $6.9 \, m \, s^{-1}$

5 a number of strides $= \dfrac{1500}{1.8} = 830$

time of stride $= \dfrac{240}{830} = 0.29 \, s$

rate of working $= \dfrac{PE \ gained}{time \ taken}$

$= \dfrac{80 \times 9.8 \times 0.15}{0.29} = 406 \, W$

c i 1200 W **ii** 50 g

6 false; false; true; false

7 velocity; current; power; force

ANSWERS TO NUMERICAL QUESTIONS

● Chapter 11 **Examination questions**

1 14 km

2 a 280 photons per second
  b about 30 photons per second, assuming about 10% reach retina

● Chapter 12 **Examination questions**

1 a i 6.5 MJ   iv 52 W, assuming steady temperature inside freezer
  b $0.018\,W\,m^{-2}\,K^{-1}$ c 3.1 MJ

2 c $\Delta E = 1.88\,eV$; transition from $-1.51\,eV$ to $-3.39\,eV$

● Chapter 13 **Marginal questions**

2 b $3.5 \times 10^{-25}\,kg$

3 a $^{2}_{1}H + {}^{9}_{4}Be \rightarrow {}^{10}_{5}B + {}^{1}_{0}n$

5 a i $1.1 \times 10^{-2}\,T$ ii $1.1 \times 10^{-6}\,T$

Examination questions

2 b i 2.2 mT

● Chapter 14 **Marginal questions**

3 a 5 K

6 a $0.11\,K\,s^{-1}$

7 a 4000 A

8 a $4.0 \times 10^{-4}\,T$ b 0.50 T   d 16 kW   e $0.73\,kg\,s^{-1}$

● Chapter 15 **Marginal questions**

4 $[M]\,[L]^{-1}\,[T]^{-1}$

5 a 1.66   b 9.64×10^{-4}

6 a 440 N   b 11 kW   c 17 kW

9 speeds up by factor 2.5

10 a $2.45 \times 10^{4}\,Pa$ b 25%

Examination questions

1 a ii m^{3}, s, m^{2}, m   iii $kg\,m^{-1}\,s^{-1}$
  b i $9.7 \times 10^{3}\,Pa$ ii 0.96 m

2 a i $180\,m^{3}\,s^{-1}$ ii $2.2\,m\,s^{-1}$

3 b i 760 N   ii 34 kW

4 b i $0.83\,m\,s^{-1}$ ii $2.3\,m\,s^{-1}$

● Chapter 16 **Marginal questions**

5 a $0.20\,mm^{-1}$, $0.043\,mm^{-1}$
  b 4.6

Examination questions

1 b 16 mm ($\mu = 0.0446\,mm^{-1}$)

2 b i 27 keV   ii $0.19\,mm^{-1}$

Index